CHEMICAL MUTAGENS

Principles and Methods for Their Detection

Volume 9

Sponsored by the Environmental Mutagen Society

CHEMICAL MUTAGENS
Principles and Methods for Their Detection
Volume 9

Edited by Frederick J. de Serres
National Institute of Environmental Health Sciences
Research Triangle Park, North Carolina

PLENUM PRESS • NEW YORK and LONDON

The Library of Congress cataloged the first volume of this title as follows:

Chemical mutagens. v. 1—
  New York, Plenum 1971—

  v. ill. 24 cm.

  "Principles and methods for their detection."
  Vols. 1—   sponsored by the Environmental Mutagen Society.
  Key title: Chemical mutagens, ISSN 0093-6855.

    1. Chemical mutagenesis—Collected works. I. Environmental Mutagen
Society.
QH465.C5C45   575.2'92   79-640909
Library of Congress   79   MARC-S

ISBN-13: 978-1-4612-9711-6   e-ISBN-13: 978-1-4613-2771-4
DOI:10.1007/ 978-1-4613-2771-4

This book was edited by F. J. de Serres
in his private capacity. No official support
or endorsement of DHHS is intended or should be
inferred.

© 1984 Plenum Press, New York

Softcover reprint of the hardcover 1st edition 1984

A Division of Plenum Publishing Corporation
233 Spring Street, New York, N.Y. 10013

Contributors

David E. Amacher, Drug Safety Evaluation, Pfizer Central Research, Groton, Connecticut 06340

Christian L. Bean, Department of Medical Genetics, University of Wisconsin-Madison, Madison, Wisconsin 53706

Leonard F. Bjeldanes, Department of Nutritional Sciences, University of California, Berkeley, California 94720

Rosalie K. Elespuru, Fermentation Program, NCI-Frederick Cancer Research Facility, Frederick, Maryland 21701. *Present address:* LBI-Basic Research Program, NCI-Frederick Cancer Research Facility, Frederick, Maryland 21701.

James S. Felton, Biomedical and Environmental Research Program, Lawrence Livermore National Laboratory, University of California, Livermore, California 94550

Martha J. Ferguson, Radiation Biology Section, Department of Radiology, University of Texas Health Science Center at Dallas, Dallas, Texas 75235

Mary Esther Gaulden, Radiation Biology Section, Department of Radiology, University of Texas Health Science Center at Dallas, Dallas, Texas 75235

Frederick T. Hatch, Biomedical and Environmental Research Program, Lawrence Livermore National Laboratory, University of California, Livermore, California 94550

Lois Jacobs, Department of Medical Genetics, University of Wisconsin-Madison, Madison, Wisconsin 53706

P. R. M. Kerklaan, Department of Radiation Genetics and Chemical Mutagenesis, State University of Leiden, 2333 AL Leiden, The Netherlands

Jan C. Liang, Department of Cell Biology, University of Texas System Cancer Center, M. D. Anderson Hospital and Tumor Institute, Houston, Texas 77030

J. T. MacGregor, United States Department of Agriculture, Western Regional Research Center, Berkeley, California 94710

Peter Maier, Institute of Toxicology, Swiss Federal Institute of Technology and University of Zurich, Schwerzenbach, Switzerland

James A. Marx, Department of Medical Genetics, University of Wisconsin-Madison, Madison, Wisconsin 53706

G. R. Mohn, Department of Radiation Genetics and Chemical Mutagenesis, State University of Leiden, 2333 AL Leiden, The Netherlands

H. F. Mower, Department of Biochemistry and Biophysics and Cancer Center of Hawaii, University of Hawaii, Honolulu, Hawaii 96822

Carmen Pueyo, Departmento de Genética, Universidad de Extremadura, Facultad de Ciencias, Badajoz, Spain. *Present address:* Departamento de Genética, Facultad de Ciencias, Universidad de Córdoba, Córdoba, Spain

Manuel Ruiz-Rubio, Departmento de Genética, Universidad de Extremadura, Facultad de Ciencias, Badajoz, Spain. *Present address:* Departamento de Génetica, Facultad de Ciencias, Universidad de Córdoba, Córdoba, Spain

L. E. Sacks, United States Department of Agriculture, Western Regional Research Center, Berkeley, California 94710

Daniel H. Stuermer, Biomedical and Environmental Research Program, Lawrence Livermore National Laboratory, University of California, Livermore, California 94550

P. A. van Elburg, Department of Radiation Genetics and Chemical Mutagenesis, State University of Leiden, 2333 AL Leiden, The Netherlands

Preface

Volume 9 of *Chemical Mutagens* consists mainly of chapters discussing the development and validation of short-term assays to detect the mutagenic effects of environmental chemicals. These chapters include an assay with the grasshopper neuroblast, a comparison of mutagenic responses of human lung-derived and skin-derived diploid fibroblasts, a forward-mutation assay in *Salmonella*, a multigene sporulation test in *Bacillus subtilis*, a specific locus assay in mouse lymphoma cells, a study of the induction of bacteriophage lambda, and the granuloma pouch assay. In addition, there are two chapters on the identification of mutagens in cooked food and in human feces.

Frederick J. de Serres

Research Triangle Park, North Carolina

Preface

Contents

Chapter 2

**Comparison of the Mutagenic Responses of Lung-Derived
and Skin-Derived Human Diploid Fibroblast Populations   67**

Lois Jacobs, James A. Marx, and Christian L. Bean

Chapter 3

**The L-Arabinose Resistance Test with *Salmonella
typhimurium*   89**

Carmen Pueyo and Manuel Ruiz-Rubio

Chapter 4
Identification of Mutagens from the Cooking of Food　111
**Frederick T. Hatch, James S. Felton, Daniel H. Stuermer, and
Leonard F. Bjeldanes**

Chapter 5

The *Bacillus subtilis* Multigene Sporulation Test for Detection of Environmental Mutagens   165

L. E. Sacks and J. T. MacGregor

Chapter 6

The L5178Y/TK Gene Mutation Assay System   183

David E. Amacher

Chapter 7

Induction of Bacteriophage Lambda by DNA-Interacting Chemicals 213

Rosalie K. Elespuru

Chapter 8

The Granuloma Pouch Assay 233

Peter Maier

Chapter 9

The Use of Multiply Marked *Escherichia coli* K12 Strains in the Host-Mediated Assay   261

G. R. Mohn, P. R. M. Kerklaan, and P. A. van Elburg

Chapter 10

The Detection of Mutagens in Human Feces as an Approach to the Discovery of Causes of Colon Cancer   283

H. F. Mower

CHAPTER 1

The Grasshopper Neuroblast Short-Term Assay for Evaluating the Effects of Environmental Chemicals on Chromosomes and Cell Kinetics

Mary Esther Gaulden, Jan C. Liang, and Martha J. Ferguson

1. Introduction

The grasshopper neuroblast (GHNb) is a newcomer to the library of tests available for evaluating the mutagenicity of environmental chemicals. Most of the current tests have been in use since the beginning of the present era of active research on the identification of environmental mutagens and carcinogens, which began to attain international momentum in the late 1960s.[28,56] Why, then, did we recently develop another assay? First, the neuroblast (Nb) of the grasshopper *Chortophaga viridifasciata* (De Geer) has been shown to be very sensitive to X rays (the effects of doses as low as 1 rad on chromosome breakage and on mitotic

Mary Esther Gaulden and Martha J. Ferguson • Radiation Biology Section, Department of Radiology, University of Texas Health Science Center at Dallas, Dallas, Texas 75235.   **Jan C. Liang** • Department of Cell Biology, University of Texas System Cancer Center, M. D. Anderson Hospital and Tumor Institute, Houston, Texas 77030.

1

rate can be detected[42,43]), so we reasoned that it might also be very sensitive to chemical mutagens. Second, the fact that the spontaneous chromosome aberration frequency in the GHNb is zero means that significant data on mutagens can be obtained with a minimum number of cells. Third, the GHNb has a short cell cycle[44] with a number of well-defined phases, and thus, much information about the effects of agents on cell progression can be obtained. This aspect of environmental mutagen action has received relatively little attention and is of considerable relevance to teratogenesis.[36] The short cell cycle (*Chortophaga*, 4 hr; *Melanoplus sanguinipes*, 2 hr, 38°C) is also advantageous for testing chemicals with short half-lives. Fourth, the GHNb is a simple, fast, reproducible, and inexpensive eukaryotic test system. No single assay developed to date is ideal for estimating the risks of environmental chemicals for humans, so a battery of systems is required, and the need to search for good ones is still with us.

Grasshopper cells have long been used for chromosome studies (see Ref. 39). Initially, germ cells were the focus of attention and were used by McClung[79] to first show that specific chromosomes determine sex. Later, J. G. Carlson, a student of McClung, undertook a study of the somatic cell chromosomes of grasshoppers, which led him eventually to work on Nbs. His first studies were done at the Cold Spring Harbor Laboratory in New York State, and *Chortophaga* was the only adult species available in the field when he arrived in early summer. This species is one of the few that is multivoltine (produces several broods a year). In other words, the embryo of this species does not have a diapause phase, i.e., a genetically determined cessation of development [128] that is usually broken by prolonged exposure to low temperature. Thus, *Chortophaga* embryos develop straight through to hatching, thereby enabling an investigator to obtain 4–6 generations a year in the laboratory.[23] Subsequently, Carlson, his colleagues, and his students have studied extensively the chromosomes and cell cycle of the living as well as the fixed Nb of *Chortophaga*, with emphasis on radiation effects. These studies provide a valuable data base as background for chemical mutagen studies.*

One attribute of the GHNb that commends it for testing is its embryonic origin. Of all the systems currently employed for mutagen testing, only one of the more widely used involves a cell of primary embryonic origin (dominant lethal test with early mammalian embryo), and it is time-consuming. In the life history of an organism, embryonic cells are among the most sensitive to ionizing radiation and probably

---

* A complete list of references for these studies is available from the senior author.

to chemical mutagens. Further, in the embryo and fetus there is evidence that different cell types have different sensitivities to mutagens, with Nbs (stem cells for the nervous system) being among the most sensitive, including those of the human.[40,45,59] A detailed rationale has been presented for the view that exposure of Nbs *in vivo* to small doses of mutagens may give rise, by chromosome aberration induction, to subtle teratogenesis of the central nervous system (CNS) in humans, resulting in functional defects.[36,40] The results obtained with a short-term mutagen test on the sensitive embryonic Nbs of the grasshopper may, therefore, have relevance to the hazards of environmental chemicals to human embryos with respect to teratogenesis as well as to mutagenesis and carcinogenesis.

The purpose of this chapter is to provide the detailed information that an investigator, unfamiliar with GHNb methods, needs in order to obtain data on chemical mutagens with a minimum of startup time. Recent work in our laboratory has shown that good rearing conditions for a grasshopper colony are essential for a constant supply of normal embryos with no spontaneous chromosome aberrations, so rearing methods are described. Previous reviews provide some of the Nb techniques[16,20,39]; details of methods pertinent to the exposure of Nbs to chemical mutagens will be given here not only for *Chortophaga*, but also for a nondiapausing strain of *Melanoplus sanguinipes*[98] that we have recently begun to study. In addition to the methods for examining chromosome aberrations and cell cycle effects, those that permit detection of other endpoints in the GHNb are also described, namely, spindle abnormalities, unscheduled DNA synthesis, and effects on normal DNA, RNA, and protein synthesis. A summary of some of the data obtained with chemicals is included.

The details given here are probably applicable with minor variations to other species of grasshopper. Grasshopper embryo development and Nb characteristics have been shown to be similar for several species, so it can be reasonably assumed that the early embryonic development in other species of grasshopper is essentially the same,[3] except perhaps for the time scale. If this is the case, the widespread distribution of grasshoppers in many parts of the world makes the GHNb technique available to investigators through the use of native species.

It should be noted that eggs of *Chortophaga* and *M. sanguinipes* survive mailing conditions quite well if they are not subjected to extreme temperatures. We will be glad to send a starter supply of eggs from our surplus to investigators who wish to initiate a colony. Dr. J. E. Henry (personal communication) tells us that he will send starter egg pods of *M. sanguinipes* when his laboratory has an excess, or that under a cooperative agreement, an investigator could be sent eggs at reasonably regular intervals.*

2. Embryo Supply

For mutagen testing, grasshopper embryos are needed year-round, so a constant supply of mature adults is required, the size of the colony being dictated by the number of embryos needed. It is therefore necessary to maintain a laboratory for rearing and maintaining egg-producing animals. With attention to a few details about food, light, temperature, and cleanliness, this can be accomplished with a minimum of time, effort, and expense.

2.1. Species

The two species we use are *Chortophaga viridifasciata* and *Melanoplus sanguinipes* (family: Acrididae; order: Orthoptera). In contrast to *Chortophaga*, few cell data on the Nb, much less other cell types, are available in the literature for *M. sanguinipes* [formerly *M. mexicanus mexicanus* (Sauss.) and *M. bilaturatus* (Walker)], which is the so-called migratory grasshopper of North America. Because of its economic importance to agriculture, *M. sanguinipes* has been much studied in other respects, e.g., embryonic development,[103] fecundity,[97] food preferences,[96] physiology,[106,131] toxic responses,[82] and sensitivity to plant growth hormones.[25] Such information is useful in establishing and maintaining a healthy colony. Of the two species, *M. sanguinipes* is the faster growing and is the more vigorous in the laboratory. Its appetite is also more voracious.

2.2. Origin of Colonies

Chortophaga viridifasciata (subfamily: Oedipodinae†) is found in the wild in eastern North America from southern Ontario to Georgia and is abundant as far west as an area bounded by a line transecting the eastern portions of Saskatchewan, Oklahoma, and Texas (approximately 50 miles east of Dallas). In the southernmost regions of its range,

---

* Dr. J. E. Henry, Rangeland Insect Laboratory, U.S. Department of Agriculture, Montana State University, Bozeman, Montana 59717.
† Two additional volumes projected by Otte[94] for a definitive treatise on North American grasshoppers will bring up to date the taxonomy of the Oedipodinae.

Chortophaga produces three generations a year, one each in spring, summer, and fall. Because of availability, we have previously used field animals for embryo supply by bringing nymphs and adults into the laboratory. Recently we have begun to establish a laboratory colony. Dr. Kenya Kawamura of the College of Agriculture in Hokkaido, Japan, informs us that he has a colony of *Chortophaga viridifasciata* in his laboratory derived from animals he obtained while in Tennessee in the late 1950s. Dr. Saralee N. Visscher of Montana State University has recently established a colony in her laboratory (personal communication), also from animals collected in Tennessee. Experience in three laboratories shows, therefore, that even though *Chortophaga* is one of the less hardy grasshoppers,[23] it can be bred satisfactorily under laboratory conditions. To avoid excessive inbreeding, we recommend occasional introduction of animals from the field to the colony.

We obtained eggs of *Melanoplus sanguinipes* (Fabricus) (subfamily: Melanoplinae) in 1980 from Drs. G. B. Staal and M. P. Pener of Zoecon Corp., whose colony was derived from the original nondiapausing strain developed by Pickford and Randell.[98] Species of *Melanoplus* in nature are univoltine; the embryos have an obligatory diapause period. Pickford and Randell had observed that in the laboatory a few eggs developed without pause to hatching after incubation at 30°C with no cooling. By selecting adults from such eggs, they were able over a period of 12 years to establish a vigorous colony of a nondiapausing strain of *M. sanguinipes*. It might be noted that Slifer and King,[119] using the same methods, had previously developed a nondiapausing strain of the much studied *M. differentialis* (Thomas). Dr. Bruce Nicklas had, to our knowledge, the only surviving colony of this strain, but he reports that it is now extinct (personal communication).

2.3. Life Cycle

The life cycle of the grasshopper consists of three phases: egg, nymph, and adult. Under the laboratory conditions for rearing grasshoppers given in Section 2.4, the durations of the egg and nymph phases of *Chortophaga* are 6 weeks each; adults survive for 6–8 weeks. The egg and nymph phases of *M. sanguinipes* are shorter, 3–4 weeks, but the life span of adults is comparable to that of *Chortophaga*.

At the time of hatching, the vermiform larva is enveloped by a membrane, thin and transparent, which serves as a provisional cuticle; it is a real cuticle in that it is acellular and chitinous. As soon as the larva reaches the soil surface, it undergoes its first molt, called the intermediate molt, and sheds the provisional cuticle, which when dry is a tiny white mass that can be found near the egg case. This first instar nymph (hatchling) can hop, but for the first few days does not move very far. For all subsequent molts the nymph grasps the screen of the cage with its jumping legs and with head down wriggles out of the old cuticle. Dry grass or sticks placed in the cage are helpful to the animals in molting. *Chortophaga* undergoes a total of seven molts for six nymph instars and the adult instar. *M. sanguinipes* has a total of five instars. Less than optimal amounts or types of food plant can increase the number of nymph instars in *M. sanguinipes*.[120]

2.4. Colony Maintenance

2.4.1. Cages

Grasshoppers are kept in screened cages to permit maximum air circulation. A variety of cages that can be hand-constructed has been described (e.g., Refs. 20, 23, 57, 78, 132); we use commercial cages that can be collapsed for autoclaving (American Biological Supply Company, Baltimore, Maryland). Small cages, 12 inches square, are preferable for rearing the nymphs. Large ones, 18 inches square, are best for adults when modified as follows. The solid metal bottom is discarded, and the cage is placed over a platform slightly larger than the base of the cage. The platform consists of a sheet of wire screen (8 mesh) tacked to a wooden frame that is supported at each corner by wooden feet 10 cm high. Two holes, 8.5 cm in diameter, are cut in the area of the screen that will be near the front of the assembled cage, so that the tops of egg-laying containers (8.2 cm × 10 cm polypropylene jars, Mallinckrodt) can be inserted level with the floor of the platform. The height of the platform feet is therefore determined by the height of the egg-laying containers. Females do not lay unless the surface of the sand is flush with the cage floor.

For field collecting, we use a cage with a wooden frame, as described previously,[20] which has a handle for ease of carrying.

Handling of the early instar nymphs causes high mortality, so hatchlings are best introduced to a nymph cage by placing an egg-laying container in the cage. When a container is left in a cage, the sand should be kept damp (not wet!). If one wishes to obtain hatchlings from egg pods that had previously been removed from egg-laying containers, such cases or even isolated eggs can be kept on top of a 5-mm-thick layer of damp, autoclaved sand in a covered petri dish (2 × 10 cm). Accumulation of water droplets on the inner surface of the lid can be prevented by placing on top of it another dish of sand. This is necessary because jumping hatchlings are trapped by the water droplets and drown. Hatchlings should not be left in closed containers for more than 24 hr, because they begin to feed 1 day after hatching and mortality is high if they do not have immediate access to food. Shallow dishes dry out quickly, which is deleterious to the unhatched eggs, so they should be left open in the cage only long enough to allow the recently hatched animals to hop out.

Overcrowding of animals in cages should be avoided, as this can reduce egg laying and result in cannibalism. About 50 males and 50 females is the maximum number of animals that should be put in an adult cage and about 75 animals in a nymph cage. Reproduction in insects is controlled not only by exogenous factors, such as light, temperature, and food, but also by endogenous factors, such as neurohormones and pheromones,[54,58] as well as by the interplay of the various factors. Although the molecular mechanisms are not always known, it has been shown that crowding in the Acrididae reduces life span and fecundity in adults and ovariole numbers in offspring.[1,2,91,122] Reduction in the rate and increase in the variability of embryonic development have been shown by Van Horn[132] to result when the number of adult pairs per cage is increased for the grasshopper *Aulocara elliotti* (Thomas). Some of the molecular events that regulate insect reproduction are being elucidated. For example, the production of vitellin, the main yolk protein of insects, has been shown in the African migratory locust, *Locusta migratoria migratorioides*, to be regulated by the juvenile hormone.[24] This hormone is also important to the normal metamorphosis of insects, so factors affecting juvenile hormone production could have profound reproductive consequences. These points emphasize the fact that investment in the maintenance of optimal rearing conditions, which requires a minimum of effort, can pay big dividends in egg and embryo yields.

2.4.2. Cleanliness; Pests

Cleanliness is an absolute necessity for the maintenance of a reproductively vigorous grasshopper colony and for optimal embryonic development. Nymph cages, which have solid floors, should be cleared of debris each day. Although access to occupied cages is limited by a surgical stockinette sleeve, debris can be easily removed through it without escape of animals if a heat lamp is aimed at the opposite side of the cage (grasshoppers are positively phototropic). Adult cages with screen floors permit droppings to fall onto a disposable sheet of paper; dead grasshoppers and food debris should, however, be removed from the cage each day. Cages should be washed and autoclaved before new groups of animals are placed in them.

Good rearing conditions are necessary to keep the spontaneous chromosome aberration frequency at zero. Neuroblasts were examined in 95 "normal", untreated embryos obtained from eggs laid by *M. sanguinipes* in cages cleaned only about twice a month and which had fresh containers of seedlings added about every 3 days. Out of 821 Nbs in late anaphase–very early telophase that were scored, five of them had a total of seven breaks (acentric chromosome fragments), i.e., 0.0085 breaks per cell. No aberrations have been found in the Nbs of normal untreated embryos from eggs laid by *M. sanguinipes* reared under optimal conditions (Section 2.4.3). In *Chortophaga* we have observed only two chromosome aberrations (acentric fragments) in thousands of untreated Nbs, and although exact rearing conditions were not documented, we suspect that they were derived from animals that did not have fresh food every day.

Containers of used seedlings should be removed from the animal room when discarded, because they provide a breeding locale for *Drosophila* and an unidentified "gnat," both of which can be pests in the grasshopper laboratory. These insects are attracted by the yeasts and microorganisms that grow in the wet seeds. To eliminate them, several strips of fly paper (with no insecticides) are hung around the growing seedlings; one strip is also hung in the 2-ft-square cage in which the seeds are initially soaked (some insect pests are small enough to enter the cage through the screen).

The ubiquitous cockroach can be a nuisance because the usual insecticide sprays that are effective against it cannot be used in a grasshopper laboratory. We have found that strategic placement of a number of flat dishes containing a dry mixture of three parts boric acid crystals to one part crystalline sugar will keep roaches under control.

2.4.3. Food, Light, and Temperature

As has been noted by Reese,[101] the best food plant for insects is one capable "of supporting growth, development and reproduction." If fed on leaves from a variety of plants, late instar nymphs and adults of *Chortophaga* collected in the field will lay fertile eggs in the laboratory for a supply of 14-day-old embryos. But the optimal food must be supplied if eggs are to be obtained that have a high hatchability rate and that yield survivable nymphs that develop into reproducing adults of maximum longevity.[23] We have found that the leaves of wheat

TABLE 1. Dry Food Mixture for Grasshoppers

Wheat germ[a]	1002 g
Cholesterol	9 g
L-Cystine	4 g
Glycine	6 g
Choline chloride	4 g
Yeast extract	55 g
Ascorbic acid	20 g
Total	1100 g

[a] Vanderzant-Adkisson special wheat germ diet for insects (ICN Nutritional Biochemicals): wheat germ, 24%; vitamin-free casein, 28%; D-sucrose, 27.75%; salt mixture, 8%; Alphacel, 12%; cholesterol, 0.05%; linseed oil, 0.2%.

seedlings are satisfactory, but that those of the seedlings of fall rye (*Secale cereale* L.) are the best food source for *Chortophaga*. Use of oat seedling leaves as a food source, according to Carothers,[23] will completely prevent reproduction of *Chortophaga*. This adverse effect of oats may possibly be caused by particular plant constituents, such as nitrogen and phosphorus[66,123] and plant growth hormones.[25,133,134] Smith[120] has reported that oats as food for *M. sanguinipes* prolongs the nymphal growth period and causes them to go through an extra instar; he attributed these effects to the fact that the animals would not eat as much oat as wheat leaves. The fecundity of *Melanoplus* is good when a wide variety of plants is provided as food,[96] but we have found that the fertility of the eggs is higher and the variability of embryo ages in a given egg case is reduced if nymphs and adults are fed rye seedling leaves. It is important that fresh food be available daily for both species; feeding *M. sanguinipes* every other day halves egg production.[121]

A dry food mixture (Table 1), modified from that described by Kreasky,[66] is supplied along with seedling leaves for both species of grasshopper. For nymphs a thin layer of this mixture is spread on the floor of the cage, so that second instar nymphs, which do not move about, can find it easily. This is especially important in the case of *Chortophaga*, because such a distribution increases considerably the survival of the second and third instar nymphs. In adult cages, the mixture is supplied in several open petri dishes (15 × 100 mm diameter).

Seedlings 7–10 days old are obtained as follows. Seeds of rye (*Secale vulgaris*), untreated with insecticides and fungicides, are available in bulk from Sexaur Seed Company, Brookings, South Dakota. Prior to use, the dry seeds are stored in closed plastic containers at 4°C or at 17°C to prevent hatching of the eggs of any insect pest with which they

**TABLE 2. Smith's Nutrient Solution for
Seedlings[120]**

Solution 1:	Water	1 liter
	H_3BO_3	2.8 g
	$MnSO_4 \cdot 4H_2O$	2.1 g
	$ZnSO_4 \cdot 7H_2O$	0.2 g
	$CuSO_4 \cdot 5H_2O$	0.08 g
	$H_2MoO_4 \cdot H_2O$	0.02 g
Solution 2:	Water	999 ml
	KH_2PO_4	0.14 g
	$MgSO_4 \cdot 7H_2O$	0.49 g
	KNO_3	0.56 g
	$Ca(NO_3)_2 \cdot 4H_2O$	1.2 g
	Solution 1	1 ml

may be infested. Seeds are soaked by placing approximately 30 ml in a wide-mouth polypropylene jar (82 mm diameter, 63 mm high, Mallinckrodt) with enough water to barely cover them. Approximately 24 hr later the germinating seeds are transferred to a 1-inch layer of damp, autoclaved sand in a similar jar with 6–10 holes (2 mm diameter) bored in the bottom and also midway around the sides; enough sand is then added to just cover the seeds and is kept damp with Smith's nutrient solution 2 (Table 2).

Watering the seedlings with the nutrient solution prolongs the life of the seedlings and enhances their nutritional value for grasshoppers; sand has a negligible amount of nutrients. After 7–10 days the seedlings are 8–10 inches tall and are ready for use. Six such dishes of fresh seedlings are kept in an adult cage and two in a nymph cage, which is enough to feed about 100 animals in each cage per day. The seedling jars are set in petri dish lids to prevent leaking water from getting on the dry food mix or on droppings. Unwilted, fresh seedlings should be used, because they serve as the main source of water for the animals.[95] Cotton soaked in water in a petri dish may also be made available to the hoppers, but it should be changed daily. *Chortophaga* is relatively intolerant to dessication and cannot survive if body water content drops below approximately 57%.[75] In other words, this species cannot survive on dry food alone even under conditions of high humidity.

Experiments in the laboratory and observations in the field by Otte and Williams[95] indicate that color phase in *Chortophaga* is primarily determined by water intake. One can, therefore, easily determine whether *Chortophaga* in the laboratory are getting optimal amounts of water; if so, there will be some green nymphs and some green adult females; if not, all animals will be brown. In the field, adult female *Chortophaga* occur in both green and brown color phases, greater proportions of green ones occurring when vegetation is predominantly green and humidity is reasonably high. Adult *Chortophaga* males in the field are almost always brown.

Light for the nymphal and adult grasshoppers and for the seedlings is automatically cycled on (16 hr) and off (8 hr) and is provided by two "Gro & Sho" plant fluorescent bulbs (General Electric, F40-PL) suspended 6–12 inches above the cages or the seedlings. In addition, fluorescent lights in the ceiling of the room are on a similar cycle.

The temperatures at which grasshoppers and their food are grown influences reproduction and longevity.[135] In our laboratory, room temperature is kept at about 27°C during the light period and at about 22°C during the dark period. Ideally, temperature during the dark period should be about 18°C, but conditions in our laboratory do not permit cycling to this lower temperature. For 8 hr of the light period, the front surface of a heat lamp (250 W infrared lamp, General Electric) is placed about 6 inches from one side of a cage to give the animals (nymphs and adults) additional optional heat, which reduces fungal infections in them. The optimal temperature produced at the side of the cage by the heat lamp is 45°C, which can be controlled by a rheostat in the line. The animals will not congregate on the screen at temperatures higher than 50°C.

The longer the light period and the higher the temperature, the greater the production of eggs, but the shorter the life of the adult. The light–dark periods and the temperatures given here allow for a high continuous rate of egg production. Decreases in the light and temperature can be manipulated to reduce egg production and lengthen the life span of the adults.

2.4.4. Egg Collection and Storage

For oviposition, the female grasshopper forces the extended length of her abdomen into soil. As she begins to withdraw slowly, she deposits a substance that she froths by agitating it with the end of her abdomen. She then lays eggs. The froth subsequently dries to form what is called an egg pod or egg case. She caps the egg pod with about a 5-mm plug of froth only. The froth not only holds the eggs together but also serves to reduce water loss through evaporation. The total time for oviposition in both *Chortophaga* and *M. sanguinipes* is 20–30 min.

To obtain hatchlings for maintenance of the colony, several containers in which eggs have been laid should be kept at room temperature and placed in nymph cages near the time estimated for the beginning of hatching. For this purpose it is best to use eggs laid about 2–3 weeks after a brood of females begins to lay. The first few pods laid are more likely to contain a higher proportion of unfertilized eggs than those laid subsequently. There is also less variability in embryonic development and a higher rate of hatching for eggs from females who are several weeks into the adult instar.[97,132]

To obtain a supply of embryos for experimental studies, we harvest egg pods three times weekly as follows. Eggs of both species are fragile for the first few days after oviposition, so they must be handled gently to prevent puncture by grains of sand. The top 2 inches of sand in the egg-laying containers is dumped, by vigorous shaking of the container, into one end of a shallow pan. To harvest *Chortophaga* pods, the sand is carefully raked with one finger or sifted through course wire to facilitate location of the pods. *Melanoplus* eggs are especially fragile, so the best method for collecting the egg pods is to apply a small stream of water from a squirt bottle to the sand while holding one end of the pan a few inches above the table. The water washes the sand to the lower end of the pan, leaving the egg pods exposed.

For storage, the egg pods are placed on top of a 5-mm layer of *damp*, autoclaved sand in petri dishes (20 × 100 mm), and the date of collection is noted on the lid. The egg pods or eggs removed from them should be kept damp but not covered with water; submersion retards development. Distilled water should be used to prevent the buildup of salts in the sand, which will alter osmotic pressure. This practice will ensure a more uniform intraegg osmotic pressure and will greatly reduce the need to adjust the osmotic pressure of the culture medium for embryos of different egg pods (Section 4.1.2).

Egg pods should be checked as frequently as necessary to ensure that they remain damp. Pods that develop fungal growth should be discarded. Rough handling of the pods during harvesting will rupture some eggs and encourage fungal and bacterial growth. Toxins produced by such organisms may affect the development of embryos of intact eggs in the affected pod and neighboring ones.

Embryonic development can be slowed by storage at 17°C (without detrimental effects) for the 3 weeks (*Melanoplus*) to 6 weeks (*Chortophaga*) it takes the embryos to reach the stage at which they are used for Nb studies. This extends the time a supply of eggs can be used for experiments. If kept solely at 17°C, *Chortophaga* embryos will continue developing but will not hatch, whereas *Melanoplus* will hatch. If *Chor-*

tophaga embryos are returned to room temperature after 2 months storage at 17°C, some of them will hatch, but the females will not lay eggs on reaching maturity. *Melanoplus* embryos similarly handled will develop into egg-laying females; we have not studied the effects on absolute fertility but have the impression that it is negligible. We recommend that those eggs that will be used for obtaining hatchlings (both species) should never be stored at temperatures below 24°C. The egg pods of both species may be stored at room temperature (25°C) or at 30°C to accelerate development.

Chortophaga eggs may be stored at 11°C to stop development completely. Storage at 11°C is deleterious to embryos at the earlier stages, but if the eggs are not put at this temperature until they have reached the stage equivalent to 11–12 days of development at room temperature, they can be stored for up to 3 months at 11°C without adverse effects on the continuation of development (to at least the 14- to 15-day stage) when the eggs are returned to room temperature. We have not examined the effects of 11°C storage on *Melanoplus* embryos.

2.5. Pathology

Two parasites are capable of wiping out a grasshopper colony: a mold, probably *Entomophaga grylli*,[89] whose occurrence we have found can be prevented by the use of a heat lamp, as noted in Section 2.4.3, and an amebic protozoan, *Malameba locustae*. These organisms usually enter a colony via the introduction of field specimens. *Malameba* can be controlled by a triple sulfonamide, thaipyrimeth.[53] It is possible that the sulfonamide might cause chromosome aberrations, but we have not used it because we have not had any problems with *Malameba* in our colony.

A virus of the poxvirus group has been described as occurring in *M. sanguinipes*.[52,68] This is an entomopoxvirus that proliferates in the fat bodies, prolongs development, and induces a "general torpor" and has a very high mortality rate.[52]

Soil nematodes (undetermined species) may occur in sand dishes containing egg pods. Although we have not studied their effects on embryonic development, it is our impression that division is slower in embryos from egg pods in dishes that have nematodes, who evidently feed on decaying eggs (unfertilized or broken). The worms can be eliminated by autoclaving the sand used for eggs.

Parasitization of *Chortophaga* eggs by the hymenopteran wasp *Scelio bisulcus* (Ashmead) has been reported for eggs laid in the laboratory by field-collected grasshoppers; whether the wasp eggs were deposited in the adult female grasshopper or directly in the grasshopper eggs could not be determined.[126] Carlson has on one occasion found a single larva of an unidentified species in several field-collected adult *Chortophaga* males (personal communication); the large larva almost filled the body cavity. We have not observed either of these parasitic larvae in our laboratory colony.

2.6. Allergy to Grasshoppers

Grasshoppers should be reared in a room dedicated to colony maintenance and one in which the air is maintained at negative pressure with respect to the surrounding rooms. This is necessary to reduce exposure of personnel to grasshopper-associated allergens, which occur mostly in dust form. The sources of these allergens include the grasshoppers themselves, the dried food mix, the droppings of the hoppers, possibly the cuticles discarded with each molt of the nymphs, the grain seeds used for food, and dried seedling leaves.

Unnecessary exposure of the respiratory track and the skin to these allergens can bring discomfort to individuals caring for the animals; reaction may become severe enough to prevent work with grasshoppers.[110] The degree of allergic response depends on the inherent sensitivity of the individual, the amount of allergen, and the length of exposure time. It is our experience that without protective devices, most people acquire with time some degree of sensitivity to grasshopper-associated allergens, the most extreme situation being the simultaneous development of skin irritation, conjunctivitis, and lung congestion that results from a contact as simple as a brief visit to a grasshopper room or the handling of one grasshopper in the field.

To avoid such undesirable complications, the worker should wear a nontoxic-particle mask, disposable surgical gloves, and a lab coat that is removed on leaving the room. Individuals with high sensitivity should use a tight-fitting dust respirator and wear disposable surgical caps in addition to gloves and lab coat.

3. Grasshopper Egg, Embryo, and Cells

3.1. The Egg Shell and Membranes

The laid eggs of *Chortophaga* and *M. sanguinipes* are easily distinguished by size and by the shape of the posterior end (Figure 1). The embryo of both species is surrounded by several layers of membranes

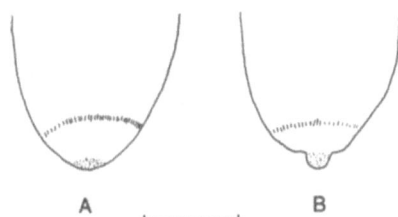

FIGURE 1. Diagrams of posterior ends of eggs of (A) *Chortophaga viridifasciata* and (B) *Melanoplus sanguinipes*. Scale: 0.5 mm.

and other material, some being formed before and some after oviposition.[51,55,113,131] The egg shell and membranes have been described for several species of grasshopper, including *Melanoplus*; although not studied in *Chortophaga*, they appear to be similar to those in *Melanoplus*.

Before the egg is laid, it is covered with a vitelline membrane that is laid down on the surface of the yolk. On top of this membrane are added two layers of a nonchitinous material called the chorion (Figure 2) that is secreted by the follicular cells in the ovarioles of the ovary. As the egg passes into the common oviduct, another layer, the thin extrachorion, is secreted onto the chorion. Note that none of these structures is of embryonic origin; the oocyte does not complete meiosis until after oviposition. About 1 week after the egg is laid, the vitelline membrane disappears and the first true embryonic membrane, the serosa, begins to appear on the surface of the yolk near the primitive

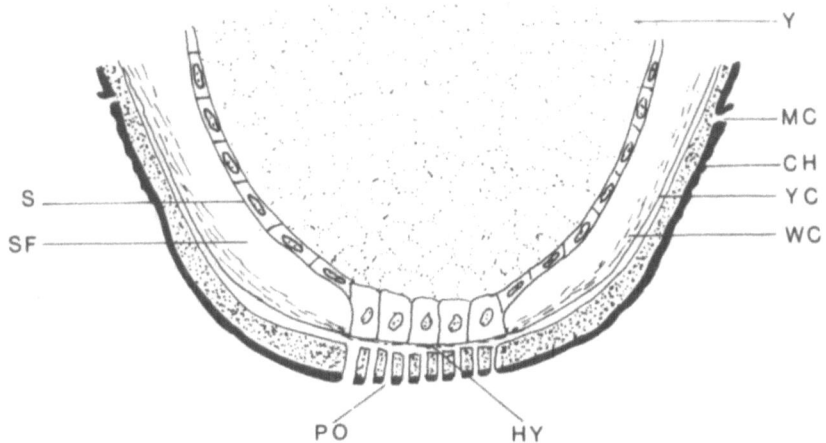

FIGURE 2. Section through posterior end of grasshopper egg. After Uvarov.[131] Y, Yolk; MC, micropyle; CH, chorion; YC, yellow cuticle; WC, white cuticle; HY, hydropyle; PO, pore; SF, serosal fluid; S, serosa.

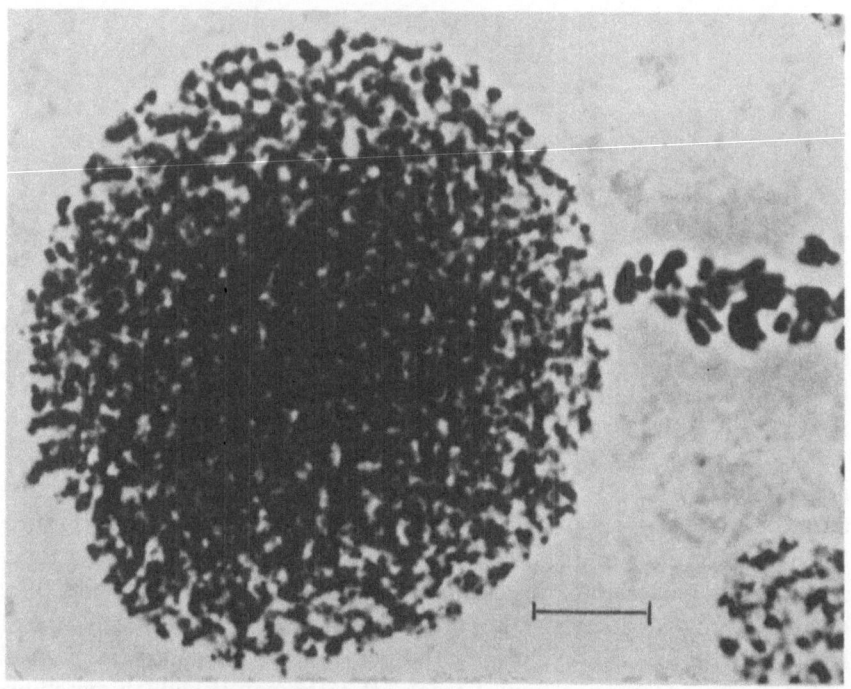

FIGURE 3. (Left) Nucleus of serosal cell and (right) metaphase chromosomes of neuroblast. Cells are somewhat flattened. *Melanoplus sanguinipes*. Scale: 30 μm.

embryo (blastoderm). Subsequently the cells of the serosa completely envelop the yolk and become highly polyploid, probably through endoreplication (Figures 3 and 4). The serosa secretes two membranes on its outer surface: a thin yellow cuticle and then a thick white cuticle consisting mainly of chitin. These cuticles lie close to the chorion[113] (Figure 2).

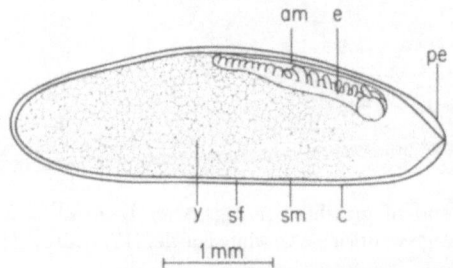

FIGURE 4. Diagram of 14-day old egg of *Chortophaga viridifasciata*. am, amnion; e, embryo; pe, posterior end; c, chorion; sm, serosal membrane; sf, serosal fluid; y, yolk. (From Gaulden and Kokomoor.[38])

Fertilization occurs at oviposition, the sperm gaining entrance to the egg through the micropyle, a ring of pores (60–65 in *Chortophaga*, 40–45 in *M. sanguinipes*) near the posterior end of the egg (Figures 1 and 2). A narrow canal that passes obliquely through the chorion connects each pore with the vitelline membrane in early development and with the serosal fluid (between serosa and the yellow cuticle) later in development.

At the posterior tip of the egg is a tiny, sievelike area consisting of many minute pores, each of which is the entrance to a canal through the chorion to a thin region of the yellow and white cuticle, called the hydropyle, that admits water to the egg[114,118] (Figure 2). That portion of the serosa underlying the hydropyle has specialized cells that probably regulate water uptake by the embryo. Waxy materials in the chorion and chitinous cuticle make most of the egg shell impermeable to water.[118] Oxygen reaches the embryo through the water that enters the egg through the hydropyle; some may also gain admission through the egg shell, but this point is still equivocal.[55,131]

The chorion can be cleared by submersion of the egg for 2 min in a 3% sodium hypochlorite solution or commercial bleach.[10,116] This permits observation of the embryo through the relatively clear cuticles and serosa. Slifer[116] reported that embryos of cleared eggs of *Melanoplus differentialis* will continue to grow and will hatch if kept in water. Whether this holds true for *Chortophaga* and *M. sanguinipes* has not been determined.

3.2. Embryonic Development

3.2.1. Normal Development

In grasshopper eggs the oocyte nucleus lies in the yolk in the posterior end of the egg beneath the micropyle.[83] Meiosis is completed approximately 4–6 hr after oviposition (25°C), and the male and female pronuclei have fused and cleavage begun by 24 hr. Cleavage nuclei migrate to the periphery of the yolk and begin forming the blastoderm at the concave side of the egg. An invagination at each edge of the blastoderm produces folds that grow to meet over the midline of the embryo. The inner layer forms the amniotic membrane, which is attached to the edge of the embryo and covers its ventral surface; the outer layer forms the serosal membrane, which envelopes the embryo and the yolk.[62] By about 5 days the embryo is detectable to the naked eye as a tiny group of cells.

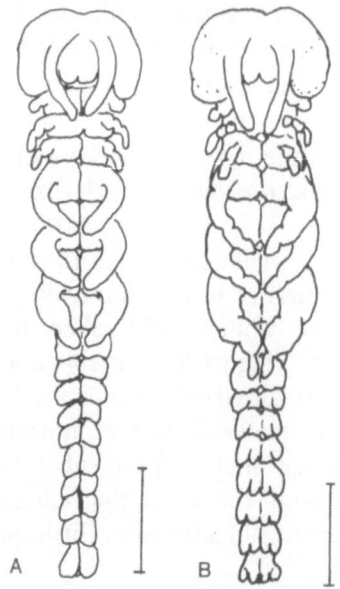

FIGURE 5. Diagrams of embryos separated from membranes. (A) Fourteen-day-old *Chortophaga viridifasciata*. Scale: 0.45 mm. (B) Nine-day-old *Melanoplus sanguinipes*. Scale: 0.42 mm.

The developmental staging of the *Chortophaga* embryo at 25°C is similar to that described by Slifer[112] for *M. differentialis*, except that the latter species enters diapause at 20 days but *Chortophaga* continues developing. The stage that is best for Nb studies in the *Chortophaga* embryo is comparable to the one midway between stages 10 and 11 of Slifer[112] and is reached at 14 days of development at 25°C (Figure 5A). The rate of embryonic development in *M. sanguinipes* is faster than in *Chortophaga*, so the optimal stage in this species is reached in 9 days at 25°C[107] (Figure 5B) or 6.5 days at 30°C[103] and is morphologically similar to a 15-day-old *Chortophaga* embryo. Hereinafter, embryo stages cited can be assumed to be those at 25°C.

Development of all the eggs in a pod is synchronous in *Chortophaga*, so that under ideal rearing conditions and with pods collected during the period of optimal fertility (Section 2.4.4), development among the 15–30 eggs in a pod will not vary by more than 0.5 day. Riegert[103] reported a similar high degree of synchrony in the nondiapausing strain of *M. sanguinipes* from which our colony is derived. We have found, however, that there can be considerable variation in development among the embryos in the 15–30 eggs in a *M. sanguinipes* egg pod, and that infertile eggs and deformed embryos may occur. The reason for this discrepancy has not been determined, but our experience indicates that some of the infertility and variability can be reduced by optimal rearing conditions. Undiagnosed pathology or the high degree of inbreeding to which this strain has been subjected may also be involved. We feel the latter possibility is an important factor for the following reasons. Riegert's finding of synchrony was made when nondiapausing animals had been reared for only 12 generations; at that time egg hatchability was close to 100%. Eight years later, in 1969, when Pickford and Randell[98] reported on the development of the completely non-diapausing strain, the animals had been through approximately 75 generations; hatchability at that time was down to about 50–70%, with nonhatching being attributed to infertility. Hatchability of the eggs in our colony of this strain of grasshoppers is in the same range. It is possible that in *M. sanguinipes* some of the variability may be related to karyotype variability (Section 3.3.1b). Whether this chromosomal variability originated before or after development of the strain is not known, but chromosomal heteromorphism of an unspecified type and extent has been reported to occur in field-collected *M. sanguinipes*.[138]

3.2.2. Isogenic Embryos

Isogenic grasshoppers can be obtained by parthenogenesis in eggs of unfertilized females, a phenomenon that occurs in many species of grasshopper, in fact, in all that have been examined for it.[131] Oviposition is the stimulus for initiation and completion of meiosis in grasshoppers, and fertilization is normally responsible for postmeiotic development of the egg.[136] Most unfertilized (haploid) eggs develop for a brief period, but only a small proportion of them attain spontaneous diploidy, probably by endoreplication, early in development and thereby are able to continue growing and finally to hatch. Of these, a few reach maturity.[65]

Isogenicity is advantageous for studies requiring a minimum of genetic variability. For example, isogenic embryos of the grasshopper *Schistocerca* have been used to study the genetic component in variability of specific neurons.[46–48] Such embryos can be obtained in the laboratory by rearing female grasshoppers in crowded conditions close to but not in direct contact with males. Pheromones from males have been proposed as being necessary for sexual maturity and oviposition by virgin female grasshoppers.[93,131]

3.2.3. Deformed Embryos

Deformed embryos are rarely observed in 14-day-old egg pods of *Chortophaga* but are sometimes seen in the 9- to 10-day old pods of *M.*

sanguinipes. For example, of 11 *M. sanguinipes* pods with a total of 238 eggs, four embryos were found to be deformed. Examination of these (and other abnormal embryos) revealed that the most prominent abnormalities include: (1) some appendages are of normal length, shape, and orientation, whereas others are short and project at abnormal angles to the body, (2) head and mouth parts are grossly underdeveloped compared to those of the thorax and abdomen, and (3) the thorax and abdomen do not lie in one plane but are small and contorted in a spiral shape.

We have examined in detail the chromosomes in some deformed embryos of *M. sanguinipes* and find various types of aberrations, such as acentric fragments, lagging chromosomes, and chromosome bridges at anaphase, as well as mosaicism for haploid/diploid cells. The most striking finding was evidence for defective spindle function: two abnormal embryos had many blocked metaphases (C-metaphases). All midmitotic cells (prometaphase, metaphase, and anaphase) were scored, and of a total of 734 small midmitotic cells, 91% were C-metaphases, and of 345 midmitotic Nbs, 78% were C-metaphases. All the cells had very short chromosomes, suggesting that they had been blocked for a long time; they were indistinguishable from those observed in embryos cultured for up to 24 hr in colchicine or Colcemid. Such blocked metaphases were not found in normal embryos from the same egg pod that contained the abnormal embryos or in the many normal embryos used as controls for mutagen testing. It is tempting to infer a causal relation of the chromosome aberrations and the blocked metaphases to the associated developmental anomalies.

These preliminary data on *M. sanguinipes* embryos are described to caution against the inclusion of deformed embryos in experimental work.

3.3. Cells

The grasshopper embryo at the stages used for chromosome and mitotic studies contains a number of different types of cells. Although the Nb has been the most studied cell type, the other cells in an embryo are useful for certain types of investigations in which homogeneity of cell type is not important. (In squashes, only the Nbs and the subesophageal cells can be unequivocally identified.)

In addition to the Nbs, the other cell types include primarily cells of: the hypodermis, a thin single-cell layer on the ventral surface of the embryo[16]; the subesophageal body, a small structure that lies under the mandibular and maxillary segments and that contains large cells

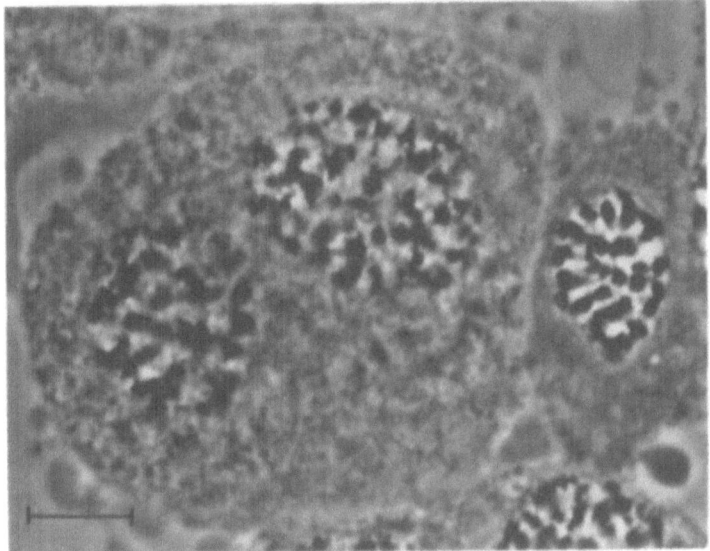

FIGURE 6. Subesophageal cell with two nuclei on left and small cell on right. *Melanoplus sanguinipes*. Scale: 5 μm.

(some have several nuclei) with a large cytoplasm to nucleus ratio (Figure 6)[64,90,105,137]; the appendages (mouth parts, legs, and abdominal appendages), whose cells are approximately 10 μm in diameter; and the ganglion cells that are the products of Nb division and that are the most numerous cell type in the body of the 14-day-old *Chortophaga* and 9-day-old *M. sanguinipes* embryos. Although we have not definitively identified primitive genital cells in either species, they are probably in the first 6–8 abdominal segments as has been reported for other species of grasshopper.[88,105,137] "Midline precursor" cells for neurons have been described in the grasshopper *Schistocerca*[9]; these cells, a little smaller than the ventral surface Nbs, are situated in the midline at the dorsal surface of the embryo. Cells similar to them have been observed in sectioned 14-day embryos of *Chortophaga*. All cells in an embryo exclusive of the Nbs and subesophageal cells will be called "small cells" in this chapter.

3.3.1. The Neuroblast

In discussing the GHNb, we will be referring to that of *Chortophaga* unless otherwise indicated. The Nb is a true blast cell and is so named because it is a neuronal precursor in insects.[49,60] It is a large cell, the

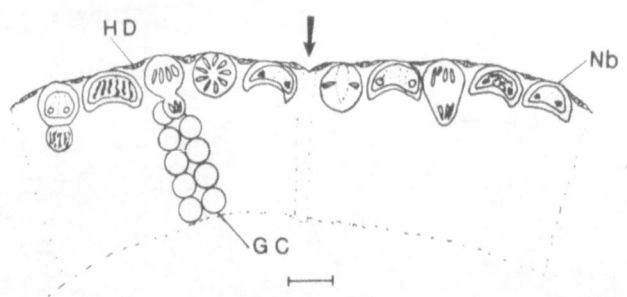

FIGURE 7. Diagram of a cross section through a thoracic segment of a 14-day-old embryo of *Chortophaga viridifasciata*. Neuroblasts from left to right are in middle telophase, late prophase, late anaphase, metaphase, interphase, metaphase, late telophase, middle anaphase, middle prophase, and interphase. Note that the convex side of the neuroblasts in interphase, prophase, and telophase is oriented toward the ventral surface of the embryo. HD, Nucleus of hypodermal cell; Nb, neuroblast; GC, ganglion cell. Arrow indicates midventral line. Scale: 30 μm.

average diameter of those in the head and thoracic segments being 25–30 μm in *Chortophaga* and 15–20 μm in *M. sanguinipes*. Its cytoplasm to nucleus ratio (volume) is small. The Nb has a definite polarity[19] (Figure 7) and divides unequally to produce a group of ganglion cells dorsally. It has a finite embryonic life span: it appears in the embryo within the first week of development and disappears in the third week.

   3.3.1a. Position. Neuroblasts lie essentially in a single layer at the surface of the head and of the ventral side of all the segments (mouth, thorax, and abdomen). They are largest in the head and smallest in the terminal abdominal segments; they divide rapidly and asynchronously. Embryos of *Chortophaga* and *M. sanguinipes* are used at the 14- and 9-day stages (Figure 5), respectively, for two reasons. First, all the Nbs from the head through the abdomen are of maximum size and are actively dividing at these stages. Second, so many ganglion cells have been produced by this time in development that the Nbs are literally bulging against the ventral surface, with the result that in many cases there is only a thin layer of hypodermal cytoplasm over the Nbs (hypodermal nuclei between Nbs) (Figure 7). Accruing from this are two assets for the investigator: when embryos are separated from yolk and membranes, Nbs have ready access to the culture medium and the contained chemicals under study, and the investigator can make microscopic observations on living Nbs *in situ* in appropriate culture preparations (Section 4.2.2).

   3.3.1b. Chromosomes. Both species of grasshopper have an X0 sex-determining mechanism; the chromosome numbers are 24,XX for the

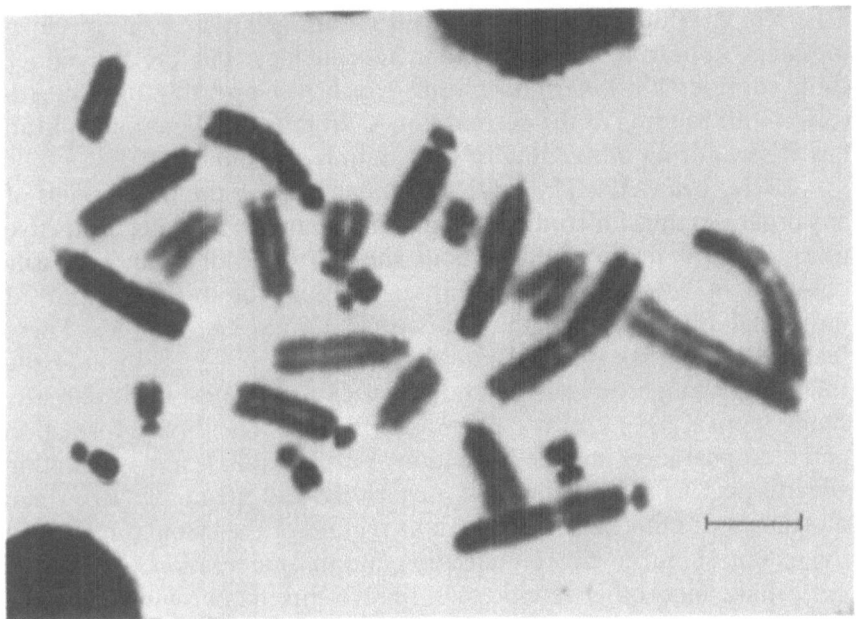

FIGURE 8. Metaphase spread of neuroblast from *Melanoplus sanguinipes* embryo. See Section 4.2.1b for method of preparation. Scale: 5.5 μm.

female and 23,X for the male. In *Chortophaga*, all of the chromosomes have terminal centromeres and vary in length at metaphase from 4 to 14 μm. There is only one pair of nucleolar organizer regions (NORs), located in a submedian position on one of the longer chromosomes; the associated secondary constriction makes this the only morphologically distinguishable pair of chromosomes in the karyotype. With the exception of some additional secondary constrictions, whose occurrence is variable, the karyotype of *Chortophaga* is invariant (M. E. Gaulden, unpublished data).

The chromosomes of *M. sanguinipes* have not previously received much attention. White[138] notes that chromosome polymorphism occurs in *M. sanguinipes*, but gives no details. Tsang *et al.*[130] have established two tissue culture cell lines from *M. sanguinipes* embryos; one culture is of "undetermined ploidy" and one is "approximately diploid." Preliminary data in our laboratories indicate that this species has chromosomes with terminal, subterminal, and submedian centromeres; chromosome lengths at metaphase vary from 2.5 to 11 μm (Figure 8). There are four nucleoli, but the loci of the NORs have not been unequivocally established. Variation observed in parts of the karyotype of *M. sangui-*

nipes Nbs precludes a single characterization: about six pairs of chromosomes appear to be invariant in morphology, but the remainder show considerable heteromorphism, which will probably only be resolved with banding of the chromosomes. Whether the heteromorphism has arisen during inbreeding of this strain is not known.

3.3.1c. Cell Cycle. The GHNb is unique among the mitotic cells of any organism thus far studied in that it has, in the living state, distinctive morphological features throughout the cell cycle that can be easily observed *in vitro* and that permit partitioning of the cycle into 22 individual phases rather than the usual G_1, S, G_2, and M. These features, which are detailed elsewhere,[15,21,22,37,41,104,129] include the following: changes in nuclear and cell sizes and shapes; nucleolar size, shape, appearance, and disappearance; nuclear membrane formation and disappearance; spindle formation at prometaphase and orientation with respect to the cell poles at metaphase and anaphase; Brownian movement of mitochondria in various regions of the cytoplasm[14]; and the extent of chromatin condensation and decondensation.

Unlike most other mitotic cells, the Nb chromatin is visible in some form during the entire cell cycle, and distinct changes in it constitute key delineations of different phases of the cycle as seen with an oil immersion objective and bright-field illumination. For example, at very late telophase, the chromatin granules, arranged in loose, linear arrangements, gradually become smaller, less numerous, and less refractile. Interphase is defined as beginning when the sparse chromatin granules appear to be randomly distributed in the nucleus, which is devoid of any other visible structures except the nucleoli. Very early prophase begins when fine fibers begin projecting from each granule; by early prophase the granules have disappeared and the nucleus is uniformly filled with thin chromatin fibers. Thus, on the basis of such detailed characteristics, the experienced observer can identify the precise phase of the cell cycle in which a living Nb is situated at any given point in time and can discern abnormalities resulting from treatment with an exogenous agent.

The cell cycle of the GHNb consists of interphase, prophase (eight subphases), prometaphase, metaphase, anaphase (three subphases), and telophase (eight subphases). DNA synthesis begins at the midpoint of middle telophase and ends at the middle of very early prophase.[35,36,80] RNA synthesis also begins in middle telophase but continues to the middle of middle prophase[109]; preliminary data on protein synthesis indicate that it begins in early telophase and ends in late prophase. Strictly speaking, therefore, the GHNb has no G_1 and G_2 as defined in mammalian cells, but as Carlson[18] has pointed out, these periods in

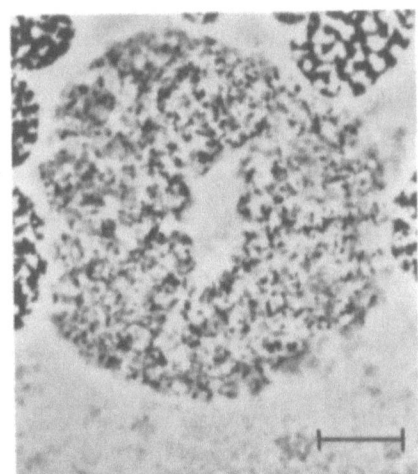

FIGURE 9. Early prophase nucleus of *Chortophaga viridifasciata* neuroblast showing the channel of cytoplasm that passes through the nucleus from the convex to the concave side of the cell. The cell was flattened by squashing, so that the channel has an oval rather than the usual round shape. Scale: 10 μm.

grasshopper, avian, and mammalian cells are not comparable morphologically, which makes comparisons difficult. If there is a G_1 period in the GHNb, it is brief. Prescott[99] has discussed evidence indicating that G_1 is not essential to progression of the cell cycle; the rapidly dividing GHNb provides additional support for this view.

The shape of the GHNb is concavoconvex (Figure 7) from late telophase until midway in late prophase when the cell rounds up. The nucleus is large (15–20 μm diameter in *Chortophaga*, 10–15 μm in *M. sanguinipes*) and shaped like a doughnut, i.e., it has a channel of cytoplasm running through the middle (Figure 9). This shape results from partial formation of karyomeres in telophase: nuclear membrane assembly in middle telophase, as revealed by electron microscopy,[124] begins around each chromosome, but by late telophase all membranes have fused to form a single one. The centrosomes, not visible in the living Nb, lie in the channel at the end near the convex surface of the cell, and at the position occupied by the pole of the spindle in the previous anaphase. At very late prophase, minutes before nuclear membrane breakdown, vestiges of karyomeres can be seen at the apolar region of the nucleus (concave side of the cell), which give the nuclear membrane a deeply scalloped appearance.

The rate at which the GHNb traverses the cell cycle at 38°C is relatively rapid. The duration of the cycle is 4 hr in *Chortophaga*[86] and is even faster in M. sanguinipes, 2 ± 0.5 hr.[44] A temperature of 38°C is used because many of the early observations made by Carlson, beginning in 1941, on GHNb *in vitro* were made at this temperature. He selected this temperature to reduce the cell cycle time (it had the additional advantage of being above summer room temperature in his non-air-conditioned laboratory in Alabama). For culture periods of less than 24 hr, this temperature has not been found to be detrimental, so we have continued to use it.

3.3.2. Ganglion Cells

The GHNb divides unequally to produce, in the position of the mother cell, a large cell that is the daughter Nb and dorsally a small cell, which is called a ganglion cell (Figure 7). Thus repeated division of the Nb produces a group of ganglion cells in columns, but the number of Nbs remains constant. In squash preparations of fixed embryos, a Nb may occasionally be found with all of its ganglion cells in position, so that the mother cell and its daughters appear as a unit.

The ganglion cell is about 10 μm in diameter, and is too deep in the embryo for microscopic observation of cellular detail with the hanging-drop culture method (Section 4.2.2) that has provided so much information about the Nb. The ganglion cells of *Chortophaga* and *M. sanguinipes* divide, but the exact number of divisions occurring before they differentiate into neurons is not known. In *Chortophaga* we have found that exposure of 14-day-old embryos to 2.5×10^{-6} M colchicine for 24 hr at 38°C is required to block progression of most or all of the approximately one dozen ganglion cells lying under each Nb (as observed in sectioned embryos). (The ganglion cells that were produced earlier in development had already become differentiated into neurons.) Similar colchicine treatment will block all of the Nbs in 4 hr. It can be inferred, therefore, that the rate of division of the ganglion cell is much slower than that of the Nb. Recently we have studied squashes of embryos exposed to Colcemid (0.5 μg/ml) for 2 hr (25°C) at the 13- to 15-day stages of *Chortophaga* and the 8- to 10-day stages of *M. sanguinipes*. Large numbers of small cells, many of which are ganglion cells, are found to be blocked in metaphase, so the ganglion cells must divide more than once at these inclusive stages of development. These findings differ from previous ones in which ganglion cells have been reported not to divide at all in *Xiphidium*[137] and in *M. differentialis*[7] or to divide once in *Schistocerca* when they are near the dorsal side of the embryo just prior to differentiation.[49]

4. Methods

4.1. Exposure

Exposure of grasshopper embryos to chemical agents can be achieved *in vivo* as well as *in vitro*. Described here are the methods we have used for exposure of grasshopper embryos *in vivo* to chemical vapors and *in vitro* to chemicals soluble in water or in organic compounds, such as dimethylsulfoxide, that are miscible with culture medium. These methods are easy, inexpensive, and yield reproducible data.

As can be deduced from the description of the grasshopper egg shell in Section 3.1, the embryo *in vivo* is accessible to ambient water as well as air. In fact, absorption of water by the grasshopper egg is maximum at developmental stages comparable to those used for mutagen testing, the mean percent weight increase through water uptake being about 50% as compared to weight at oviposition.[108] Slifer[117] has shown the eggs of these stages to be the most susceptible to the toxic effects of small molecules such as iodine and potassium iodide dissolved in water; she demonstrated that these substances reached the embryos. These data suggest that it may be possible to expose grasshopper embryos *in vivo* to agents dissolved in water, but it remains to be seen whether moderate- or high-molecular-weight compounds in solution would be admitted to the embryo by the hydropyle cells.

In this chapter the expression "exposure of Nbs" means their exposure *in situ*, i.e., in the embryo *in vivo* (vapors) or *in vitro* (chemicals dissolved in medium). Methods for isolating and culturing single Nbs have not yet been perfected.

4.1.1. *In Vivo* Exposure to Vapors of Volatile Liquids

Grasshopper embryos may be treated *in vivo* by exposure of eggs to the vapor of a volatile liquid in a sealed chamber. Several (3–5) intact eggs are placed on wet filter paper in a small dish suspended in a Coplin jar (approximately 100-ml volume). The desired amount (usually 0.01–0.5 ml) of the liquid agent to be tested is placed on the flat bottom of the jar (approximately 1 inch square) before the screw cap is tightly sealed with petroleum jelly. At the incubation temperature of 24°C, these small volumes of volatile liquids usually evaporate very quickly. The vapors permeate the air surrounding the eggs and reach the embryo through respiration avenues. Duration of treatment depends on the potency and the amount of the compound in the jar, but it usually ranges from 2 to 16 hr. Embryos exposed to air alone in sealed jars serve as untreated controls. Two to three chemicals (3–4 doses of each) plus a zero dose can be tested concurrently. The mitotic arresting effects of several volatile liquids have been demonstrated with this method,[73] including two common solvents, benzene and toluene, and four common anesthetics, chloroform, halothane, ether, and ethyl chloride.

If the density d in mg/liter and the molecular weight (mol wt) of an agent are known, the atmospheric level in ppm may be estimated from the formula

$$\text{ppm} = \frac{(\text{mg/liter}) \times 24{,}470}{\text{mol wt}}$$

which assumes that conditions are relatively close to standard temperature and pressure (STP: 25°C, 760 mm mercury pressure).[127] The constant 24,470 in the numerator represents the gram molecular volume in milliliters at STP. A constant derived from the perfect gas law for the prevailing temperature and pressure must be substituted when conditions vary widely from STP. If the liquid completely evaporates, the mg/liter of vapor per jar can be calculated by the formula

$$\frac{d \times \text{liquid volume}}{\text{jar volume}}$$

where liquid volume is the amount of the volatile compound put in the jar. If not all of the liquid in the jar evaporates, the volume of the unevaporated liquid must be subtracted from the liquid volume. A more sophisticated system must be developed so that atmospheric levels of vapors can be more accurately determined.

4.1.2. *In Vitro* Exposure to Soluble Chemicals

Before *Chortophaga* embryos could be used for *in vitro* testing of chemicals, it was necessary to determine (1) how long the Nbs continue dividing at a normal rate *in vitro*, and (2) whether the response of Nbs *in vitro* to mutagen action is the same as that *in vivo*. For this purpose we used X rays, which can be administered with equal ease to embryos both *in vivo* and *in vitro* and whose dose can be reliably measured under both conditions. It was found that Nbs divide in embryos *in vitro* for 20 hr at 38°C (five cell cycles) at the same rate as those *in vivo*,[70] and that the dose–response of acentric chromosome fragments in embryos exposed to 8, 16, and 32 rad of X rays *in vitro* does not differ from that *in vivo*. These data indicated the validity of observations made on Nbs exposed to mutagens *in vitro*.

In view of the facts that the Nbs are covered by a thin hypodermal cell layer and that a few lie deep in the embryo, one might question whether chemicals reach them *in situ* at a concentration near or equal to that in the medium. Although we have no quantitative data on this point, there is indirect evidence that chemicals penetrate the embryo very quickly. For example, spheration of nucleoli caused by 4-nitroquinoline-1-oxide (4NQO) or spindle disorganization caused by colchicine occurs about 2–5 min after the embryo is first exposed to the chemical. The few Nbs that lie 60–90 μm below the hypodermal cells show the same effects no more than 2–3 min later than the Nbs right at the ventral surface, i.e., a total of about 5 min after the embryo is placed in medium containing the chemical.

Many of the materials and methods previously described[20] for preparing *in vitro* hanging-drop cultures of individual grasshopper embryos are applicable to the *in vitro* methods now used for exposing them to chemicals. This information is briefly reviewed here together with pertinent details of current methods. Unless noted, all materials are sterile, the dishes, pipettes, etc., being autoclaved rather than dry-heat sterilized.

For a typical experiment, at least 5–10 embryos are exposed to each of 3–5 doses of a chemical, including a zero dose, so more eggs are required than can be obtained from a single egg pod. All eggs are removed from several pods, the unfertilized ones (small) are discarded, and the eggs are cleared of sand particles with forceps, care being taken not to puncture them. The eggs from each pod are placed on clean, moist filter paper in a separate dish. Equal numbers of eggs from the different pods are then combined in as many groups as there are doses of chemicals to be tested.

The eggs with which one dose of a chemical will be tested are placed in 70% ethyl alcohol for 1 min to sterilize their surfaces and then transferred to a petri dish containing a sheet of filter paper. When the egg surfaces are dry (about 1 min), the eggs are placed in 9 ml of culture medium: Mark's M-20 insect medium (Gibco) with 7.5% fetal calf serum (Gibco), and 100 units penicillin/ml and 100 μg streptomycin/ml. A Syracuse watch glass (67-mm diameter) is used for dissection under a stereoscopic microscope equipped with a dust shield.[20] Forceps and dissecting knives, made from surgical needles, are sterilized by insertion in a platform (with dust cover) over an ultraviolet box (15-W germicidal lamp) situated behind the microscope (Figure 10).[20] The embryo of each egg is removed from the chorion and then separated from the serosal membrane and yolk. The amnion is slit open over the embryonic midline and over the head. This ensures that the Nbs, which lie near the surface of the head and the ventral plane of the thoracic and abdominal segments, have full access to the treatment medium. With practice and sharp knives, the embryo can be separated from all egg materials in less than 1 min, so that ten embryos can be prepared within 15 min after the eggs are put in the dish of culture medium. The embryos are then transferred with a minimum of medium by a

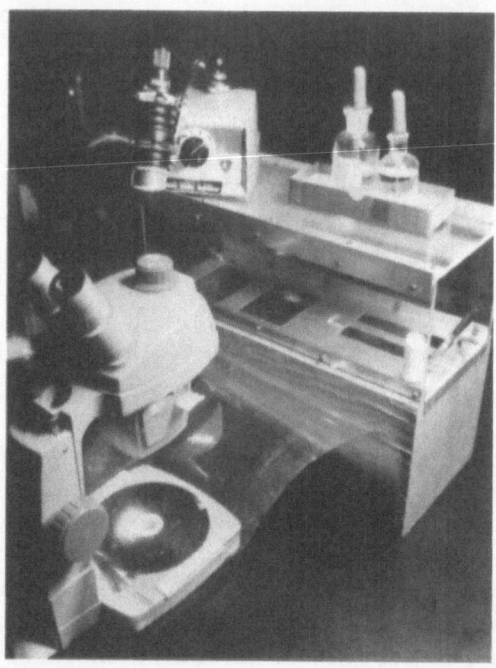

FIGURE 10. Dissecting microscope with a dust shield covering the stage. A glass plate on the stage extends to the ultraviolet sterilization box, so that two dishes of medium can be accommodated during the handling of dissected embryos. The sterilization box is equipped with an ultraviolet sterilamp and covered with a slotted platform for sterilization of slides, cover glasses, forceps, and quartz pipettes. A Lucite cover over the box serves as a shield against dust and prevents exposure of the operator to ultraviolet radiation.

Pasteur pipette to a dish (35 mm × 10 mm plastic petri dish) containing 3 ml of medium and the desired concentration of the chemical under study. The dish is placed in a humidified incubator at 38°C. At the end of the exposure period, the embryos are washed twice (1 min each) by transfer to dishes of 3 ml culture medium with no chemical. They may then be processed for cell analysis immediately or left in culture medium for a period of recovery before processing (Section 4.2).

Neuroblasts are sensitive to change in the osmotic pressure of culture medium; medium hypotonic to them will retard or completely inhibit their division. If distilled water is used to dampen the sand in which the eggs are laid and stored (Section 2.4.4), little or no increase in osmotic pressure of the culture medium will be required. The osmotic pressure is best adjusted with dextrose or dextrose plus sodium chloride rather than sodium chloride alone. Usually *Chortophaga* embryos do not require any adjustment of the complete medium, but *M. sanguinipes* embryos may require some adjustment, especially those from eggs laid by older females. To determine whether the osmotic pressure of the culture medium needs to be modified (see Ref. 20), several embryos should be placed in the medium for 1 hr and then examined for the number of mitotic cells. For this purpose an embryo can be prepared in a hanging-drop culture (Section 4.2.2) or, alternatively, squashes of embryos held in culture medium (Section 4.2.1a) can be compared to those made of embryos taken directly from eggs.

4.1.3. Acute and Chronic Exposure

Analysis of chromosome aberrations requires that cells be in metaphase or anaphase, so only those concentrations of a chemical that are not completely inhibitory to cell cycle progression can be studied. For higher doses of chemicals, an "acute" exposure of 1 hr followed by 3 hr (1 + 3) for recovery (38°C) is a good treatment regimen for *Chortophaga* embryos.[71,72] A recovery period is especially important for those agents whose action is S phase dependent, because the cells that are in the S phase during treatment must be allowed to proceed to metaphase and anaphase. The 1- plus 3-hr regimen for chromosome aberration analysis was arbitrarily selected in the first experiments because the Nb cell cycle in *Chortophaga* averages 4 hr at 38°C. Although a 1-hr exposure causes some mitotic inhibition, there is enough recovery in 3 hr to allow at least some cells to reach midmitosis (prometaphase, metaphase, and anaphase) in 3 hr. Preliminary *in vitro* experiments with *M. sanguinipes* indicate that acute exposures to higher doses of a chemical, e.g., 10^{-5} M mitomycin C, of 0.5–1 hr must be followed by as much as 3 hr recovery to obtain a sufficient number of cells for chromosome analysis, even though the cell cycle duration in untreated Nbs of this species is about 2 hr. This observation suggests that mitosis in *M. sanguinipes* Nbs may be more sensitive to chemicals than that of *Chortophaga*, but more work is needed to verify this inference.

Studies on the mutagenic action of environmental chemicals should include long-term or "chronic" exposures to very low doses, because chronic exposure of a large number of people to a low dose may have a greater impact on a population than an acute exposure of a small number of people to a high dose. For chronic exposure of *Chortophaga* embryos *in vitro*, we have used either a 4-hr or an 8-hr treatment (38°C), which produces more chromosome aberrations than are observed with acute exposure. For example, with 4NQO, 17 chromosome breaks (acentric fragments) per 100 cells were found with acute exposure to 5 μM, as compared with 29 breaks per 100 cells after chronic exposure to only 0.5 μM.[71] This difference may be explained in part by the fact that the lower dose causes minimal inhibition of mitosis, so more cells with aberrations reach anaphase, the stage at which they are scored.

4.1.4. Activation Systems

The young grasshopper embryo does not have the enzymes necessary for converting promutagenic compounds into active mutagens; the development of such an activation system occurs later in embryonic development. To use the GHNb for testing promutagens, it is necessary therefore to use exogenous activation systems. We have examined in detail the efficacy of two systems for activating promutagens to induce chromosome aberrations in the GHNb,[72] namely, S12 mixture (S12 fraction + cofactors) prepared from phenobarbital-induced rats, and primary cultures of hepatocytes from uninduced rats.

For the S12 fraction, male Sprague-Dawley rats (150–250 g, Charles River Breeding Laboratory) were injected intraperitoneally once a day for 4 days with phenobarbital in 0.9% saline (80 mg/kg body weight) and were starved for 18 hr before removal of the liver. The methods of Remmer et al.[102] used for preparing the liver microsomes (S12 fraction) can be briefly summarized as follows. Four volumes of a sucrose solution (0.25 M sucrose + 10 mM Tris hydrochloride, pH 7.25) is added to one volume of liver, which is then minced; the tissue is disrupted with a Teflon pestle in a Potter-Elvejhem homogenizer. Centrifugation of the homogenate is carried out at $12,000 \times g$ 10 min, and 2-ml aliquots of the supernate (S12) are stored in tightly stoppered vials at $-80°C$. One batch of the S12 fraction is sufficient for a large number of experiments.

Fresh S12 mixture is prepared immediately before an experiment according to the method of Au et al.:[5] 0.3 ml of the S12 fraction is added to 0.7 ml of a cofactor solution (11.4 mM $MgCl_2$, 47 mM KCl, 7.1 mM glucose 6-phosphate, 5.7 mM NADP, and 140 mM K_2HPO_4–KH_2PO_4 buffer at pH 7.4). For the activation medium, 0.1 ml of the S12 mix per ml of culture medium containing the chemical being tested is used. The amount of microsomal protein in the S12 fraction used for all of our experiments is 0.071 mg/ml of medium as determined by the method of Masters et al.[77] for measuring the NADPH-cytochrome c reductase activity. We use this concentration of S12 mix because it is the average one used for S9 or S12 mix in the more common mutagen assay tests.[29]

Primary cultures of rat hepatocytes are prepared according to the method of Fry et al.[32]; aseptic conditions are maintained through all steps. Uninduced male Sprague-Dawley rats are used. Excised liver (80–100 g) is immediately placed in phosphate-buffered saline (PBS) and approximately 0.5-mm-thick pieces are sliced by hand. The slices are washed twice in 10 ml PBS by shaking for 10 min in a 37°C water bath (approximately 100 oscillations/min), followed by two 10-min shakings in 10 ml PBS containing 0.5 mM ethylene-glycol-bis-(β-aminoethyl ether) N,N-tetraacetic acid (EGTA). The liver slices are then digested for 1 hr (shaking bath, 37°C) in an enzyme solution containing 5 mg collagenase, 10 mg hyaluronidase, and 0.2 ml 250 mM $CaCl_2$ in 10 ml Hanks balanced salt solution adjusted to pH 7.5 with 2.8% $NaHCO_3$. The cloudy suspension obtained is put through a single layer of bolting cloth ("Nitex" monofilament nylon screen, 125 μm pore size: Tetko, Inc., Elmsford, New York), and the filtrate obtained is centrifuged at 50 × g for 1 min. For washing, the cell pellet is twice resuspended in PBS and centrifuged. Cell yield and viability (trypan blue exclusion) are determined on the final cell pellet that is resuspended in 10 ml of culture medium.

Hepatocytes with ≥80% viability are used at a concentration of 2 × 10^5 cells/ml, which represents 0.076 mg microsomal protein/ml or 0.38 mg microsomal protein/10^6 hepatocytes. A range of hepatocyte concentrations was tested for activation: 2 × 10^5 to 2 × 10^6 per ml. A concentration of 2 × 10^5 cells/ml was chosen because it does not by itself significantly affect the mitotic index of GHNbs but does activate chemicals, e.g., cyclophosphamide (CPhos), to induce chromosome aberrations. Hepatocytes alone do not cause aberrations.

It should be pointed out that the S12 mix and the hepatocyte concentration used for activation were not selected because they had the same microsomal protein content; the similarity was a chance occurrence. That microsomal protein content is not the only factor involved in activation is indicated by the fact that significant differences in the yield of chromosome aberrations in the GHNb were found with the two activation systems for the same concentration of CPhos (Figure 11).[72] This observation supports the view of Bigger et al.[11] that activation is enhanced by the inherent intracellular location and environment of the relevant enzymes.

The use of freshly isolated hepatocytes in suspension as an activating system for the short-term exposure of GHNbs to promutagens eliminates the time-consuming step of having to allow cells to attach before being used for activation (see Ref. 8). Thus the time required to test an indirect-acting mutagen with the GHNb is reduced considerably. Another possible advantage of using freshly isolated hepatocyes is that their activating enzymes are probably at maximum concentration and

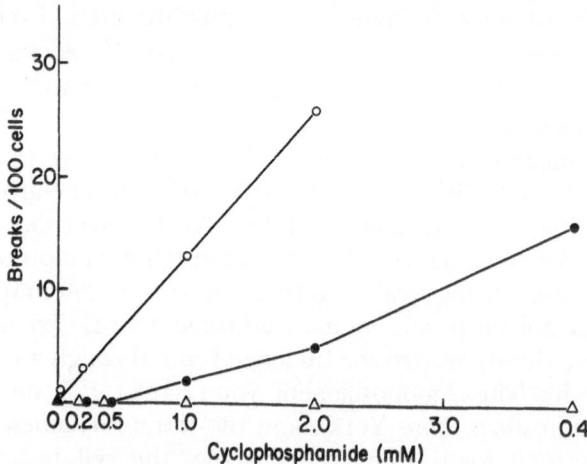

FIGURE 11. Dose–response of chromosome break frequency (acentric fragments) in *Chortophaga* neuroblasts. Embryos were exposed *in vitro* to cyclophosphamide (CPhos) with and without S12 mix or freshly isolated rat hepatocytes in suspension. A 1-hr exposure was followed by a 3-hr recovery at 38°C. The concentration of S12 mix was 0.1 ml/ml of medium and that of hepatocytes was 2×10^5 cells/ml. (From Liang and Gaulden.[72]). (O) With rat hepatocytes, (●) with S12 mix, (△) without hepatocytes or S12 mix, (▲) combined data for no CPhos.

function; hepatocytes rapidly lose their ability to synthesize cytochrome P-450 when grown in tissue culture.

4.1.5. Special Safety Precaution

All operations involving solutions of mutagens or potential mutagens, needless to say, are carried out with the necessary precautions. Because embryos must be transferred into and out of mutagen solutions with pipettes, the possibility of aerosol formation should be emphasized. It is well in this instance to remember that high-speed photography has demonstrated the production of an aerosol consisting of 15,000 droplets (most under 10 μm in diameter) when the last drop of fluid in the tip of a pipette is expelled with moderate force![67]

4.2. Preparation of Embryos for Cell Analysis

4.2.1. Fixed Cells

4.2.1a. Squashes of Whole Embryos. After exposure of embryos to chemicals, they are made into squash preparations for cytogenetic analysis. The endpoints to be analyzed determine the type of squash preparation made. Two types of squashes are made: The direct squash method is used for scoring acentric chromosome fragments at late anaphase and very early telophase (LA-VET) and for determining the anaphase to metaphase ratio as well as the mitotic index of fixed cells. The hypotonic-pretreated embryo squash is used to determine the occurrence of aneuploidy and polyploidy.

Methods for making direct embryo squashes have been described previously.[39] Briefly, embryos are fixed in 3:1 methanol–glacial acetic acid for at least 1 min, after which the embryos can be stained and squashed, or they can be left in the fixative indefinitely. After fixation, embryos are stained in acetocarmine (0.5–0.75% carmine in 45% glacial acetic acid) for 3–5 min; gentle agitation of the dish will facilitate submersion of the embryo, which tends to float when first dropped in the stain. Each embryo is placed on an individual slide, covered with a drop of acetocarmine–Karo mixture and a cover glass, and then squashed ɔ a near monolayer of cells. The mixture consists of three parts acetocarmine, three parts distilled water, and two parts colorless corn (*Zea mays*) syrup (Light Karo). Evaporation of water at the edge of the cover glass results in a semipermanent seal by the Karo residue. No cells are lost with the direct squash method.

In making direct squashes, it is helpful to keep in mind that acetic acid softens tissue and facilitates separation of the cells. The grasshopper embryo, at the stages used, has no tough connective tissue. The durations of embryo submersion in the fixative and in the stain must therefore be balanced so that the embryo does not fall apart or break into pieces before being transferred to the slide. The procedure described for making direct squashes emphasizes speed of preparation, and the short time in the acetocarmine, which weakly stains the chromosomes, is sufficient if the cells are viewed with phase optics. If, instead, bright-field optics is used, a longer staining period is required. For this, the 3:1 fixation is made with ethanol rather than methanol. Fixation for 1 min and staining for 20 min are recommended. Ethanol hardens the embryo, so that a longer time in the acetocarmine is required for softening, which at the same time results in a deeper staining of chromosomes. Treatment with some chemicals, e.g., mitomycin C, may make embryos fragile, so ethanol–acetic acid may be preferable for their fixation.

In hypotonic-pretreated embryo squashes[73] the individual cells are more separated than in the direct squash. Embryos are placed in 0.075 M KCl for 20 min, fixed for 1 min in 3:1 methanol–glacial acetic acid, stained 3 min in acetocarmine, and then put on slides, one embryo per slide. A drop of 1 : 1 glacial acetic acid–distilled water is placed on an embryo to soften the tissue for about 1 min. A few more drops of the 50% acetic acid are added plus a cover glass, and the embryo is squashed to a monolayer of cells. The edges of the cover glass are quickly sealed with Permount (Fisher Scientific Company). Chromosomes in such squashes gradually lose stain but can still be viewed with phase optics.

Squashing must be done carefully so as to prevent movement of the cover glass, which will produce rolls of cells that cannot be analyzed. Squashing may be done with a finger or with an instrument designed to apply equal pressure over the embryo.[39] Prior to squashing, a piece of blotting paper is placed over the slide in such a way that the tip of one corner of the cover glass can be held in place with a finger to prevent movement. The blotting paper will prevent the mounting medium expelled by the squashing from running back under the cover glass and raising it. A little practice will quickly establish the pressure needed for optimal squashing.

Slides may be examined with a low power, nonimmersion objective within a few minutes after squashing; an oil immersion lens may be used several hours later when the medium at the edges of the cover glass has dried. The Karo mounting medium is termed "semipermanent" because in an air-conditioned laboratory with low humidity the cells remain unaltered for several years. For long-term storage, it is recommended that slides be kept in a horizontal position and not on their sides.

4.2.1b. Separated Cells. A suspension of single cells from fixed grasshopper embryos can be made with a minimum of effort. Dropping such a suspension onto slides in much the same manner as that used for lymphocytes[31] results in well-spread chromosomes with little or no overlying cytoplasm (Figure 8). This type of preparation is especially useful for analyzing metaphase chromosomes.

About ten embryos are placed in 3 ml of medium containing Colcemid (0.5 μg/ml) for 30 min at 38°C (1 hr at 25°C). They are then transferred to Ohnuki's[92] hypotonic medium: 4.0 g KCl in 1000 ml distilled water, 2.3372 g $NaNO_3$ in 500 ml water, and 0.9023 g $NaC_2H_3O_2$ in 200 ml water. The three solutions are made separately and then combined for a total of 1700 ml. The embryos are left in Ohnuki's medium (3 ml) for 20 min at 38°C, followed by 10 min in 0.8% sodium citrate at 38°C. The embryos are fixed in three changes of 3 : 1 methanol–glacial acetic acid for a total of 5 min and then placed in 3 ml of 45% acetic acid in a 5 ml siliconized conical centrifuge tube for a 5-min centrifugation at 250 × g. The supernatant is discarded and the pellet resuspended in 3 ml of fixative by gently tapping the tube. This suspension is centrifuged for 5 min at $250 \times g$ and all but about 0.5 ml of the supernatant is discarded. The cell pellet is then resuspended in the 0.5 ml supernatant by drawing it in and out of a siliconized Pasteur pipette several times, care being taken not to bubble air into the suspension.

A slide, previously acid-cleaned and thoroughly rinsed in distilled water, is taken from 95% ethanol in which it is stored, dipped once in ice-cold distilled water, and then placed on a platform at a 45° angle. One drop of the fixed cell suspension is immediately dropped on the slide from a height of 2–4 inches. The slide is then placed flat on the table and several drops of cold fixative are added; it is then set on end to drain the fixative and to air dry. Slides are stained by submersion in a 1:50 Giemsa (R66 Improved Gurr)–distilled water solution for 3–6 min, then air dired, and a cover glass applied with Permount (Fisher Scientific Company).

4.2.1c. Sectioned Embryos. For some types of investigations in which certain Nbs observed in a living embryo preparation must be reexamined in the fixed state, frontal serial sections of paraffin-embedded embryos are useful. Serial thin sections of grasshopper embryos for electron microscopy can also be used.[124] In frontal sections prepared by either method, the relative positions of the Nbs in the segments are maintained, so that previously identified cells can be easily relocated. Details for making 1- to 10-μm-thick paraffin sections are given by Carlson and Gaulden.[20] Such methods are obviously more laborious and time-consuming than those for making squashes, but they are particularly advantageous for examining by autoradiography the uptake of labeled precursors at individual parts of the cell cycle.[35,37,80,109,125] The relocation of Nbs in serial sections can be helpful, for example, in determining the stage-dependent effects of chemicals on the synthesis of nucleic acids and proteins and in detecting mutagen-induced unscheduled DNA synthesis.

When frontal sections are required, it is helpful to fix the embryo on the cover glass used for the hanging-drop preparation (Section 4.2.2). The embryo will adhere to the glass through the dehydration and embedding procedures so that the ventral surface remains flat and parallel to one surface of the paraffin block. This facilitates the cutting of true frontal sections.

4.2.2. Hanging-Drop Cultures of Embryos

Observations on living cells are invaluable for studying some aspects of mutagen action. All cell organelles, except the centrosomes, can be

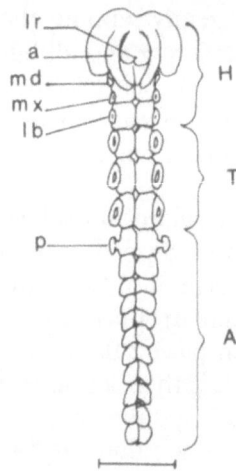

FIGURE 12. Diagram of 14-day-old *Chortophaga viridifasciata* embryo with mouth parts (except labrum) and appendages removed. H, Head; T, thorax; A, abdomen; lr, labrum; a, antenna; md, mandible; mx, maxilla; lb, labium; p, pleuropodium. Scale: 0.45 mm.

seen in living Nbs, so the effects of mutagens on the cell cycle can be followed and described with considerable detail. The location of Nbs near the ventral surface of the 14-day-old *Chortophaga* and the 9-day-old *M. sanguinipes* embryos makes it possible to observe them *in situ* with high-power optics. A method for making hanging-drop cultures of a portion of the embryo was devised for this purpose by Carlson[16] and is still used, but with a different culture medium.

Details of the materials and methods needed for preparing such cultures are described by Carlson and Gaulden,[20] the main features of which are as follows. An egg's surface is sterilized by submersion in 70% ethanol for 1 min. The embryo is removed from the egg in culture medium (Section 4.1.2), and the amniotic membrane is slit over the thorax and pulled away. The head, cut off at the first maxillary segment, and the abdomen, cut at the second segment (Figure 12), are pushed aside. These two cuts not only reduce the size of the tissue for culture purposes, but also prevent the partially cut amnion from forcing the embryo into a contorted form. The mouth parts and legs are cut away close to the body, leaving the pleuropodia as the only appendages; they are useful for correct designation of the segment numbers when the preparation is viewed with an oil immersion lens. (The pleuropodia are small protuberances on either side of the first abdominal segment as seen in Figures 12 and 13; they secrete enzymes that help the vermiform larva to get out of the egg at hatching.[63,115] This piece of embryo is transferred to a dish of "hanging-drop" medium of slightly lower osmotic pressure than that of the dissecting medium (attained by the

Right

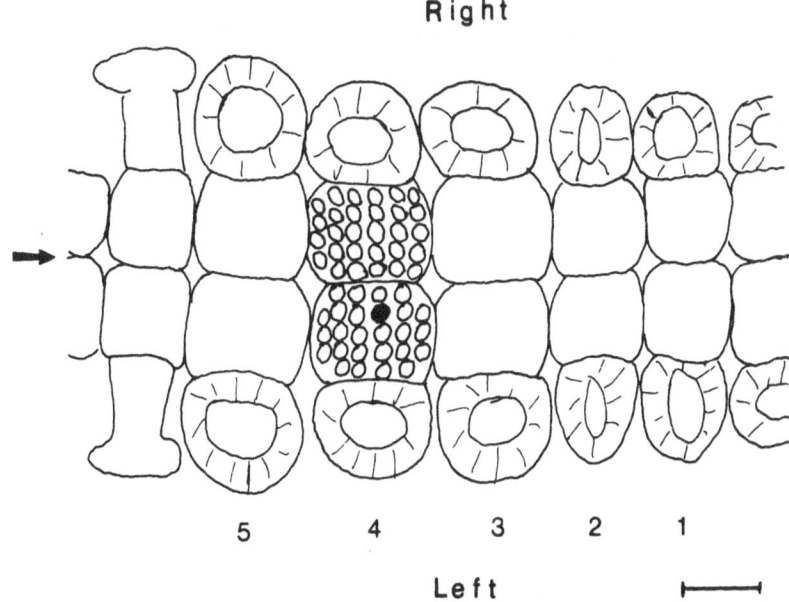

5   4   3   2   1

Left   ├────────┤

FIGURE 13. Diagram of hanging-drop culture of 14-day-old *Chortophaga viridifasciata* embryo as seen with 10× ocular of compound microscope. The uncut pleuropodia on the first abdominal segment (next to segment 5) help orient the viewer. As the mandibular segment is injured when the head is severed from the body, the maxillary segment is called segment 1. The more or less linear arrangement of neuroblasts at right angles to the midventral line (arrow) permits the "mapping" of their positions within each half segment. The solid circle indicates the neuroblast in position 4L42 (see Section 4.3.3b). Scale: ~80 μm.

addition of a small amount of water, about 0.1 ml/10 ml medium). The embryo piece is then transferred, ventral side down, to a small drop of hanging-drop medium in the center of a 22-mm-square cover glass of #1 thickness, previously "greased" with a bit of yolk. The embryo is positioned so that the midventral line is parallel to two opposite edges of the cover glass. Addition of yolk to the medium on the cover glass is not necessary with Mark's M-20 + fetal calf serum medium. The cover glass is inverted over a small depression in a slide with the abdomen to the observer's right (in the compound microscope it appears to be on the left); the edges are sealed with a minimum of mineral oil so that the cover glass cannot move.

Evaporation of water from the very small drop of medium into the air in the depression results in a slight increase in osmotic pressure of the medium surrounding the embryo to bring it to the optimal pressure. Caution must be exercised to prevent the cover glass from becoming warmer than the slide, e.g., by breathing on it or by setting the slide on a cold object. This causes water vapor in the depression to condense on the slide, which in effect removes water from the culture medium in the hanging drop and increases the osmotic pressure of the medium. Hanging-drop cultures are usually observed with a microscope enclosed, except for the eyepieces, in a box maintained at 38°C.[20] If the slide is first placed on the solid part of the microscope stage for 30 sec, it will quickly warm up to prevent condensation in the depression when a drop of immersion oil (38°C) is applied to the cover glass. If more than one depression slide preparation is being observed in a day's experiment, the slides should be stored in a metal box, preferably copper, that has feet to ensure that it is surrounded by the warm air in the incubator.

The Nbs, which now are just beneath the cover glass (Figure 13), can be observed with an oil immersion lens and bright-field illumination; the microscope light must be equipped with a heat filter (glass or solution,[20] as the cells are sensitive to heat. The embryo is too thick for observation with phase optics.

The time required to make a hanging-drop preparation includes: 1 min for egg surface sterilization; 3 min for separation of the embryo from the egg shell, membranes, and yolk, and cutting away of the unwanted parts; and about 1 min to put the embryo piece on a cover glass and seal it over a depression slide.

4.3. Analysis of Mutagen Effects

4.3.1. Chromosome Aberrations at Late Anaphase–Very Early Telophase

The prevalent, and often the only, type of chromosome aberration observed in the GHNb at low doses of mutagens is the acentric fragment. It is scored as single or double, rather than being classified as chromatid or chromosome, for the following reasons. A single fragment (Figure 14A) may be derived from either a chromatid break or from a chromosome break in which the broken ends of the two acentric pieces have rejoined. In other words, a single fragment may result from a break that occurred after or before chromosome replication. Because the two sister chromatids repel each other at the beginning of anaphase in the GHNb,[12] a rejoined chromosome fragment that through self-repelling forces has become rod-shaped at anaphase is indistinguishable from a chromatid fragment. A double fragment is a pair of fragments exactly the same length that lie close to each other (Figures 14B and 14D); they represent one chromosome break. That single and double

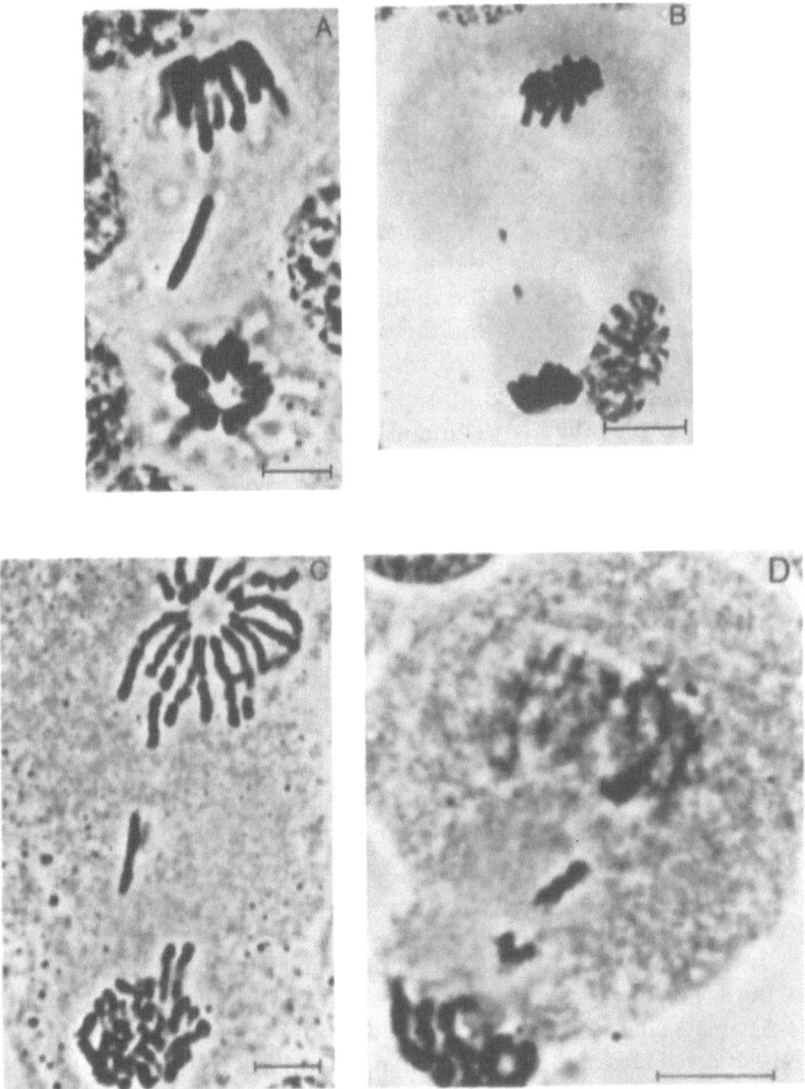

FIGURE 14. Neuroblasts with acentric fragments; oriented with ganglion cell chromosome group at the bottom of the figure. (A) Late anaphase with a single fragment; ganglion cell chromosomes viewed from the centromere ends. Embryo exposed *in vitro* to 100 μM mitomycin C (MMC) 1 hr + 3 hr recovery, 38°C. (B) Late anaphase with a pair of fragments (one break). Embryo exposed *in vivo* to 3 R of X rays (3 R/min); fixed after 176 min at 38°C. (From Gaulden and Read.[42]) (C) Very early telophase with a pair of fragments, only one of which is entirely in the focal plane. Embryo exposed *in vitro* to 1 μM MMC for 1 hr + 3 hr recovery, 38°C. Note that the chromosomes at the poles have begun telophase decondensation but the acentric fragment is still in the late anaphase condition. (D) Early telophase with a pair of fragments in late anaphase condition. Embryo exposed *in vivo* to 16 rad of γ rays (64 rad/min) and fixed 170 min (38°C) later. (A–C) *Chortophaga*; (D) *Melanoplus sanguinipes*. Scales: 10 μm.

fragments both arise from single breaks is confirmed by their strictly linear dose–response to X rays.[13,42,70] Their frequencies are usually combined and expressed as the number of breaks per 100 cells or 100 chromosomes, the latter being a little more accurate because the number of chromosomes for the two sexes differs by one (Section 3.3.1b).

Acentric fragments as well as bridges and rings are easily detected in Nbs at late anaphase (LA) (Figures 14A and 14B) and very early telophase (VET) (Figure 14C) in squash preparations. Acentric fragments usually lie at or near the center of the spindle, i.e., some distance from the centric fragments and the unbroken chromosomes at the poles of this large cell (Figure 14). Scoring of chromosome aberrations in the GHNb at these stages is rapid because (1) the unequal division of the Nb makes it immediately distinguishable from the other cell types in the embryo, (2) acentric fragments, bridges, and rings can be unequivocally identified, and (3) easy recognition of the large Nb and aberrations permits viewing with a low-power objective (25 × or 40 ×), which speeds the scanning of slides. A cell appearing to have an aberration with low-power optics is always examined with a 100 × lens for confirmation before it is scored. In a Nb with mutagen-induced constrictions that are unusually long, the portion of the chromosome distal to the constriction will sometimes extend beyond the anaphase group of chromosomes. This may be mistaken by the novice for an acentric fragment or a lagging chromosome when viewed with low-power optics, but can be resolved with higher power lenses. Further, some small cell inclusions or debris may be mistakenly identified by phase optics as chromosome fragments, so they should also be viewed with bright-field optics to determine whether or not their staining properties are identical to those of the centric chromosomes.

Inclusion of Nbs in VET for the scoring of aberrations is justified for the following reasons. This stage is traversed at 38°C in 3 min (as is LA) and is characterized as beginning with the simultaneous apparent completion of cleavage furrow formation and loss of the smooth outline of the anaphase chromosomes. The latter represents the beginning of the decondensation of chromosomes. Previously it has been shown in the GHNb that decondensation of chromatin lying outside the nuclear region, i.e., in the cytoplasm, at VET (Figure 14C) and even early telophase (ET) (Figure 14D) begins later than it does in the chromosomes lying within the nuclear region.[33] In other words, at VET and ET the acentric chromosome fragments in the cytoplasm are in the anaphase condition, whereas the centric fragments and unbroken chromosomes have already begun to decondense. Acentric fragments can therefore be clearly identified in VET and also at ET. We do not score ET cells, because squashing often separates the newly formed daughter ganglion cell from the Nb at this phase; as separation may possibly cause loss of any fragments lying near the cleavage furrow, spurious data could be recorded.

Note that analysis of cells at LA-VET eliminates the need for colchicine or Colcemid treatment. This reduces the time of prefixation procedures and also enables an investigator to detect effects of a chemical that may be expressed in metaphase, e.g., spindle malfunction (Section 5.3.2), while at the same time scoring for chromosome aberrations at LA-VET.

There are approximately 500 Nbs in an embryo; in untreated embryos about 5% of them are in LA-VET at any given time at 38°C. The minimum number of cells used for testing a single concentration of a chemical is usually 100. With 23 chromosomes in the cells of male embryos and 24 in those of females, 100 cells represents an average of 2350 chromosomes. This number of cells can usually be obtained with 5–10 embryos if mitotic activity has not been reduced by more than 50%. In those cases in which only one or two acentric fragments are observed in 100 cells for a given concentration of a chemical, an additional 100 cells is usually scored. One hundred cells can be scored quickly and is a sufficient number for obtaining statistically significant results because of the zero frequency of spontaneous chromosome aberrations in the GHNb. We routinely include a group of untreated embryos in every experiment to check for spontaneous aberrations.

For chemical testing it is advisable to include periodically positive and negative controls as well as untreated control embryos. Bleomycin is a good positive control chemical for the GHNb, because it induces a relatively high frequency of chromosome breaks at very low doses (0.125 μM)[71] with a minimum of mitotic inhibition (Figure 15). Alcohol can serve as a negative control substance. Sucrose is not recommended for negative controls, because it may increase the osmotic pressure of the culture medium, which, if high enough, can cause chromosome stickiness with resulting breakage at anaphase (M. E. Gaulden, unpublished data).

The time required to perform and obtain dose–response data on chromosome breakage with one experiment is relatively short and can be estimated from the following. Approximately 20–30 min is required to clean the eggs from three pods. As noted above (Section 4.1.2), ten embryos can be removed from the eggs for exposure in 15 min. Exposure plus recovery takes 4 hr with *Chortophaga* and 2–3 hr with *M. sanguinipes*. Fixation, staining, and squashing of ten embryos can be completed in 30 min. Embryos can be exposed, therefore, to 4–5

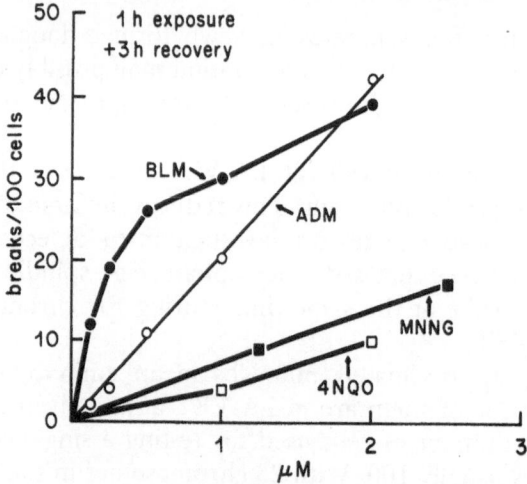

FIGURE 15. Dose–response of chromosome break frequency (acentric fragments) to bleomycin (BLM), Adriamycin (ADM), N-methyl-N'-nitro-N-nitrosoguanidine (MNNG), and 4-nitroquinoline-1-oxide (4NQO). 1-hour exposure plus a 3-hr recovery at 38°C. (From Liang and Gaulden.[71])

concentrations of a chemical and made into slides in 1 day, and the scoring can be completed in 2 days.

4.3.2. Mitotic Apparatus Effects

When an agent acts in some manner on the microtubules of dividing cells, it interferes with spindle function and arrests cells at metaphase and/or causes anaphase abnormalities. Complete disorganization of the spindle results in an appearance similar to that of cells arrested by colchicine (C-mitosis), i.e., the metaphase chromosomes are randomly distributed, their lengths being related to the time of exposure (the longer in colchicine, the shorter the chromosomes). A partially blocked metaphase in a squashed cell has an appearance intermediate between that of a normal metaphase and a typical C-mitosis. Grasshopper embryos are excellent material with which to demonstrate the C-mitotic effects of chemicals and to obtain quantitative values with which to compare different chemicals.[73] The several million cells (of all types) in an embryo can be squashed to a monolayer, so the accumulation of C-mitoses can be easily observed in a direct squash.

For quantitation of C-mitotic effects in squashes, the mitotic index (MI) and the anaphase–metaphase ratio (A/M) can be used to measure the potency and toxicity of chemicals. When an agent completely arrests mitosis at metaphase, the A/M ratio is zero and the MI is elevated (if the agent does not prevent interphase or prophase cells from reaching metaphase). The MI in each embryo is determined by counting the number of metaphases and anaphases in 3000 cells scanned in an embryo; the A/M ratio is determined by counting the number of anaphases present for every 100 metaphases scanned. The values of MI and A/M for treated and untreated embryos can then be analyzed for statistical significance.

To detect aneuploidy and polyploidy, embryos treated with mitotic poisons are allowed to recover for varying periods of time at room temperature, the last 2 hr of which Colcemid (0.16 μg/ml) is added to the medium. The embryos are then made into hypotonic-pretreated squashes (Section 4.2.1a).

Mitotic apparatus effects can also be detected with observations on living cells, as described in the next section.

4.3.3. Cell Cycle Effects

The many morphologically definable phases of the GHNb cell cycle and its short duration (Section 3.3.1c) make the Nb a superior cell with which to determine the effects of chemicals on cell progression. Such effects can be determined in several ways with both the fixed cells of squashes and the living cells of hanging-drop cultures.

4.3.3a. Fixed Cell Counts. The simplest but least definitive, method, is that of counting the number of Nbs in midmitosis (prometaphase, metaphase, anaphase, very early telophase) in 5–10 embryo squashes for a dose of a chemical. The relative mitotic index (MI) is used and is defined as the average number of midmitotic Nbs in the treated embryos expressed as a percentage of the average number in the untreated embryos. This method provides a quick measure of the relative mitotic effects of different doses of a chemical at the time the cells were fixed and is useful for establishing the upper limits of concentration and exposure times that can be used to obtain sufficient numbers of cells for chromosome analysis.[70-72] This fixed-cell method, however, is time-consuming and therefore is not useful for observations at the short post-treatment intervals that are required to obtain detailed information on cell progression effects.

4.3.3b. Living Cell Counts. Observations on living Nbs are best for determining cell cycle effects. In the hanging-drop preparation, the Nbs lie in a single layer at the ventral surface with only a thin, single-cell layer of hypodermal cells between them and the cover glass (Figure 7). It is possible, therefore, to observe Nbs *in situ* with a 100× objective (oil immersion). All the cell structures except the centrosomes can be distinguished in the living Nb with bright-field illumination. In the segments of the body, the Nbs lie more or less in rows at right angles to the midventral line (Figure 13). As their relative positions and their numbers do not change with division (Section 3.3.2), it is possible to designate ("map") the position of individual Nbs. For example, a cell at position 4L42 is in the fourth segment, on the left side of the body, in the fourth row, and is the second cell in that row (Figure 13). Information on the effects of chemicals on the cell cycle can be obtained with one or a combination of three types of data collection methods with living Nbs.

First, selected mapped cells can be followed by *repeated observations* over a period of time. This type of observation enables the investigator to record the exact times at which a treated cell moves from one phase to another, so that even small departures from the normal progression rate of specific phases of the cycle can be determined. At the same time, abnormalities occurring in specific cell structures can be noted. Such detailed information can be recorded for about 6–10 cells at a time, the precise number being determined by the length of the interval between observations; rapidly occurring events will limit the number of cells on which precise data can be obtained. For example, anaphase is defined as beginning when the centromere regions of the chromosomes begin to move apart. This happens within a matter of seconds at 38°C, and the sister chromatids complete separation and move to the poles of the spindle in 3 min.[19] To follow such events requires quick observations made at short intervals on a relatively small number of cells, so, at most, the investigator is limited to no more than two embryos at a time, one treated and one untreated. We use double depression slides so that the observer can quickly move back and forth between two embryos for observations.

The second living cell method, called *mitotic counts*, can be used when a chemical does not affect the duration of midmitosis. It utilizes the Nbs in the three thoracic segments. The number of midmitotic Nbs (prometaphase, metaphase, anaphase) is recorded at intervals equal to the duration of midmitosis, which at 38°C is 22 min for *Chortophaga* and 15 min for *M. sanguinipes*. This is a period in the cycle of almost unvarying duration. If the treatment does not affect this period, counts made every 22 (or 15) min will include every cell that passes through midmitosis. The ratio of the number of midmitotic cells in the treated embryos to the number in the untreated at each count, over a period of time equal to that of the cell cycle, provides a good measure of the rate at which cells are reaching midmitosis. This method does not, of course, reveal the phase or phases of the cycle that are retarded or accelerated by an agent.

It is possible to expose Nbs to a chemical in a hanging-drop preparation, but because they are so close to the cover glass, they may not have as much access to the chemical as when the embryo is free in a dish of medium. Routinely, therefore, embryos are dissected as for a hanging-drop culture and then exposed in dishes as described for whole embryos (Section 4.1.3). At the end of the exposure period the embryo is washed and transferred to hanging-drop medium from which it is made into a hanging-drop culture. The depression slide holding the culture is placed in the microscope incubator box (38°C) 21 min (*Chortophaga*) or 14 min (*M. sanguinipes*) after the end of exposure; the first count is made 1 min later. For making mitotic counts, the embryo is positioned so that the pleuropodium (on first abdominal segment) on the left side of the body is centered in the field of a $10 \times$ objective. When the oil immersion lens is shifted into place, the observer is oriented and can immediately move to the left half of the third thoracic segment (segment 5 in the culture: Figure 13) and begin counting midmitotic cells. For rapid counting, the order in which the segments are scanned is: 5L, 4L, 3L, 3R, 4R, 5R. By using a hand counter, the observer only has to keep track mentally of which half-segment is being scanned. Counting in this manner can be completed in 1 min.

The time required for the preculture procedures and for the counting limits the number of embryos on which data can be obtained at one sitting to six *Chortophaga* embryos in about 6 hr and to four *M. sanguinipes* embryos in about 4 hr, when exposure is only 30–60 min. On the average there are about 20 Nbs per half segment that lie close or next to the cover glass, so the number of cells surveyed in the three thoracic segments is about 120. If untreated embryos are included with each group of six or four embryos examined, statistically reliable results can be obtained when counts on a total of 10–12 embryos (1200–1400 Nbs) per treatment are accumulated.

The third method for obtaining cell progression data on living Nbs is used if a treatment affects the duration of midmitosis. It is a modification of the second method and is called the *mapped count* method. At the time of each count, the position of every Nb in midmitosis is noted on an outline of the thoracic segments with a small circle and the number of the count is entered in the circle (Figure 16). If midmitosis in a cell is prolonged, the map will show two or more successive counts for that cell. An accelerated midmitotic period can be detected by the routine checking of the positions of cells observed in very early or early telophase in order to detect cells that were not in midmitosis at the

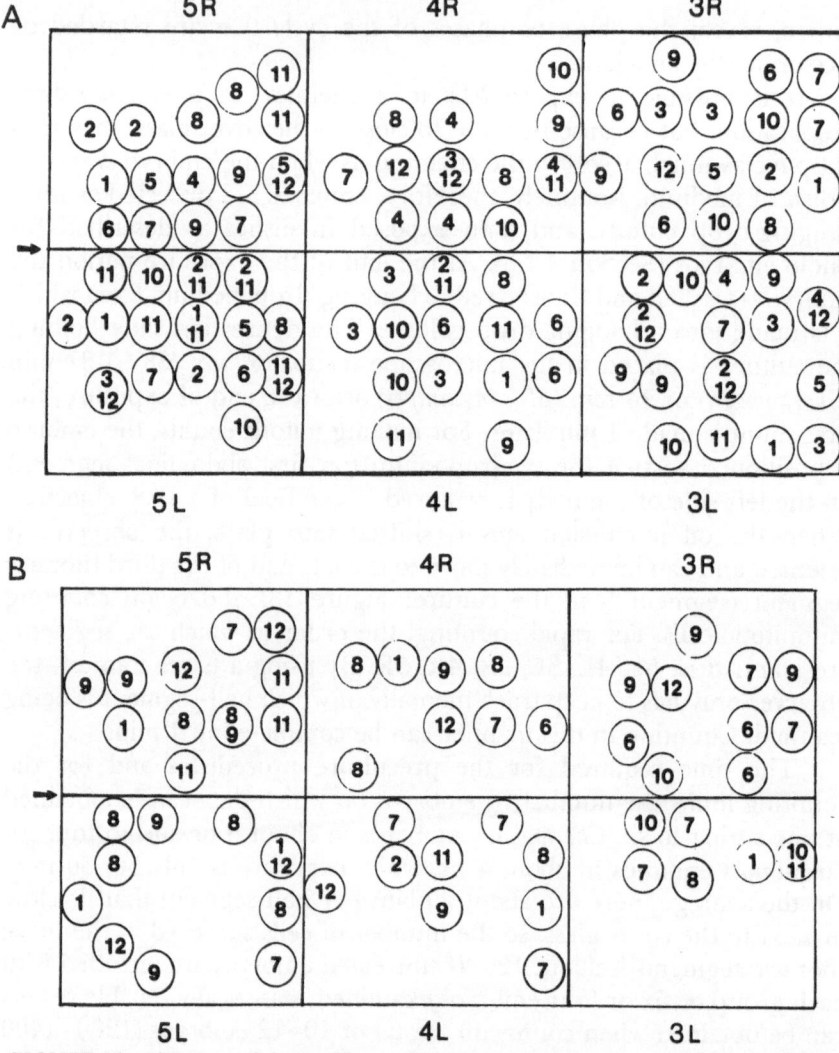

FIGURE 16. Diagrams of the positions of *Melanoplus sanguinipes* neuroblasts observed to be in midmitosis at varying times after preparation of hanging-drop cultures (mapped count method). (A) Untreated embryo. (B) Embryo exposed *in vivo* to 64 rad of γ rays (64 rad/min). Counting intervals: 15 min; count 1 made 15 min after irradiation. The number in a cell represents the counting period at which it was observed to be in midmitosis (prometaphase, metaphase, or anaphase). Cells with two numbers were in midmitosis either at two successive counts, i.e., this period was a bit longer than 15 min, or at two widely spaced counts, i.e., the cell passed through midmitosis twice in the 3 hr of observations (38°C). In the latter cases, the difference between the counts equals the duration of the cell cycle, e.g., a cell in midmitosis at counts 5 and 12 means that it traversed the cycle in 1.5 hr. Note that 11 cells in the untreated embryo went through midmitosis twice; one cell was observed in two successive counts. Only one cell was observed to divide twice in the irradiated embryo; two cells were observed in successive counting periods. See Section 4.3.3b.

previous count. In the same manner, cells in late prophase can be noted to detect unusual prolongation of this phase.

The number of cells first observed to be in midmitosis at each count can be tallied to provide data comparable to those obtained by the "mitotic count" method. Cells observed only at very early or early telophase are scored as being in midmitosis at the time of the previous count.

This method takes more time than the one with which only total numbers of midmitotic cells are recorded at each counting interval. It takes about 3 min to record the positions of 12–15 cells, the maximum number usually observed in the three thoracic segments at one count. Practically speaking, data on no more than four *Chortophaga* or two *M. sanguinipes* embryos can be obtained at one sitting with the mapped count method.

4.3.4. Autoradiography

Autoradiography of sections or squashes of embryos can be made to determine localized uptake of labeled precursors of DNA, RNA, and proteins.[34,35,69,80,81,109,129] Details of techniques have been previously described[20,129] for the exposure of identified cells to a radioactive label at a known phase or phases of the cycle and reidentification of them in the autoradiographs of squashes as well as of sectioned embryos. This method has been used to determine not only the periods of normal uptake in untreated GHNb, but also the induction of unscheduled DNA synthesis (UDS) by radiation.[80,129] It should also prove useful for detecting UDS induced by chemicals, because small, statistically significant differences in grain counts can be distinguished in Nbs in squash preparations in which the cells are flattened to a thickness of 1–2 μm.

4.3.5. Other Methods

Banding of chromosomes is of great value in localizing specific regions of chromosomes that are particularly susceptible to breakage by mutagens. Preliminary work in our laboratories with the conventional methods used for banding mammalian chromosomes has failed to band GHNb chromosomes. Reports of efforts at banding insect chromosomes have recently been reviewed[39]: in most cases, only bands typical of C-bands in mammals are obtained even when G-banding methods are used, i.e., large blocks of chromatin are stained. Greilhuber[50] has postulated that the lack of banding in plant chromosomes is caused by the large amount of DNA per unit length of chromosome, which necessitates a high degree of coiling and which in turn prevents resolution of bands. Anderson *et al.*[4] have recently compared, however, the lengths and volumes of some plant and animal chromosomes and found no differences, so another explanation must be sought. We have proposed[39] that the occurrence of spontaneous banding in insects, including the GHNb, suggests that failure to obtain pretreatment-induced G-bands may not be due to an inherent characteristic of insect chromosomes, but rather may result from a flaw or deficiency in the techniques thus far used. Further work is obviously needed.

The GHNb chromosomes have not yet been examined with methods that reveal sister chromatid exchange, but the short cell cycle of Nbs and *in vitro* cultures of the embryos are advantageous for the use of this technique for detecting mutagen action.

5. Response of the Grasshopper Neuroblast to Mutagens

Space does not permit a complete review of the GHNb mutagen data, but a brief discussion of some aspects of the Nb response to radiation and to certain chemicals will show the types of information obtainable.

5.1. Reproducibility of Data

For the testing of chemicals and for comparisons of the mutagenic activity of different chemicals, it is desirable to use a system that gives highly reproducible results. That the GHNb has this feature was first demonstrated by X-ray data. Carlson in 1941[13] reported a linear dose–response curve for chromosome aberrations in Nbs of *Chortophaga* when embryos were exposed *in vivo*. Forty years later when we repeated his experiments at the lower exposures (8, 16, 32 R) for comparison with embryos irradiated *in vitro*, we obtained frequencies of acentric chromosome fragments similar to his.[70] The acentric fragment is the predominant type of aberration observed at low doses of X rays.

Reproducibility of chemical mutagen data with the GHNb has also been demonstrated.[71] In five separate experiments with a 1-hr *in vitro* embryo exposure to 1 μM 4NQO + 3 hr recovery, 100 cells in each experiment were scored for acentric fragments. The numbers of chromosome breaks/100 cells in successive experiments were found to be 4, 3, 4, 3, and 5, or an average of 3.8. Each of these experiments utilized the eggs from a single egg pod. We can conclude, therefore, that the induced frequencies of breaks were not significantly different in the Nbs of the five egg pods. In an experiment requiring the use of more than one egg pod, one can expect all the embryos to respond alike to a given concentration of a chemical.

5.2. Radiation

From the standpoint of comparing the mutagenic efficacy of chemical exposures, a standard test agent is desirable, the ultimate aim of such a comparison being assessment of human risks associated with environmental chemicals. It was initially proposed by a group of mutagen researchers[26,30] that ionizing radiation serve as the standard mutagen and that the "REC" or rem-equivalent-chemical (rem: radiation-equivalent-man) be used as the unit of reference. A REC was defined as the dose (concentration × time) of a chemical that induces genetic damage equivalent in amount to that induced by 1 rem of chronic irradiation. Critical evaluation of the REC almost immediately raised concern about its use (e.g., Ref. 6), but the initially perceived need for it or a similar value has not diminished. As the complexity of chemical mutagen action in different cell and organ systems has become increasingly clear with additional research, it has become obvious that unqualified comparison of a chemical with radiation in human risk assessment is difficult if not impossible at the present time for a number of reasons. For example, "dose" is not simply the product of the concentration administered multiplied by the time of treatment, but may involve different factors in different systems, such as detoxification or elimination in whole animals, or varying half-lives of a chemical in different culture media. Recently Murthy[87] has redefined REC as radiation-equivalent-chemical and has suggested that its validity be examined thoroughly within given test systems before comparison to other systems, much less humans, be attempted. We have begun studies with the GHNb to determine whether this approach will be useful.

Much work has been done on the response of the GHNb to a wide range of doses of γ- and X-radiation. Acentric chromosome fragment induction is a sensitive index of damage, because the effects of 1 R of X rays (3 R/min)[42] can be quantitated, as shown in Figure 17. The overall radiation sensitivity of the GHNb is 0.011 break/cell per R, which is 4–5 times higher than pollen mother cells of *Tradescantia*,[76] ten times higher than human lymphocytes[74] or Chinese hamster cells,[85,100] and 15 times higher than mouse erythroblasts.[61]

The linear dose–response of acentric fragment frequency from 128 R down to 1 R (Figure 17) indicates that "calibration" of the GHNb response to a chemical can be roughly made by comparing the frequency

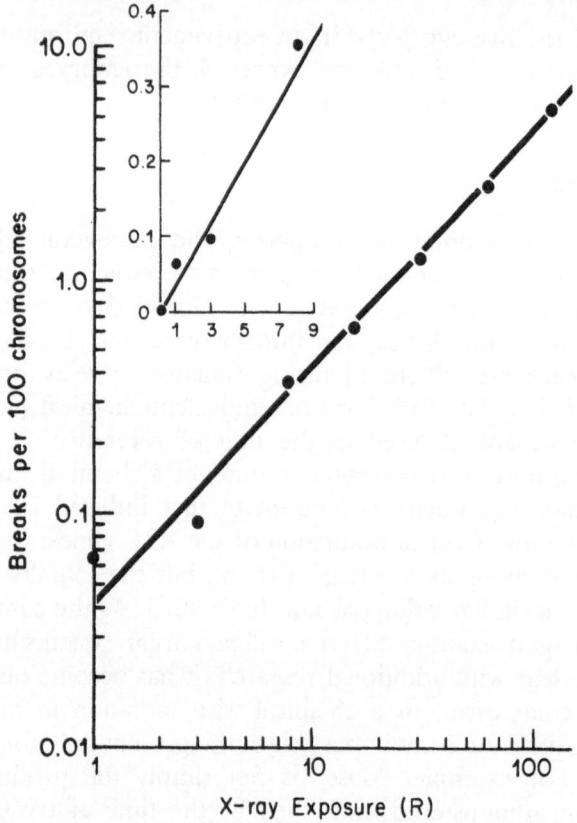

FIGURE 17. Linear dose–response of chromosome breaks (acentric fragments) after *in vivo* exposure of *Chortophaga* neuroblasts to X rays. (From Gaulden and Read.[42])

obtained at a given dose of a chemical to the same frequency of aberrations induced by *x* rad (or R) of X rays. For example, an 8-hr exposure to 0.25 μM of 4NQO produces 0.14 break/cell (Figure 18),[71] which is equivalent to the number induced by exposure *in vitro* to 13–14 R of X rays delivered at 192 R/min.[42] Such a comparison does not, however, take into account the rate at which the chemical enters the cell, the total amount that accumulates in the cell, the portion of the cell cycle traversed during exposure, or the extent of mitotic inhibition. These factors are being investigated. Needed also are data on chromosome aberration frequency in the GHNb after chronic exposure to radiation at very low dose rates.

Confidence in the linearity of the dose–response of GHNb chromosomes to X rays is provided by the zero frequency of spontaneous chromosome aberrations, a unique feature of the GHNb. In fact, none

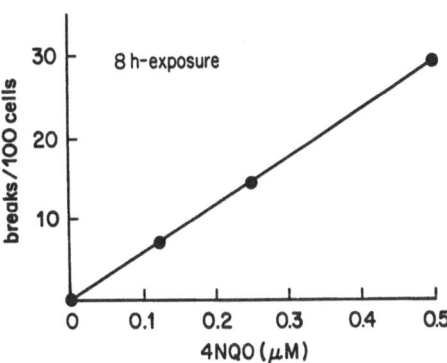

FIGURE 18. Dose–response of chromosome breaks in *Chortophaga* neuroblasts exposed *in vitro* to low concentrations of 4-nitroquinoline-1-oxide (4NQO) for 8 hr at 38°C (no recovery period). (From Liang and Gaulden.[71])

were observed in tens of thousands of untreated Nbs over a 50-year period in *Chortophaga* embryos obtained from the eggs of field-collected grasshoppers in either the laboratory of Dr. J. G. Carlson or that of the senior author. In the 4 years since we began to rear *Chortophaga* in the laboratory, we have observed two acentric fragments in two Nbs out of approximately 10,000 scored. This suggested that the circumstances of rearing might be responsible. Although we have not systematically investigated the effects of every aspect of rearing, data obtained recently indicate that a constant supply of fresh food and the absence of debris in the cages reduce "spontaneous" aberrations to zero in *Chortophaga* and in *M. sanguinipes* (Section 2.4.2). Fresh food appears to be the more important factor. If so, this suggests an intriguing relation of nutrition to mutagenesis in the grasshopper.

A paradox encountered with the GHNb that defies interpretation at present is its high sensitivity to all mutagens thus far tested in the face of its essentially zero spontaneous chromosome aberration frequency.

5.3. Chemical Mutagens

5.3.1. Chromosomes and Cell Cycle

The GHNb is a sensitive cell for the detection of chromosome aberrations induced by chemicals, both direct- and indirect-acting, as shown in Figures 11, 15, and 18. From these and data on other chemicals to be reported elsewhere, two general facts relevant to mutagen action on chromosomes have emerged. First, chromosomes respond in a variety of ways to mutagens, i.e., there are effects other than chromosome aberrations. Some of these effects are detectable at phases of the cell cycle outside the midmitotic ones, and they are

associated with abnormal progression rates. Second, continuous observations on individual living Nbs show that the cells in midmitosis at any
given point during recovery from a short exposure to a chemical
represent cells that were in several different phases of the cycle at the
time of treatment. For example, mapped Nbs observed in a hanging-
drop culture to be in middle prophase, very early prophase, and
interphase at the beginning of a 1-hr treatment with 5 μM N-methyl-
N'-nitro-N-nitrosoguanidine (MNNG) were all in midmitosis 4 hr later
(38°C).

Detection of chromosome aberrations depends of necessity on the
presence of cells in metaphase and anaphase. If cells containing
aberrations reach these stages sooner or later than the time of fixation,
the effects of a chemical may be underestimated. Conversely, an agent
that inhibits or retards the progression of a given phase but produces
no aberrations in that phase may result in a larger proportion of cells
with aberrations that reach midmitosis; this would cause an overesti-
mation of the number of aberrations induced. In other words, the
number of aberrations observed at midmitosis may differ considerably
from the number actually induced. It is the latter quantity that is of
greatest value in attempts to extrapolate from data obtained with test
systems to risk estimates for exposed human populations. Thus, knowledge of the phases of the cell cycle affected by a chemical can help not
only in establishing the fixation times that yield the highest aberration
frequencies, but also in the interpretation of the interaction of the cells
with the chemical. Such knowledge can be unequivocally obtained with
the GHNb.

Some of the chromosome abnormalities observed that are not the
usual aberrations are as follows. In fixed Nbs that had been exposed
to MNNG at concentrations above 2.5 μM for 1 hr, some anaphases
were observed in which the chromatids behaved as though they were
sticky.[71] At early anaphase the sister chromatids appeared to adhere
to each other except at the centromere regions. With observations on
living cells it was found that these centromere regions move to the
poles of the spindle at an unaltered rate, with the result that often the
remainder of the chromatids are left near the spindle equator. In some
cases the centromere region is severed and in others a thin strand of
chromatin connects it with the remainder of the chromatid (Figure 19).
Repeated observations on living Nbs reveal that such anaphase config-
urations do not occur until about 3 hr after 1-hr exposure (38°C) and
are not seen in all of the Nbs reaching anaphase at this or later times.[36]
From this it can be inferred that the phenomenon arises from events
induced in specific phases of the cell cycle (to be determined) somewhat

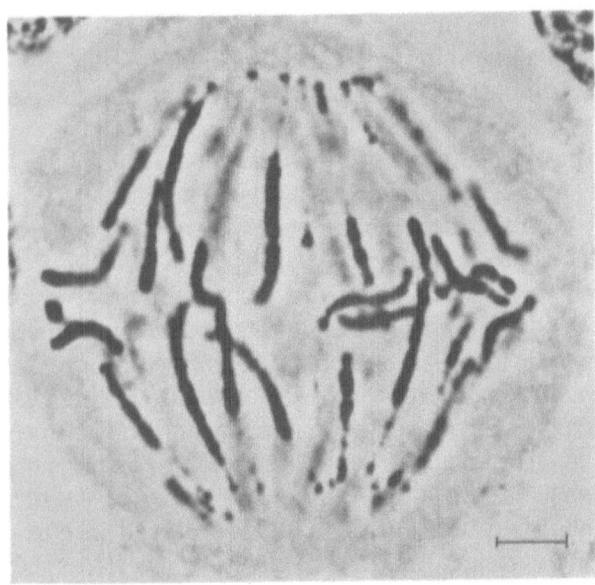

FIGURE 19. *Chortophaga* neuroblast in middle anaphase after a 1-hr *in vitro* exposure of the embryo to 5 μM *N*-methyl-*N'*-nitro-*N*-nitrosoguanidine followed by a 3-hr recovery (38°C). The centromere region has essentially become severed from the remainder of the chromosome in a number of chromosomes. Observations on living neuroblasts reveal that this appears to result from chromatid stickiness except at the centromere. Scale: 5 μm. (From Liang and Gaulden.[71])

removed in time from the succeeding midmitotic period. In cells displaying this phenomenon at middle anaphase, the number of acentric fragments observed when the cells move into late anaphase–very early telophase (LA-VET) are usually too numerous for accurate counting.[71]

One chromosome abnormality observed in prophase cells is the development of C-mitotic chromosomes within a late prophase nucleus; this effect is induced by exposure to doses of mitomycin C as low as 10^{-8} M (4 hr, no recovery) or 10^{-6} M (1 hr + 3 hr recovery). It was first observed in squashes of fixed embryos made to determine chromosome aberration frequencies in LA-VET cells. Many C-metaphases were found, which was puzzling in view of the presence of normal metaphases and anaphases. Observations on living Nbs showed that cells reaching very late prophase remain at this point in the cycle for up to several hours; breakdown of the nuclear membrane is delayed but the chromosomes continue condensing and assume a C-mitotic appearance.[44] In squashes, the cells appear to be in a blocked metaphase state, but are actually in very late prophase. When the nuclear membrane eventually breaks down, formation of the spindle is normal and the cell traverses midmitosis at the normal rate.

Another abnormality observed in prophase is the bleomycin-induced reversion of the chromatin in Nbs at midprophase to a state resembling that of very early prophase, and even interphase, chromatin. It occurs with exposure to amounts as low as 0.125 μM for 1 hr. During the recovery period the chromatin begins prophase condensation again and moves slowly to midmitosis. At amounts of 0.5 μM and above, the time required for recovery increases. The prophase reversion induced by bleomycin is similar in all respects to that induced by radiation,[17,18,21] and it results in a considerable retardation of cell progression.

One puzzling feature of the GHNb response to chemical mutagens is the high frequency of acentric fragments and the absence of two-hit aberrations, such as dicentrics (anaphase bridges) or ring chromosomes. The deficiency of two-hit aberrations is not due to difficulties in detecting them, because they are readily observed in Nbs exposed to doses of X rays greater than about 25 rad. Several explanations are possible: (1) The concentrations of chemicals that do not drastically inhibit cell progression may not be high enough to induce many two-hit aberrations. (2) The progression of cells with two-hit aberrations may be more retarded than those with one-hit aberrations, and consequently do not reach midmitosis at the fixation times used. Contradicting this perhaps is the fact that Nbs are observed with two or more acentric fragments, i.e., the cells have sustained a total of more than one "hit" and have reached midmitosis. (3) The low observed frequency of two-hit aberrations with X rays or chemicals may be related to the large size of the Nb nucleus, i.e., the spacing of the chromosomes is not conducive to the rejoining of broken ends that is necessary for the formation of rings and dicentrics. (4) Perhaps S-dependent chromosome breakage by chemicals is a factor. The real explanation awaits further experiments.

5.3.2. Mitotic Apparatus Effects

Interference with the function of the mitotic apparatus of the GHNb has been demonstrated to occur when embryos are exposed *in vitro* to ultraviolet radiation,[21,104] colchicine,[37] or sodium hydrosulfite.[111] Observations on living cells reveal the dose–response (38°C). At lower doses of these agents, there is a reduction in size of the spindle and abnormalities in cytokinesis, and at higher doses the spindle can be completely disorganized so that the chromosomes are scattered at random in the cell (C-metaphase). At very high doses of colchicine, such as $25–50 \times 10^{-6}$ M, cells in anaphase at the beginning of treatment form a restitution tetraploid nucleus as a result of spindle disorganization. During recovery from colchicine treatment, the cells in C-metaphase reorganize the spindle and complete division. Prolonged exposure for 24 hr to a low concentration of colchicine (0.1×10^{-6} M) does not cause disorganization of the spindle but does affect its function so that metaphase is retarded. More relevant for mutagen studies is the occasional occurrence of a single chromosome at one pole of a metaphase spindle, which may lead to aneuploidy.[37] These data are helpful in evaluating similar effects produced by environmental chemicals.

Recently, C-mitotic effects have been demonstrated to result from exposure of grasshopper embryos *in vivo* to the vapor of several volatile liquids, including halothane, benzene, toluene, and chloroform.[73] Halothane only induces complete C-mitotic block at amounts of 0.2 ml or more per jar ($\geq 463,000$ ppm); the MI was not elevated until the amount of halothane used was 0.5 ml ($\sim 1,160,000$ ppm) or more per jar. Benzene at 0.05 ml per jar ($\sim 138,000$ ppm) was shown to cause a complete metaphase block in two of the three embryos tested, and the average MI in treated embryos was nine times that of the controls. Toluene at 0.05 ml per jar ($\sim 115,000$ ppm) induced a complete C-mitotic block; however, no increase in the MI was observed. Failure of toluene to increase the MI in embryos may be due to its toxicity or its preventing cells from reaching metaphase. Chloroform was the most potent mitotic arrestant among the four agents that gave positive results with the grasshopper embryo assay: 0.05 ml per jar ($\sim 152,000$ ppm) induced a complete C-mitotic block in all embryos tested, and the average MI was 11 times that of the controls.

The importance of studying chemicals that disturb mitotic apparatus function relates to possible carcinogenic action and heritable effects. Polyploid and aneuploid cells, both of which can result from exposure to mitotic arrestants, occur in tumor cells, and aneuploidy is a significant cause of germ cell-derived genetic disorders in humans. Although a cause–effect relation of these chromosome abnormalities to tumor incidence in humans has not been firmly established, there are strong suggestive data in humans occupationally exposed to industrial solvents, such as benzene,[27,84] that spur us to examine the effects of solvents on cells.

6. Advantages of the Grasshopper Neuroblast for Testing Chemicals

The grasshopper neuroblast (GHNb) has a number of features that make it a simple, fast, reproducible, and inexpensive eukaryotic test system. The principal ones can be summarized as follows.

1. No spontaneous chromosome aberrations occur when rearing conditions are optimal.
2. The GHNb is very sensitive to mutagen-induced chromosome damage, so the effects of chronic exposure to very low doses can be quantitated.
3. Mutagen-induced acentric chromosome fragments, the prevalent aberration at low doses, can be rapidly and unequivocally scored in late anaphase–very early telophase Nbs.
4. The frequency of chromosome aberrations induced by a given dose of a mutagen is highly reproducible.
5. Use of late anaphase–very early telophase Nbs for chromosome analysis eliminates the need for Colcemid and permits the simultaneous detection of other mutagen effects, such as spindle disorganization.
6. The short cell cycle (4 hr in *Chortophaga*, 2 hr in *M. sanguinipes*), the large number of Nbs per embryo, the simple and quick squash method, and the ease of scoring make for fast collection of data. The short cell cycle is also advantageous for testing chemicals with short half-lives.
7. The cell cycle of the GHNb has 22 distinct phases, so observations on living cells reveal the specific portions of the cycle affected by a mutagen and provide accurate quantitation of effects on cell progression.
8. Large numbers of grasshopper embryos (eggs) can be obtained seasonally from field-collected animals or year-round from a laboratory colony, whose maintenance is simple and inexpensive.

ACKNOWLEDGMENTS

The authors are grateful to N. B. Weber and C.-K. Chao for technical assistance, C. S. Starnes and J. Cheek for manuscript preparation, Dr. J. G. Carlson for critical review of the manuscript, Drs. B. Nicklas, D. Wise, and S. N. Visscher for advice on grasshopper colony maintenance, and Altrusa International for a predoctoral fellowship for J. C. L. This research was supported by NIH grant ES03077 and by a Southwestern Medical Foundation grant.

7. References

1. F. O. Albrecht, Facteurs internes et fluctuations des effectifs chez *Nomadacris septemfasciata* (Serv.), *Bull. Biol. Fr. Belg.* 93, 414–461 (1959).

2. F. O. Albrecht, M. Verdier, and R. E. Blackith, Détermination de la fertilité par l'effet de group chez le criquet migrateur (*Locusta migratoria migratorioides* R. and F.), *Bull. Biol. Fr. Belg. 92*, 350–427 (1958).

3. D. T. Anderson, The development of hemimetabolous insects, in: *Developmental Systems: Insects* (S. J. Counce and C. H. Waddington, eds.), Vol. 1, pp. 95–163, Academic Press, New York (1972).

4. L. K. Anderson, S. M. Stack, and J. B.' Mitchell, An investigation of the basis of a current hypothesis for the lack of G-banding in plant chromosomes, *Exp. Cell Res. 138*, 433–436 (1982).

5. W. Au, M. A. Butler, S. E. Bloom, and T. S. Matney, Further study of the genetic toxicity of gentian violet, *Mutat. Res. 66*, 103–112 (1979).

6. C. Auerbach, The effects of six years of mutagen testing on our attitude to the problems posed by it, *Mutat. Res. 33*, 3–10 (1975).

7. V. Baden, Embryology of the nervous system in the grasshopper, *Melanoplus differentialis* (Acrididae; Orthoptera), *J. Morphol. 60*, 159–188 (1936).

8. H. Bartsch, T. Kuroki, M. Roberfroid, and C. Malaveille, Metabolic activation systems *in vitro* for carcinogen/mutagen screening tests, in: *Chemical Mutagens. Principles and Methods for Their Detection*, Vol. 7 (F. J. de Serres and A. Hollaender, eds.), pp. 95–161, Plenum Press, New York (1981).

9. C. M. Bate, Embryogenesis of an insect nervous system. I. A map of the thoracic and abdominal neuroblasts in *Locusta migratoria, J. Embryol. Exp. Morphol. 35*, 107–123 (1976).

10. D. Bentley, H. Keshishian, M. Shankland, and A. Toroian-Raymond, Quantitative staging of embryonic development of the grasshopper, *Schistocerca nitens, J. Embryol. Exp. Morphol. 54*, 47–74 (1979).

11. C. A. H. Bigger, J. E. Tomaszewski, A. Dipple, and R. S. Lake, Limitations of metabolic activation systems used with *in vitro* tests for carcinogens, *Science 209*, 503–505 (1980).

12. J. G. Carlson, Mitotic behavior of induced chromosomal fragments lacking spindle attachments in the neuroblasts of the grasshopper, *Proc. Natl. Acad. Sci. USA 24*, 500–507 (1938).

13. J. G. Carlson, An analysis of X-ray induced single breaks in neuroblast chromosomes of the grasshopper (*Chortophaga viridifasciata*), *Proc. Natl. Acad. Sci. USA 27*, 42–47 (1941).

14. J. G. Carlson, Protoplasmic viscosity changes in different regions of the grasshopper neuroblast during mitosis, *Biol. Bull. 90*, 109–121 (1946).

15. J. G. Carlson, Microdissection studies of the dividing neuroblast of the grasshopper, *Chortophaga viridifasciata* (De Geer), *Chromosoma 5*, 199–220 (1952).

16. J. G. Carlson, The grasshopper neuroblast culture technique and its value in radiobiological studies, *Ann. NY Acad. Sci. 95*(2), 932–941 (1961).

17. J. G. Carlson, A detailed analysis of X-ray-induced prophase delay and reversion of grasshopper neuroblasts in culture, *Radiat. Res. 37*, 1–14 (1969).

18. J. G. Carlson, X-ray-induced prophase delay and reversion of selected cells in certain avian and mammalian tissues in culture, *Radiat. Res. 37*, 15–30 (1969).

19. J. G. Carlson, Anaphase chromosome movement in the unequally dividing grasshopper neuroblast and its relation to anaphases in other cells, *Chromosoma 64*, 191–206 (1977).

20. J. G. Carlson and M. E. Gaulden, Grasshopper neuroblast techniques, *Meth. Cell Physiol. 1*, 229–276 (1964).

21. J. G. Carlson and A. Hollaender, Mitotic effects of ultraviolet radiation of the 2250 A region, with special reference to the spindle and cleavage, *J. Cell. Comp. Physiol.* *31*, 149–174 (1948).

22. J. G. Carlson and R. D. McMaster, Nucleolar changes induced in the grasshopper neuroblast by different wavelengths of ultraviolet radiation and their capacity for photorecovery, *Exp. Cell Res.* *2*, 434–444 (1951).

23. E. E. Carothers, Notes on the taxonomy, development and life history of certain Acrididae (Orthoptera), *Trans. Am. Entomol. Soc.* *49*, 7–24 (1923).

24. T. T. Chen and G. R. Wyatt, Juvenile hormone control of vitellogenin synthesis in *Locusta migratoria*, in: *International Conference on Regulation of Insect Development and Behavior (Sci. Pap. Inst. Organ. Phys. Chem. Wroclaw Tech Univ.,* No. 22), Part II, pp. 535–566, Wroclaw, Poland (1981).

25. A. Chrominski, S. N. Visscher, and R. Jurenka, Exposure to ethylene changes nymphal growth rate and female longevity in the grasshopper *Melanoplus sanguinipes, Naturwissenschaften 69*, 45–46 (1982).

26. Committee 17, Environmental mutagenic hazards, *Science 187*, 503–514 (1975).

27. B. J. Dean, Genetic toxicology of benzene, toluene, xylenes and phenols, *Mutat. Res.* *47*, 75–97 (1978).

28. F. J. de Serres, The utility of short-term tests for mutagenicity in the toxicological evaluation of environment agents, *Mutat. Res. 33*, 11–15 (1975).

29. F. J. de Serres and J. Ashby, eds., *Progress in Mutation Research*, Vol. 1, *Evaluation of Short-Term Tests for Carcinogens*, Elsevier/North-Holland, New York (1981).

30. J. W. Drake, Environmental mutagenesis: Evolving strategies in the USA, *Mutat. Res. 33*, 65–72 (1975).

31. U. Francke and N. Oliver, Quantitative analysis of high-resolution trypsin-Giemsa bands on human prometaphase chromosomes, *Hum. Genet. 45*, 137–165 (1978).

32. J. R. Fry, C. A. Jones, P. Wiebkin, P. Bellemann, and J. W. Bridges, The enzymic isolation of adult rat hepatocytes in a functional and viable state, *Anal. Biochem. 71*, 341–350 (1976).

33. M. E. Gaulden, Telophase behavior of extranuclear chromatin and its bearing on telophase changes in chromosomes, *Experientia 10*, 18–20 (1954).

34. M. E. Gaulden, DNA synthesis and X-ray effects at different mitotic stages in grasshopper neuroblasts, *Genetics 41*, 645 (1956).

35. M. E. Gaulden, in: *Mitogenesis* (N. S. Ducoff and C. F. Ehret, eds.), pp. 38–39, University of Chicago Press (1959).

36. M. E. Gaulden, Chromosome aberrations as a cause of subtle teratogenesis and use of the grasshopper neuroblast to test potential mutagens and teratogens, *Cytogenet. Cell Genet. 33*, 114–118 (1982).

37. M. E. Gaulden and J. G. Carlson, Cytological effects of colchicine on the grasshopper neuroblast *in vitro* with special reference to the origin of the spindle, *Exp. Cell Res.* *2*, 416–433 (1951).

38. M. E. Gaulden and K. L. Kokomoor, Influence of yolk on mitotic rate in untreated and x-rayed grasshopper neuroblasts *in vitro*, *Proc. Soc. Exp. Biol. Med. 90*, 309–314 (1955).

39. M. E. Gaulden and J. C. Liang, Insect cells for testing clastogenic agents, in: *Cytogenetic Assays of Environmental Mutagens* (T. C. Hsu, ed.), pp. 107–135, Allenheld, Osmun & Co., Totowa, New Jersey (1982).

40. M. E. Gaulden and R. C. Murry, Medical radiation and possible adverse effects on the human embryo, in: *Radiation Biology in Cancer Research* (R. E. Meyn and H. R. Withers, eds.), pp. 277–294, Raven Press, New York (1980).

41. M. E. Gaulden and R. P. Perry, Influence of the nucleolus on mitosis as revealed by ultraviolet microbeam irradiation, *Proc. Natl. Acad. Sci. USA 44*, 553–559 (1958).

42. M. E. Gaulden and C. B. Read, Linear dose–response of acentric chromosome fragments down to 1 R of x-rays in grasshopper neuroblasts, a potential mutagen test system, *Mutat. Res. 49*, 55–60 (1978).

43. M. E. Gaulden, M. Nix, and J. Moshman, Effects of oxygen concentration of X-ray-induced mitotic inhibition in living *Chortophaga* neuroblasts, *J. Cell. Comp. Physiol. 41*, 451–470 (1953).

44. M. E. Gaulden, M. J. Ferguson, and N. B. Weber, A eukaryotic cell with a 2-hour cell cycle (38°C) for determining effects of mutagens on chromosomes and cell cycle kinetics, *Environ. Mutagen. 5*, 375–376 (1983).

45. R. E. Gibson and S. M. D'Ambrosio, Differing levels of excision repair in human fetal dermis and brain cells, *Photochem. Photobiol. 35*, 181–185 (1982).

46. C. S. Goodman, Neuron duplications and deletions in locust clones and clutches, *Science 197*, 1384–1386 (1977).

47. C. S. Goodman, Isogenic grasshoppers: Genetic variability in the morphology of identified neurons, *J. Comp. Neurol. 182*, 681–705 (1978).

48. C. S. Goodman, Isogenic grasshoppers: Genetic variability and development of identified neurons, in: *Neurogenetics: Genetic Approaches to the Nervous System* (X. O. Breakefield, ed.), pp. 101–151, Elsevier, New York (1979).

49. C. S. Goodman, Embryonic development of identified neurons in the grasshopper, in: *Neuronal Development* (N. C. Spitzer, ed.), pp. 171–212, Plenum Press, New York (1982).

50. J. Greilhuber, Why plant chromosomes do not show G-bands, *Theor. Appl. Genet. 50*, 121–124 (1977).

51. J. C. Hartley, The shell of acridid eggs, *Q. J. Microsc. Sci. 102*, 249–255 (1961).

52. J. E. Henry, B. P. Nelson, and J. W. Jutila, Pathology and development of the grasshopper inclusion body virus in *Melanoplus sanguinipes*, *J. Virol. 3*, 605–610 (1969).

53. J. E. Henry and E. A. Oma, Sulphonamide antibiotic control of *Malameba locustae* (King and Taylor) and its effect on grasshoppers, *Acrida 4*, 217–226 (1975).

54. K. C. Highnam and P. T. Haskell, The endocrine systems of isolated and crowded *Locusta* and *Schistocerca* in relation to oocyte growth and the effects of flying upon maturation, *J. Insect Physiol. 10*, 849–864 (1964).

55. H. E. Hinton, *Biology of Insect Eggs*, Vol. 1, Pergamon Press, Oxford, (1981).

56. A. Hollaender, A history of attempts to quantify environmental mutagenesis, in: *Environmental Mutagens and Carcinogens* (T. Sugimura, S. Kondo, and H. Takebe, eds.), pp. 21–36, Alan R. Liss, New York (1982).

57. P. Hunter-Jones, Rearing and Breeding of Locusts in the Laboratory, Pamphlet of Anti-Locust Research Centre, Wrights Lane, London, W.8., England (1966).

58. *International Conference on Regulation of Insect Development and Behavior* (*Sci. Pap. Inst. Organ. Phys. Chem., Wroclaw Tech. Univ.*, No. 22), Parts I and II, Wroclaw, Poland (1981).

59. L. J. Jacobs, J. A. Marx, and T. R. Grey, Comparison of the mutagenic responses of lung-derived and skin-derived human diploid fibroblast populations, *Environ. Mutagen. 4*, 373 (1982).

60. M. Jacobson, *Developmental Neurobiology*, 2nd ed., Plenum Press, New York (1978).

61. D. Jenssen and C. Ramel, Factors affecting the induction of micronuclei at low doses of x-rays, MMS and dimethyl-nitrosamine in mouse erythroblasts, *Mutat. Res. 58*, 51–65 (1978).

62. O. A. Johannsen and F. H. Butt, *Embryology of Insects and Myriapods*, McGraw-Hill, New York (1941).

63. B. M. Jones, Endocrine activity during insect embryogenesis. Control of events in development following the embryonic moult (*Locusta migratoria* and *Locustana pardalina*, Orthoptera), *J. Exp. Biol. 33*, 685–696 (1956).

64. R. G. Kessel, Cytological studies on the subesophageal body cells and pericardial cells in embryos of the grasshopper, *Melanoplus differentialis* (Thomas), *J. Morphol. 109*, 289–319 (1961).

65. R. L. King and E. H. Slifer, Insect development. VIII. Maturation and early development of unfertilized grasshopper eggs, *J. Morphol. 56*, 603–619 (1934).

66. J. B. Kreasky, A growth factor in romaine lettuce for the grasshoppers *Melanoplus sanguinipes* (F.) and *M. bivattatus* (Say), *J. Insect Physiol. 8*, 493–504 (1962).

67. Laboratory Safety Monograph, Supplement to the NIH Guidelines for Recombinant DNA Research, U. S. Department of Health, Education and Welfare, Public Health Service, National Institutes of Health, Bethesda, Maryland, (January 2, 1979), p. 5.

68. W. H. R. Langridge and D. W. Roberts, Structural proteins of *Amsacta moorei*, *Euxoa auxiliaris*, and *Melanoplus sanguinipes* entomopoxviruses, *J. Invertebr. Pathol. 39*, 346–353 (1982).

69. W. M. Leach, The thymidine pool in grasshopper neuroblasts during mitosis, *J. Cell Biol. 36*, 282–286 (1968).

70. J. C. Liang and M. E. Gaulden, The neuroblast of the grasshopper embryo as a new mutagen test system. I. *In vitro* radiosensitivity, *Mutat. Res. 93*, 401–408 (1982).

71. J. C. Liang and M. E. Gaulden, Neuroblast of the grasshopper embryo as a new mutagen test system. II. Chromosome breakage induced by *in vitro* exposure of embryos to the direct-acting mutagens 4NQO, MNNG, Adriamycin and bleomycin, *Environ. Mutagen 4*, 279–290 (1982).

72. J. C. Liang and M. E. Gaulden, The neuroblast of the grasshopper embryo as a new mutagen test system. III. Chromosome breakage induced by cyclophosphamide is greater with activation by rat hepatocytes than by S12 mix, *Mutat. Res. 119*, 71–77 (1983).

73. J. C. Liang, T. C. Hsu, and J. E. Henry, Cytogenetic assays for mitotic poisons: The grasshopper embryo system for volatile liquids, *Mutat. Res. 113*, 467–479 (1983).

74. D. C. Lloyd, R. J. Purrott, G. W. Dolphin, D. Bolton, and A. A. Edwards, The relationship between chromosome aberrations and low LET radiation dose to human lymphocytes, *Int. J. Radiat. Biol. 28*, 75–90 (1975).

75. D. Ludwig, The effect of different relative humidities on respiratory metabolism and survival of the grasshopper *Chortophaga viridifasciata* De Geer, *Physiol. Zool. 10*, 342–351 (1937).

76. T. H. Ma, Micronuclei induced by X-rays and chemical mutagens in meiotic pollen mother cells of *Tradescantia*, a promising mutagen test system, *Mutat. Res. 64*, 307–313 (1979).

77. B. S. S. Masters, C. H. Williams, Jr., and H. Kamin, The preparation and properties of microsomal TPNH-cytochrome *c* reductase from pig liver, in: *Methods in Enzymology* (R. W. Estabrook and M. E. Pullman, eds.), Vol. 10, pp. 565–572, Academic Press, New York (1967).

78. P. C. Mazuranich, Construction of a metal-framed cage for studies with grasshoppers, *Acrida 4*, 151–154 (1975).

79. C. E. McClung, The accessory chromosome—Sex determinant? *Biol. Bull. 3*, 43–84 (1902).

80. R. A. McGrath, X-ray-induced incorporation of tritiated thymidine into deoxyribonucleic acid of grasshopper neuroblast chromosomes, *Radiat. Res. 19*, 526–537 (1963).

81. R. A. McGrath, W. M. Leach, and J. G. Carlson, Cell stages refractory to thymidine incorporation induced by x-rays, *Exp. Cell Res. 37*, 39–44 (1965).

82. K. S. McKinlay and W. K. Martin, Effects of temperature and piperonyl butoxide on the toxicity of six carbamates to the grasshopper *Melanoplus sanguinipes* (=*M. bilituratus*), *Can. Entomol. 99*, 748–751 (1967).

83. J. W. McNabb, A study of the chromosomes in meiosis, fertilization, and cleavage in the grasshopper egg (Orthoptera), *J. Morphol. Physiol. 45*, 47–95 (1928).

84. F. Mitelman, P. G. Nilsson, L. Brandt, G. Alimena, R. Gastaldi, and B. Dallapiccola, Chromosome pattern, occupation, and clinical features in patients with acute nonlymphocytic leukemia, *Cancer Genet. Cytogenet. 4*, 197–214 (1981).

85. W. Moore, Jr. and M. Colvin, Chromosomal changes in the Chinese hamster thyroid following x-irradiation *in vivo*, *Int. J. Radiat. Biol. 14*, 161–167 (1968).

86. G. A. Mueller, M. E. Gaulden, and W. Drane, The effects of varying concentrations of colchicine on the progression of grasshopper neuroblasts into metaphase, *J. Cell Biol. 48*, 253–265 (1971).

87. M. S. S. Murthy, Radiation equivalence of genotoxic chemicals. Validation in cultured mammalian cell lines, *Mutat. Res. 94*, 189–197 (1982).

88. O. E. Nelsen, The segregation of the germ cells in the grasshopper, *Melanoplus differentialis* (Acrididae; Orthoptera), *J. Morphol. 55*, 545–575 (1934).

89. D. R. Nelson, W. D. Valovage, and R. D. Frye, Infection of grasshoppers with *Entomophaga* (=*Entomophthora*) *grylli* by injection of germinating resting spores, *J. Invertebr. Pathol. 39*, 416–418 (1982).

90. S. Neumann-Visscher, The embryonic diapause of *Aulocara elliotti* (Orthoptera, Acrididae), *Cell Tiss. Res. 174*, 433–452 (1976).

91. M. Norris, Reproduction in the desert locust (*Schistocerca gregaria* Forsk.) in relation to density and phase, *Anti-Locust Bull. London 13* (1952).

92. Y. Ohnuki, Demonstration of the spiral structure of human chromosomes, *Nature 208*, 916–917 (1965).

93. O. Okelo, Studies on the Reproductive Physiology of the Female Grasshopper, *Schistocerca vaga* Scudder, Ph.D. Thesis, University of California, Berkeley (1975).

94. D. Otte, *The North American Grasshoppers. Acrididae: Gomphocerinae and Acridinae*, Vol. I, Harvard University Press, Cambridge, Massachusetts (1981).

95. D. Otte and K. Williams, Environmentally induced color dimorphisms in grasshoppers, *Syrbula admirabilis, Dichromorphia viridis*, and *Chortophaga viridifasciata, Ann. Entomol. Soc. Am. 65*, 1154–1161 (1972).

96. R. Pickford, Observations on the reproductive potential of *Melanoplus bilituratus* (Wlk.) (Orthoptera: Acrididae) reared on different food plants in the laboratory, *Can. Entomol. 90*, 483–485 (1958).

97. R. Pickford, Survival, fecundity, and population growth of *Melanoplus bilituratus* (Wlk.) (Orthoptera: Acrididae) in relation to date of hatching, *Can. Entomol. 92*, 1–10 (1960).

98. R. Pickford and R. L. Randell, A non-diapause strain of the migratory grasshopper, *Melanoplus sanguinipes* (Orthoptera: Acrididae), *Can. Entomol. 101*, 894–896 (1969).

99. D. M. Prescott, The cell cycle and the G1 period, in: *Cell Growth* (C. Nicolini, ed.), pp. 305–314, Plenum Press, New York (1980).

100. R. J. Preston, J. G. Brewen, and K. P. Jones, Radiation-induced chromosome aberrations in Chinese hamster leukocytes, A comparison of *in vivo* and *in vitro* exposures, *Int. J. Radiat. Biol. 21*, 397–400 (1972).

101. J. C. Reese, Insect dietetics: Complexities of plant–insect interactions, in: *Current Topics in Insect Endocrinology and Nutrition* (G. Bhaskaran, S. Friedman, and J. G. Rodriguez, eds.), pp. 317–335, Plenum Press, New York (1981).

102. H. Remmer, H. Griem, J. B. Schenkman, and R. W. Estabrook, Methods for the elevation of hepatic microsomal mixed function oxidase levels and cytochrome P-

450, in: *Methods in Enzymology* (R. W. Estabrook and M. E. Pullman, eds.), Vol. 10, pp. 703–708, Academic Press, New York (1967).

103. P. W. Riegert, Embryological development of a nondiapause form of *Melanoplus bilituratus* Walker (Orthoptera: Acrididae), *Can. J. Zool. 39*, 491–494 (1961).
104. H. S. Roberts, The mechanism of cytokinesis in neuroblasts of *Chortophaga viridifasciata* (De Geer), *J. Exp. Zool. 130*, 83–105 (1955).
105. M. L. Roonwal, Studies on the embryology of the African migratory locust, *Locusta migratoria migratorioides* Reiche and Frm. (Orthoptera, Acrididae). II. Organogeny, *Philos. Trans. R. Soc. London B 227*, 175–244 (1937).
106. J. G. Saha, R. L. Randell, and P. W. Riegert, Component fatty acids of grasshoppers (Orthoptera: Acrididae), *Life Sci. 5*, 1597–1603 (1966).
107. R. W. Salt, A key to the embryological development of *Melanoplus bivittatus* (Say), *M. mexicanus mexicanus* (Sauss.), and *M. packardii* Scudder, *Can. J. Res. D 27*, 233–235 (1949).
108. R. W. Salt, Water uptake in eggs of *Melanoplus bivittatus* (Say), *Can. J. Res. D 27*, 236–242 (1949).
109. S. O. Schiff, Ribonucleic acid synthesis in neuroblasts of *Chortophaga viridifasciata* (de Geer), as determined by observations of individual cells in the mitotic cycle, *Exp. Cell Res. 40*, 264–276 (1965).
110. S. S. Sekhon, Eleanor H. Slifer: An appreciation, *J. Morphol. 168*, 3–4 (1981).
111. E. I. Shaw, Protection by sodium hydrosulfite against x-ray-induced mitotic inhibition in grasshopper neuroblast, *Proc. Soc. Exp. Biol. Med. 92*, 232–236 (1956).
112. E. H. Slifer, Insect development. IV. External morphology of grasshopper embryos of known age and with a known temperature history, *J. Morphol. 53*, 1–21 (1932).
113. E. H. Slifer, The origin and fate of the membranes surrounding the grasshopper egg; together with some experiments on the source of the hatching membrane, *Q. J. Microsc. Sci. 79*, 493–506 (1937).
114. E. H. Slifer, The formation and structure of a special water-absorbing area in the membranes covering the grasshopper egg, *Q. J. Microsc. Sci. 80*, 437–459 (1938).
115. E. H. Slifer, A cytological study of the pleuropodia of *Melanoplus differentialis* (Orthoptera, Acrididae) which furnishes new evidence that they produce the hatching enzyme, *J. Morphol. 63*, 181–206 (1938).
116. E. H. Slifer, Removing the shell from living grasshopper eggs, *Science 102*, 282 (1945).
117. E. H. Slifer, Variations, during development, in the resistance of the grasshopper egg to a toxic substance, *Ann. Entomol. Soc. Am. 42*, 134–140 (1949).
118. E. H. Slifer, Diapause in the eggs of *Melanoplus differentialis* (Orthoptera, Acrididae), *J. Exp. Zool. 138*, 259–282 (1958).
119. E. H. Slifer and R. L. King, The inheritance of diapause in grasshopper eggs, *J. Hered. 52*, 39–44 (1961).
120. D. S. Smith, Utilization of food plants by the migratory grasshopper, *Melanoplus bilituratus* (Walker) (Orthoptera: Acrididae), with some observations on the nutritional value of the plants, *Ann. Entomol. Soc. Am. 52*, 674–680 (1959).
121. D. S. Smith, Fecundity and oviposition in the grasshoppers *Melanoplus sanguinipes* (F.) and *Melanoplus bivittatus* (Say.), *Can. Entomol. 98*, 617–621 (1966).
122. D. S. Smith, Crowding in grasshoppers. I. Effect of crowding within one generation on *Melanoplus sanguinipes*, *Ann. Entomol. Soc. Am. 63*, 1775–1776 (1970).
123. D. S. Smith and F. E. Northcott, The effects on the grasshopper, *Melanoplus mexicanus mexicanus* (Sauss.) (Orthoptera: Acrididae), of varying the nitrogen content in its food plant, *Can. J. Zool. 29*, 297–304 (1951).

124. B. J. Stevens, The fine structure of the nucleolus during mitosis in the grasshopper neuroblast cell, *J. Cell Biol. 24*, 349–368 (1965).

125. R. E. Stephens and J. G. Carlson, Action of actinomycin D on grasshopper neuroblasts in culture, *Exp. Cell Res. 74*, 42–50 (1972).

126. R. E. Stephens, M. B. Cole, Jr., A. A. Cole, and J. G. Carlson, Parasitization of *Chortophaga viridifasciata* by larvae of *Scelio bisulcus*, *Association of Southeastern Biologists Bull. 15*, 55 (1968).

127. I. Sunshine, ed., *CRC Handbook of Analytical Toxicology*, p. 673, Chemical Rubber Co., Cleveland, Ohio (1969).

128. C. A. Tauber and M. J. Tauber, Insect seasonal cycles: Genetics and evolution, *Annu. Rev. Ecol. Syst. 12*, 281–308 (1981).

129. J. Thornton and M. E. Gaulden, Relation of X-ray-induced thymidine uptake to chromosome reversion at prophase in grasshopper neuroblasts, *Int. J. Radiat. Biol. 19*, 65–78 (1971).

130. K. R. Tsang, F. A. Freeman, T. J. Kurtti, M. A. Brooks, and J. E. Henry, New cell lines from embryos of *Melanoplus sanguinipes* (Orthoptera: Acrididae), *Acrida 10*, 105–112 (1981).

131. B. Uvarov, *Grasshoppers and Locusts, A Handbook of General Acridology*, Cambridge University Press, London (1966).

132. S. N. Van Horn, Studies on the embryogenesis of *Aulocara elliotti* (Thomas) (Orthoptera, Acrididae). II. Developmental variability and the effects of maternal age and environment, *J. Morphol. 120*, 115–134 (1966).

133. S. N. Visscher, Regulation of grasshopper fecundity, longevity and egg viability by plant growth hormones, *Experientia 36*, 130–131 (1980).

134. S. N. Visscher, Effects of abscisic acid in animal growth and reproduction, in: *Abscisic Acid* (F. T. Addicott, ed.), pp. 553–579 Praeger, New York (1983).

135. S. N. Visscher, R. Lund, and W. Whitmore, Host plant growth temperatures and insect rearing temperatures influence reproduction and longevity in the grasshopper, *Aulocara elliotti* (Orthoptera: Acrididae), *Environ. Entomol. 8*, 253–258 (1979).

136. D. F. Went, Egg activation and parthenogenetic reproduction in insects, *Biol. Rev. 57*, 319–344 (1982).

137. W. M. Wheeler, A contribution to insect embryology, *J. Morphol. 8*, 1–160 (1893).

138. M. J. D. White, *Animal Cytology and Evolution*, 3rd ed., Cambridge University Press, Cambridge (1973).

CHAPTER 2

Comparison of the Mutagenic Responses of Lung-Derived and Skin-Derived Human Diploid Fibroblast Populations

Lois Jacobs, James A. Marx, and Christian L. Bean

1. Introduction

We have established a series of pairs of lung- and skin-derived fibroblast cultures from human embryonic tissues. Each pair of skin and lung populations was established simultaneously from the same individual using carefully controlled, identical conditions. These paired populations provide a unique opportunity to study differences between normal diploid fibroblast populations from skin and lung without the confusion of genetic differences between donors. They also permit the study of differences among individuals without the confusion of variables such as different ages of donors, different biopsy sites, different conditions for initiating cell populations, and different cell population doubling levels. Our goals have been (1) to analyze the variance in cell killing

Lois Jacobs, James A. Marx, and Christian L. Bean • Department of Medical Genetics, University of Wisconsin-Madison, Madison, Wisconsin 53706.

and mutability among normal individuals in the human population and (2) to analyze the variance in mutability between fibroblast cultures derived from two different human tissues.

1.1. Human Cells *in Vitro* as a Model Test System

The direct evaluation of changes occurring in the somatic or germ cells of humans who have been accidentally exposed to known or suspected carcinogens or mutagens has not been technically or ethically possible. As an alternative, various *in vivo* and *in vitro* models have been developed to substitute for the intact human. Unfortunately, no single model or combination of models has thus far been completely satisfactory. Extensive basic research and technical innovation will be required before model systems can be adequately calibrated for accurately predicting risks of hereditary damage to the somatic or germ cells of humans.

We have proposed that systematic, quantitative investigations of mutagenic changes occurring *in vitro* in normal, diploid human cells derived from different tissues in the human body and from different individuals in the human population should contribute significantly to the future interpretation of other models as they relate to human cancer and hereditary disease. Sound quantitative data describing the range of mutagenic responses of cells from normal individuals in the human population are needed as baseline control values for further investigation of high-risk groups, for studies of the mechanisms of mutagenesis and the role of somatic mutation in human cancer, and for evaluation of the validity of short-term test systems for making risk estimates in screens for potential mutagens and carcinogens.

As a basic research tool, cultured diploid human fibroblasts offer several advantages, which have been discussed in greater detail elsewhere.[17,19] Of greatest importance, the cells are of human origin and they are karyotypically stable. Skin fibroblast cultures can be established from biopsies taken from most individuals in the human population, including patients and their relatives in cancer-prone families. Normal human fibroblast cultures have highly stable, diploid karyotypes that can be studied in detail using chromosome banding techniques. In contrast to other mammalian cell lines, primary cultures of diploid human fibroblasts do not transform spontaneously to established permanent lines. The characterization of the normal diploid human fibroblast as an *in vitro* genetic organism is now quite extensive and is increasing rapidly. Data are available describing *in vitro* growth characteristics and populations kinetics, detailed karyotype analyses, DNA repair, DNA replication patterns, gene mapping, and genetic analysis of phenotypic expression utilizing somatic cell fusion.

Possible technical limits or special requirements of cultured diploid human fibroblasts for mutagenesis studies need not be seen as insurmountable disadvantages (for further discussion of these points, see Ref. 19). Although fibroblasts are neither the germ cells that transmit hereditary disease nor the epithelial cells believed to be most frequently involved in human cancers, one of the major values of work with human fibroblasts may be the development of technical capabilities and insight for future studies of mutagenesis in other human cell types. With available tissue culture techniques, most other human tissues do not consistently produce vigorously proliferating cultures with sufficient clone-forming ability and *in vitro* doubling potential to permit the detection, isolation, and characterization of spontaneous and induced variants. We have, however, developed procedures for routinely establishing and maintaining human diploid fibroblast cultures with cloning efficiencies and doubling potentials sufficient to perform detailed mutagenesis experiments.[16,17,19]

Experiments using diploid human fibroblasts can be carried out over long periods of time in spite of the fact that the cultures have a finite *in vitro* life span. An almost unlimited supply of the same cell population can be made available by storing a pooled population of viable cells from a single biopsy in identical aliquots in liquid nitrogen. One ampule of the same pooled cell population can be used to initiate each experiment, removing the possibility of shifts in population characteristics that may occur in continuously propagated permanent cell lines.

With present methods, human fibroblasts are not the organism of choice for prescreening programs that require quick, easy, and inexpensive evaluation of large numbers of compounds, because of the cells' growth requirements and their somewhat longer doubling time (approximately 18–22 hr). For the basic study of genetic change in normal human somatic cells, however, experiments can be performed readily; the information gained should be worth the additional effort.

1.2. Choice of the Genetic Marker: The *hpt* Locus

Quantitative mutagenesis determinations are dependent on the availability of selectable markers. The most widely used method of detecting single-locus mutation in diploid human fibroblasts is the selection for 8-azaguanine- or 6-thioguanine-resistant cells.[43] Resistance to 8-azaguanine (AG) or 6-thioguanine (TG) results from deficiencies of hypoxanthine phosphoribosyltransferase (HPRT, E.C. 2.4.2.8)[11] activity, coded for by the X-chromosomal *hpt* locus (reviewed in Ref. 7). HPRT is a salvage enzyme that ordinarily converts hypoxanthine, guanine, and xanthine to their respective nucleotides. Both AG and TG are innocuous to cells at 10^{-6}–10^{-4} M concentrations until they are converted to a toxic ribonucleotide by HPRT. The proliferation of normal cells that have HPRT is inhibited in medium containing AG or TG, but because HPRT is dispensible under ordinary culture conditions, HPRT-deficient mutant cells can proliferate to form resistant colonies.

All forward mutations that sufficiently reduce HPRT activity, even to complete absence, can be detected as viable mutants, making the TG-selection system a sensitive detection method. This is in contrast to ouabain resistance[30] or α-aminitin resistance[3] systems that detect only certain mutations in the structural genes (probably missense) because the enzymes are vital: complete absence of the enzyme activity inhibits cell proliferation. Because the *hpt* locus is X-linked,[23,26,33] single mutations, dominant or recessive, are phenotypically expressed both in female cells and male somatic cells.

We have demonstrated that large numbers of independent analogue-resistant colonies can be isolated and characterized from independent mutagenesis experiments, thus verifying their mutant nature.[8,17,18] Thioguanine-resistant cultures established from biopsies taken from spontaneously occurring mutant persons in the human population (the Lesch–Nyhan syndrome) are the *in vivo* standard to which cell variants arising *in vitro* are compared.[2,8,33]

A potential limitation of studies of mutation using thioguanine selection is that the *hpt* locus represents only a small fraction of the entire human genome. Mutagenic responses of the *hpt* locus may not be representative of other X-linked or autosomal loci. However, recent detailed studies of mutation in human diploid fibroblasts at the autosomal *APRT* locus (APRT: adenine phosphoribosyltransferase, E.C. 2.4.2.70) using our mutagenesis protocol have resulted in mutagenic responses similar to that predicted by our earlier *hpt* data.[41]

1.3. The Mutagen

N-Methyl-*N'*-nitro-*N*-nitrosoguanidine (MNNG) is an alkylating agent that induces mainly AT to GC transitions.[27] *In vivo*, MNNG probably reacts with thiol groups to liberate the methylating species.[25] It does not require metabolic activation.

MNNG is mutagenic in bacteria,[29] higher plants,[32] fungi,[28] and mammalian cells in culture,[4,12,15,22,34,35] apparently acting primarily during the period of DNA synthesis;[6,13] it is carcinogenic in rats,[37,38,42] dogs,[40] and hamsters.[24] Repair of MNNG-induced damage is thought to be X-ray mimetic (short patch).[5]

2. Materials and Methods

2.1. Derivation, Maintenance, and Storage of Cell Cultures

2.1.1. Acquisition of Tissues

Lung and skin tissues are obtained from fetal material from therapeutic abortions. Fetal tissues are being used in this study because we can easily obtain different tissues from the same individual in a highly reproducible manner. It is very difficult to obtain tissues that are as well matched for age and biopsy site from other sources. The cultures established from fetal tissues for this study proliferate rapidly, they have high cloning efficiencies (40–70%[16]), and they have high population doubling potential (70–90 population doublings[16]), characteristics that are critical to the success of our mutagenesis protocol. Abortuses included in this study are not preselected by us in any way. We receive the tissues from a large, urban medical center; the population is of diverse racial composition. Our sample size is obviously small, but our study is the first extensive effort to generate data of these kinds.

2.1.2. Primary Cultures

Biopsies are immersed immediately in FCS-F10. Primary cultures are prepared by transferring the tissue with sterile forceps to a tissue culture dish containing a few drops of growth medium (FCS-F10), where it is minced as finely as possible with sterile iris scissors. The fine pieces are immobilized under a sterile 25-mm round glass coverslip, or, alternatively, for larger tissue samples, the fine pieces are evenly distributed over the surface of a 60-mm glass culture dish and allowed to attach for 1 hr in a 37°C incubator. Medium is added and the dishes are incubated undisturbed for 1 week. Medium is replaced at the end of the first week and changed twice the second and subsequent weeks. Biopsies vary, but fibroblast outgrowth is usually apparent in 3–7 days for fetal lung and skin biopsies. Monolayers of cells are dispersed with 0.25% trypsin (Gibco). All stock cultures, survival determinations, and mutagenesis experiments are grown in 60 × 15 mm plastic dishes (Lux). Cultures are kept in a 37°C incubator with a humidified atmos-

TABLE 1. Human Diploid Fibroblast Populations

Cell strain	Repository number	Description
S132	AG4392	Embryonic skin, female
L132	AG4393	Embryonic lung, female
S133	AG4431	Embryonic skin, female
L133	AG4432	Embryonic lung, female
S136	AG4433	Embryonic skin, male
L136	AG4434	Embryonic lung, male
S140	AG4449	Embryonic skin, male
L140	AG4450	Embryonic lung, male
S144	AG4451	Embryonic skin, female
L144	AG4452	Embryonic lung, female
S146	GM6111	Embryonic skin, female
L146	GM6112	Embryonic lung, female
S158	AG6557	Embryonic skin, male
L158	AG6558	Embryonic lung, male
S160	AG6561	Embryonic skin, female
L160	AG6562	Embryonic lung, female

phere containing 5% CO_2 and 95% air. Cultures are routinely tested for mycoplasma using broth and agar cultures and immunoprecipitation.[1,31]

Primary cultures from each tissue sample are expanded to at least 2×10^7 cells. Cultures of this cell population are pooled and identical aliquots of 10^6 cells are stored in liquid nitrogen. The cell populations are at population doubling levels of three to five because the expansion requires only one or two subcultures at a one to six split ratio. A new ampule is thawed for each survival or mutagenesis experiment to ensure uniformity of cell population characteristics. At least 20 ampules of each strain are frozen, providing material for extensive investigation if any given strain is found to be of special interest. All cell populations used in these studies have normal diploid karyotypes, determined by Giemsa banding.

Paired skin and lung cultures derived from eight different abortuses were used in the experiments described here (Table 1). Skin and lung cultures from the same abortus are assigned the same number; the appropriate tissue of origin is designated by a prefix of S or L. Six of these pairs of cultures are available from the National Institutes of Aging's Cell Repository and the National Institutes of General Medical Sciences Human Genetic Mutant Cell Repository at the Institute for Medical Research in Camden, New Jersey; the corresponding repository numbers are listed in Table 1.

2.1.3. Media

For routine cell growth, 85 volumes of Ham's F10[14] (Gibco, hypoxanthine-free) containing 100 units/ml penicillin and 100 μg/ml streptomycin are combined with 15 volumes of fetal bovine serum (Sterile Systems) to constitute FCS-F10. The FCS-F10-TG contains 3.0 × 10^{-5} M 6-thioguanine. Mutagen is applied to cells in serum-free F10 buffered to pH 6.8 with 0.02 M HEPES (N-2-hydroxyethylpiper-izine-N'-2-ethanesulfonic acid). Antibiotics are used in all procedures described here because risk of contamination is increased in mutagenesis experiments that require repeated handling of hundreds of cultures over long periods of time. The concentrations of penicillin and strep-tomycin that we use do not reduce the growth rate or the cloning efficiency of the fibroblasts and do not mask the presence of myco-plasma.

2.1.4. Mutagen Preparation

N-Methyl-N'-nitro-N-nitrosoguanidine (MNNG, mol wt 147.1, Ald-rich, Milwaukee, Wisconsin) is dissolved in double distilled water at a concentration of 10^{-3} M (147 μg/ml) by rapid stirring for 10–20 min at room temperature. This solution is sterilized by filtration through a 0.22-μm Millipore filter and is stored in aliquots at −20°C. Immediately before use, aliquots of MNNG are thawed and diluted to the desired concentrations with sterile, double distilled water.

2.2. Characterization of the Cell Population

2.2.1. Cloning Procedures

The cloning efficiency is given by

$$\text{Cloning efficiency} = \frac{\text{clones formed}}{\text{cells plated}} \times 100$$

Cloning efficiency is independent of the number of cells in the inoculum.[18] Freshly confluent cultures are trypsinized and diluted in FCS-F10 to approximately $(0.5-1.0) \times 10^5$ cells/ml. The cells are counted with a hemacytometer, which permits observation of the cells and of the suspension; a good suspension has fewer than 5% of its cells in clumps of two or more cells. For most strains, final cloning suspensions are adjusted to give an average of 50–100 cells per 60-mm dish, resulting in about 20–40 clones per dish. When the cloning efficiency is reduced by lethal treatments, the inoculum may be increased to 10^3 cells per dish and more dishes may be used to give enough clones for statistical accuracy. Seven days after plating, the medium is changed. Colonies are stained and counted when they range from 1 to 5 mm in diameter, usually 10–14 days after inoculation.

For staining, the dishes are emptied, rinsed once with 5 ml of 0.9% saline to remove serum residue, and fixed for 10 min in 95% ethanol. The ethanol is then removed and the dishes are stained for 3 min with 0.2% (w/v) methylene blue in 0.1 N citrate buffer, rinsed in distilled water, and air dried. Magnified images of the colonies are projected onto white paper using an overhead projector. Individual cells can be seen this way and only clones of at least 100 cells are counted. However, the majority of clones counted for determining the cloning efficiencies contain thousands of cells.

2.2.2. Chromosome Preparations

Metaphases are collected from actively dividing 48-hr cultures that are approximately 80% confluent, using 0.1 μg/ml Colcemid for 60 min. Cells are harvested using 1:5000 w/v EDTA (ethylenediamine tetraacetic acid) solution and 0.15% trypsin solution to remove cells from the growth surface. Hypotonic solution, 0.075 M KCl, is applied for 15 min at 37°C to swell the cells and disperse the chromosomes in the cytoplasm. Fixative, three parts absolute methanol to one part glacial acetic acid, is applied and changed three times prior to making slides (air dried). The slides are aged by a 2-hr exposure to UV (long-wave mineral light) on a slide warmer at 56°C.

The G-banding is accomplished using Wright's strain (MCB, Norwood, Ohio) by a modification of the method described by Yunis and Chandler.[44] Slides are dipped through the following solutions:

1. Hanks buffered salt solution, Ca^{2+}- and Mg^{2+}-free (HBSS) (1–5 sec)
2. 50% methanol (three to five dips)
3. 100% methanol (three to five dips)

The methanol is blotted from the edges of the slide and the slide is dried on a slide warmer. Slides are stained horizontally on a staining rack using a freshly mixed solution of three parts Sorensen's phosphate buffer pH 6.8 (0.07 M) to one part Wright's stock for 2–4 min. Slides are rinsed 10 sec in gently running water and blotted dry with filter paper.

This method of staining has the advantage that the stain is heavy enough along the length of the chromosomes to permit accurate breakage analysis to be done on the same preparations used to prepare banded karyotypes.

2.3. Survival Determinations

Many strains of human diploid fibroblasts have now been treated with MNNG using the protocol we have developed; all have been sensitive to the same range of MNNG concentrations.[18,41]

The observation of too few colonies is a recurrent problem in the analysis of survival or mutagenesis data. Since sample variance is related inversely to sample size, data points based on the observation of only a few clones are subject to large variance and the significance of the difference between two data points will be difficult to determine. Therefore, previous MNNG survival results are used to estimate the number of cells required at each dose to give at least 50–100 surviving colonies. Approximately 10–14 days after treatment, colonies are fixed, stained, and counted. Only colonies containing 100 or more cells are scored.

2.4. Mutagenesis Determinations

The mutagenesis protocol used for these experiments is summarized in the following paragraphs; a more detailed description is given in Ref. 19.

For each experiment, an ampule of cells is thawed and grown through an increase of three to four population doublings. Seventy-two hours prior to mutagen treatment, a minimum of four cultures are pooled and used to inoculate 60-mm dishes at 10^4 to 3×10^4 cells per dish. The number of cells inoculated per dish is chosen so that approximately 10^5 cells per dish will be present 3 days later. The inoculation density is modified as needed to allow for the different proliferation rates of individual strains. When cell populations exceed 10^5 per 60-mm dish at the time of treatment, the killing effects of the mutagen are reduced; the required cell population density is twofold to fourfold lower than that observed previously for a foreskin-derived strain.[18] Plating cell populations 3 days rather than 24 hr prior to exposure to mutagens yields a more reproducible mutant frequency for a given dose and a given cell strain, probably by minimizing differences in proliferation rates at the time of exposure to the mutagen. The lack of reproducibility when cells are plated 24 hr prior to exposure may arise because early fluctuations in proliferation rates alter the extent of repair or the length of the appropriate expression time.

Immediately prior to treating the cells with mutagen, a fresh aliquot of MNNG is thawed. The MNNG is dissolved in distilled water and is stored as concentrated aliquots at $-20°C$, where it is stable for as long as 7 months.[18] Appropriate dilutions of mutagen are added as 0.05-ml aliquots to each dish containing a final 5 ml volume of F10 buffered to pH 6.8 with 2.0 M HEPES. Full dose–response experiments include six mutagen doses and one control. Partial dose–response experiments include two or three mutagen doses and one control. Mutagen-containing cultures are incubated at 37°C in a CO_2 incubator for 4 hr. The mutagen-containing medium is then removed and fresh F10 with 15% fetal bovine serum is added. Cell counts are made to determine the exact number of cells present at the time of treatment.

After 1 hr of incubation in serum-containing medium (to inactivate the mutagen[18]), cloning cultures are prepared to determine cell survival. For each dose, three dishes are suspended and pooled; dilutions are made to inoculate survival dishes using a number of cells per dish and a number of dishes that will result in at least 50–100 surviving clone-formers for each treatment. Nonselective medium (FCS-F10) is replaced on mutagenesis plates every second day during the following 7-day phenotypic expression period. Cultures are subcultured when they approach confluence to ensure that the populations maintain maximal rates of proliferation. Medium is replaced on survival dishes 7 days after they are plated; they are fixed and stained 10–14 days after plating. The survival of cells as a function of MNNG is the same for cells that are treated and allowed to proliferate *in situ* (mutagenesis cultures) as for cells that are treated and then subcultured immediately following mutagen treatment (survival cultures).

Seven days after treatment, all mutagenesis plates for one dose are trypsinized and pooled. The suspensions are counted and the number of cells per dish at the time of replating are recorded. This suspension is used to inoculate dishes for three determinations: (1) the replating efficiencies of the mutagenized populations in the absence of thioguanine (50–100 cells/60-mm dish in FCS-F10), (2) reconstruction dishes to monitor recovery of the TG phenotype in the presence of TG (10^4 mutagenized, normal cells and 50 TG^r cells/60-mm dish in FCS-F10-TG), and (3) TG-selection dishes to measure the induced mutation frequency (10^4 mutagenized normal cells/60-mm dish in FCS-F10-TG). Selective medium (FCS-F10-TG) is replaced on the mutagenesis and reconstruction plates and nonselective medium (FCS-F10) is replaced on the replating efficiency dishes 7 days after replating. The survival dishes are fixed, stained, and counted 10–14 days after treatment. The replating efficiency, reconstruction, and mutagenesis plates are fixed, stained, and counted when the colonies are 3–5 mm in size; most would be expected to contain 10^3–10^4 cells per colony.

As in survival determinations, the observation of too few colonies is a recurrent problem in the analysis of mammalian cell mutagenesis data. Dose points based on the observations of only a few colonies from a small sample of surviving cells are subject to large variances and the significance of the difference between two data points will be difficult to determine. The number of dishes is modified so that for each dose at least 20 mutant colonies are observed and 10^5 viable cells are examined (a viable cell is one that proliferates into a colony of 100 or more cells, as established by replating efficiencies). Suitable numbers of dishes for reconstruction, replating efficiency, and survival determinations usually bring the total to 1000 dishes for a seven-point dose–response experiment. Mutagenesis and survival dose–response experiments are performed in triplicate for each cell strain.

2.5. Quantification of Mutation Frequencies

During mutagenesis experiments with human fibroblasts many complex factors can influence the final observed mutant frequency: the stability of the mutagen during treatment, the cell population density, the metabolic condition of the cells at the time of treatment, the amount of cell killing, the effect of the mutagen on the proliferation of surviving cells, the extent of cross-feeding between mutants and nonmutants, the time needed for phenotypic expression of the newly induced genotype, and the effectiveness of the selective agent to select a spectrum of mutant phenotypes.

To attempt to minimize experimental variance, we are following a mutagenesis protocol that rigorously defines the preparation of cell populations prior to treatment, the preparation and application of the mutagen, and the conditions for phenotypic expression and selection of mutants. Further details regarding the development of this protocol are given in Ref. 19 and 21.

The calculation of the incidence of induced mutants per viable cell involves several correction factors. The procedure used to make these corrections and calculate the mutant incidence is described in detail in Ref. 19.

2.6. Safety Precautions

The most convenient way to minimize chemical risk is to prepare and to apply the mutagen using plastic disposable pipets, test tubes,

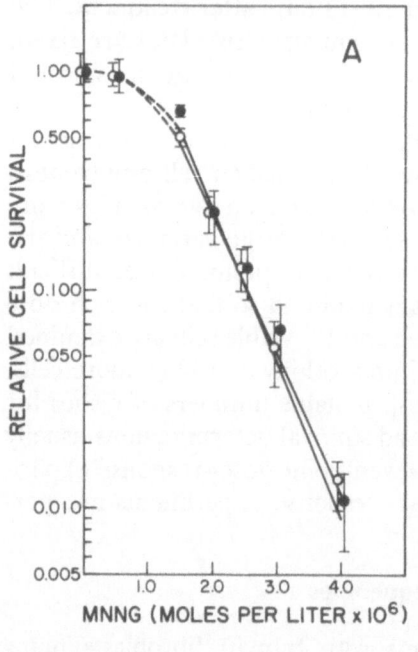

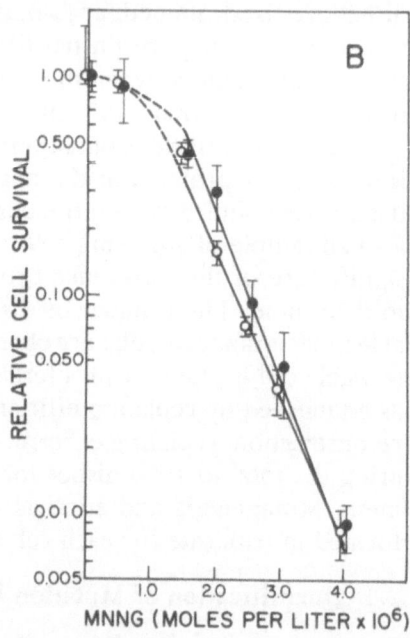

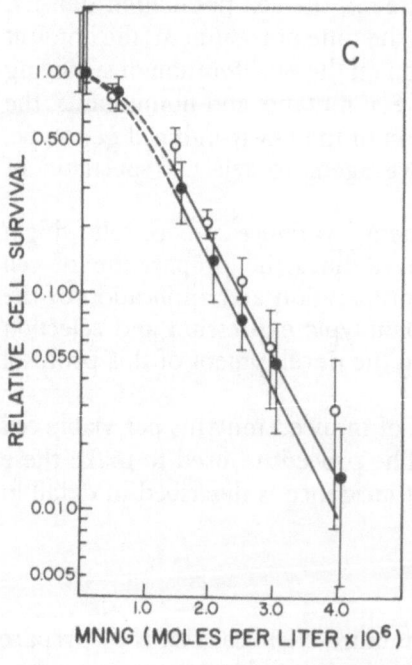

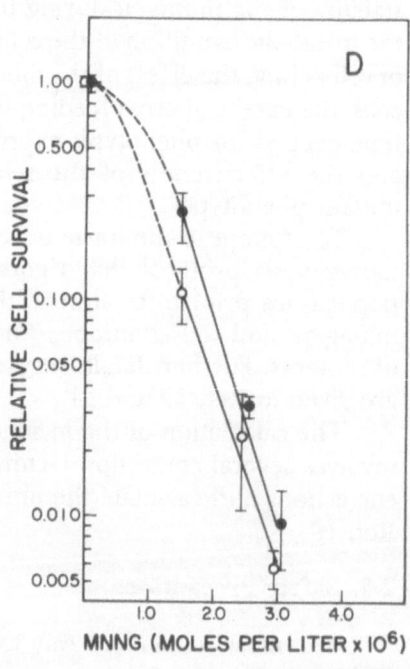

and flasks in an area protected by plastic disposable underpads. Persons handling the mutagens wear disposable plastic gloves and face masks. No pipetting is done by mouth. The mutagen-containing medium is removed from cultures with a peristaltic pump system that minimizes the production of aerosols. All mutagen-containing fluids and all paper and plastic wastes are disposed of by the University Chemical Waste Disposal Service.

3. Differences in Survival Responses among Skin- and Lung-Derived Populations

Of the 12 lung and skin strains that were studied in detail, S132, L132, S133, L133, S136, and L136 are very similar in their survival following exposure to a range of MNNG concentrations (Figures 1A–1C). Differences in the magnitude and slope of the survival curves among these three individuals and between the skin and lung populations from each individual are minor. At 2.0×10^{-6} M MNNG, survival for the six strains is approximately 15–30%; at 4.0×10^{-6} M their survival is 0.7–3%.

In contrast, S140, L140, S144, L144, S146, and L146 cultures demonstrate greater cytotoxic sensitivity and, additionally, the skin cultures are more sensitive than the lung cultures. Compared to strains from 132, 133, and 136, L146 (Figure 1D), L140 (Figure 2A), and L144 (Figure 2B) are as much as tenfold more sensitive and S146, S140, and S144 are as much as 20- to 30-fold more sensitive. Experiments with the 140 and 144 pairs included more dose points than experiments with 146, but the regression curves are similar for all three pairs. At 2.0×10^{-6} M MNNG, survival for the three skin strains is approximately 3–4% (compared to 15–30% for pairs 132, 133, and 136); at 4.0×10^{-6} M their survival is 0.03–0.1% (compared to 0.7–3% for pairs 132, 133, and 136).

←

FIGURE 1. Relative cell survival after exposure to a range of MNNG concentrations. (A) S132 (O) and L132 (●). (B) S136 (O) and L136 (●). (C) S133 (O) and L133 (●). (D) S146 (O) and L146 (●). Each point is the average of three independent determinations representing 30–2000 observed clones and a minimum of 30 60-mm plates. Standard errors of the mean are plotted. Concentrations of MNNG at exposure were identical for the different strains, although some points have been slightly offset for visual clarity. Linear regression lines were determined using individual data points from the three independent experiments for MNNG concentrations of 1.5×10^{-5} M and higher. Survivals were determined simultaneously with mutagenesis determinations presented in Figure 3–5.

Clearly, individuals 140, 144, and 146 are more sensitive to the cytotoxic effects of MNNG than individuals 132, 133, and 136, with the sensitivity being most pronounced for skin cultures.

4. Differences in Mutagenic Responses among Skin- and Lung-Derived Populations

Although the survival responses of skin and lung cultures from individuals 132, 133, and 136 are not significantly different, skin cultures from individuals 132, 133, and 136 yield significantly lower frequencies of induced TG^r mutants than the corresponding lung cultures (Figures 3A–3C). At 4.0×10^{-6} M MNNG (approximately 0.7–3.0% relative survival) the lung cultures yield average frequencies of $(1.0–1.4) \times 10^{-3}$ induced mutants per clonable cell, whereas the corresponding skin cultures yield frequencies of $(4.0–9.0) \times 10^{-4}$ induced mutants per clonable cell. Average values are graphed in Figure 3; however, the higher mutant yield for lung compared to skin was observed in each of the nine individual experiments for 132, 133, and 136.

Apparent differences in mutagenic responses among strains might result from differences in survival. However, for pairs 132, 133, and 136 the survivals for lung and skin are not significantly different. The minor differences that have been observed (Figure 1) are not consistent for lung compared to skin: regression lines for the survival response of skin and lung 132 are indistinguishable; the regression line for S136 reflects a slightly increased sensitivity compared to L136; and in contrast, the regression line for S133 reflects a slightly decreased sensitivity compared to L133. Data plotted as a function of survival in Figure 5 show that the survival differences do not account for the observed differences in mutagenic response for these three pairs.

Apparent differences in mutagenic responses might also result from differences in the length of the optimal phenotypic expression times for lung and skin. The skin- and lung-derived cell populations differ in their growth rates, population doubling potentials, cloning efficiencies, and clonal morphologies;[16] therefore they could also differ in their optimal expression times following exposure to mutagens.

We have carefully examined several of the lung and skin cultures for differences in the length of the optimal phenotypic expression time. Our data provide strong evidence that the 7-day expression period in our standard mutagenesis protocol is optimal for both lung- and skin-derived cultures over a range of MNNG doses.[21] The differences in

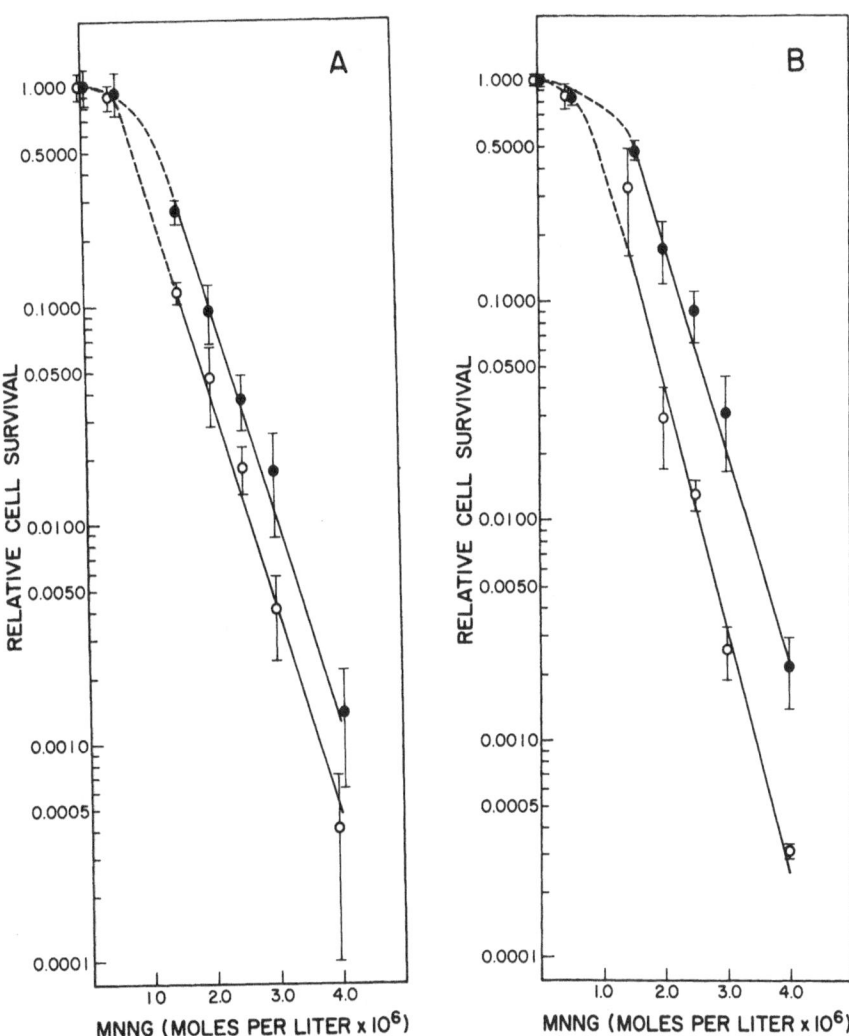

FIGURE 2. Relative cell survival after exposure to a range of MNNG concentrations. (A) S140 (○) and L140 (●). (B) S144 (○) and L144 (●). Data characteristics and analyses were as in Figure 1.

magnitude of induced mutant frequencies between skin and lung for individuals 132, 133, and 136 are not resulting from different optimal phenotypic expression times for the skin and lung populations.

The differences in mutant frequencies between lung and skin cultures derived from the same individual are not always observed (Figure 3D). The S144 and L144 are very similar in response throughout the mutagen dose range, with skin being most similar to lung at the

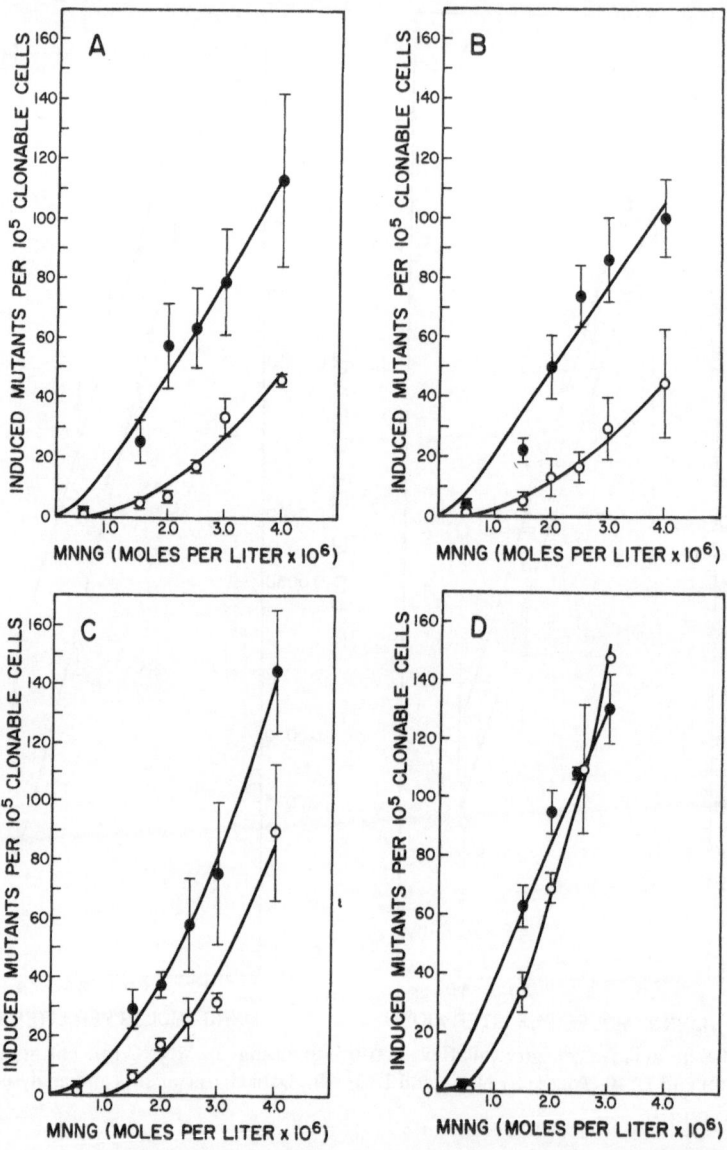

FIGURE 3. Frequency of induced mutants per 10^5 clonable cells as a function of MNNG concentration. (A) S132 (O) and L132 (●). (B) S133 (O) and L133(●). (C) S136 (O) and L136 (●). (D) S144 (O) and L144 (●). Each point is the average of three independent determinations, representing more than 3×10^6 cells (4×10^5 to 1.6×10^6 clonable cells) from a minimum of 300 60-mm plates. The point at 3.0×10^{-6} MNNG for S144 is from a single experiment; the total number of surviving cells in the other two experiments was too low to be included. The number of colonies observed per point ranged from ten for the lowest doses to more than 700 for the highest doses. Standard errors of the mean are plotted. Concentrations of MNNG were identical for the different strains, although some points have been slightly offset for visual clarity. Regression lines were fit to the second-degree polynomial equation $y = a + a_1x + a_2x^2$, using individual data points from the three independent experiments for each strain.

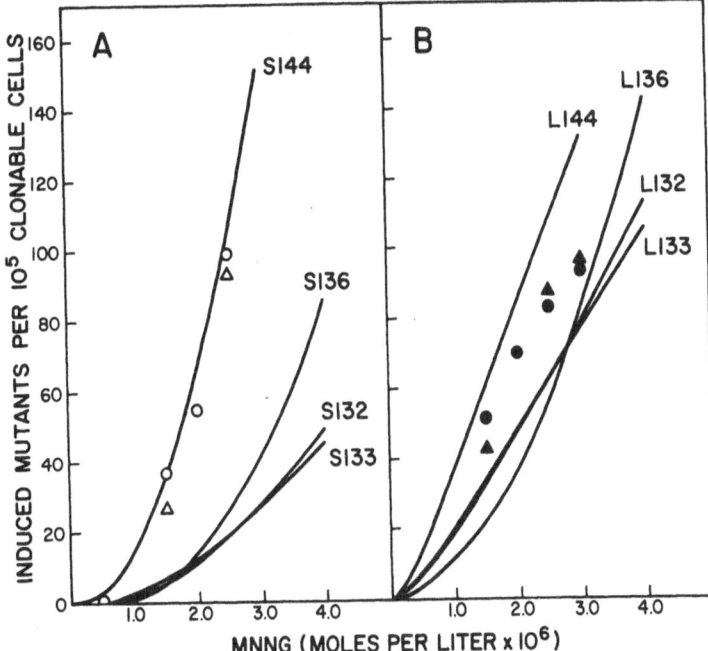

FIGURE 4. Frequency of induced mutants per 10^5 clonable cells as a function of the concentration of MNNG. The regression curves in Figure 3 are used to compare (A) the responses of skin from different individuals and (B) the responses of lung from different individuals. (A) S132, S133, S136, S144, as labeled; S140 (O); S146 (△). (B) L132, L133, L136, L144, as labeled; L140 (●); L146 (▲). Data for pairs 140 and 146 were determined in experiments using too few MNNG concentrations to permit regression; each point is the average of three independent experiments. The numbers of observed clones and dishes per point are in the ranges given for the experiments in Figure 3. Error bars have been omitted for clarity, but they are similar in magnitude to those in Figure 3.

highest MNNG concentrations. A similar relationship occurs for pairs 140 and 146 (circles and triangles, Figures 4A, 4B).

At similar MNNG concentrations, skin cultures derived from individuals 140, 144, and 146 yield significantly greater numbers of mutants than individuals 132, 133, and 136 (Figure 4A). Fewer dose points were evaluated for pairs 140 and 146, so points rather than a regresion line are plotted for these strains in Figures 4 and 5. At 3.0 × 10^{-6} M MNNG, S140, S144, and S146 yield a fourfold greater number of induced mutants than S132, S133, and S136 (Figure 4A). In parallel, L140, L144, and L146 yield higher induced mutant frequencies than L132, L133, and L136 although the magnitude of the differences is less pronounced for the lung cultures than for the skin cultures (Figure 4B).

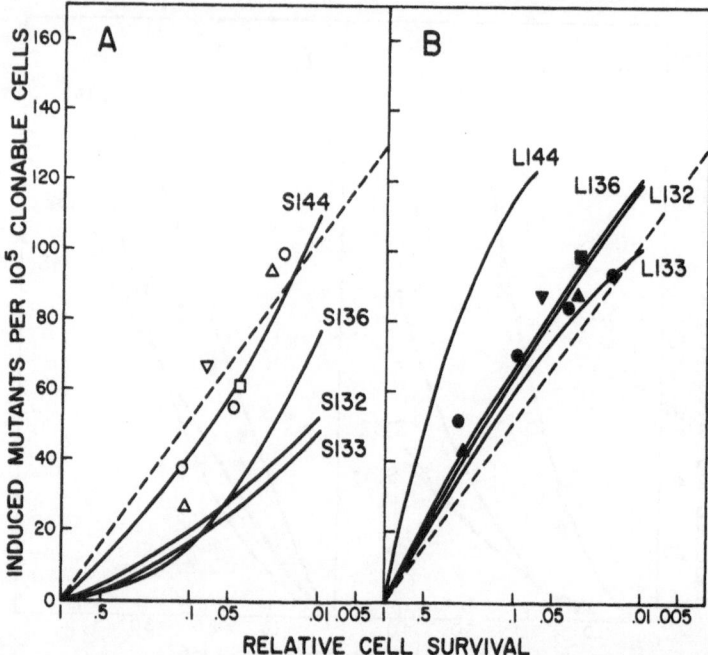

FIGURE 5. Frequency of induced mutants per 10^5 clonable cells as a function of relative cell survival. Data are as described in Figures 3 and 4, except that regression curves were fit using survival as the x coordinate. The dashed line is a point of reference for comparing (A) the responses of skin cultures to (B) those of lung cultures. (A) S158 (∇); S160 ($\square$). (B) L158 ($\blacktriangledown$); L160 ($\blacksquare$). Data for pairs 158 and 160 are averages from two independent experiments. The numbers of observed clones and dishes per point are in the ranges given for experiments in Figure 3.

In order to analyze the possible differences in mutagenic response without the confusion of differences in cell survival, the data are plotted as a function of relative cell survival in Figure 5. The 132, 133, and 136 curves are virtually unchanged because of their very similar cell survivals. The S140, S144, and S146 samples are still more mutable, although the magnitude of the increased mutagenic sensitivity is diminished (Figure 5A). Only L144 remains distinctly sensitive among the lung cultures (Figure 5B). The strains most sensitive to the cytotoxic effects of MNNG yield the highest mutant frequencies, even when the data are corrected for survival differences.

Two additional pairs (158 and 160), for which duplicate experiments were run, show skin responses similar to S144 (Figure 5A) and lung responses similar to the majority of lung cultures (Figure 5B). The skin and lung cultures from individuals 158 and 160 yield similar mutant frequencies at equivalent doses of mutagen.

5. Discussion

In the strains studied so far, the skin populations from different individuals have demonstrated average mutagenic responses differing by as much as sixfold for the same MNNG dose (e.g., S133 and S144). The magnitude of the experimental variation in these determinations necessitates multiple measurements to permit accurate interpretation of observed differences. Using the most widely diverse data from individual experiments in our study, the difference observed between S133 and S144 would have appeared to have been 30-fold, rather than the sixfold difference obtained from triplicate determinations.

Within the range of responses for skin cultures, the strains are divided into two sets: S140, S144, S146, S158, and S160 yield high numbers of mutants compared to S132, S133, and S136. Earlier mutagenesis data for foreskin-derived fibroblasts were in the same range as the skin cultures in this study.[18] In contrast, the lung cultures fall into a much tighter data distribution. After correction for relative survival differences, L144 is the only lung population that is clearly different from the others.

The greatest difference in mutagenic responses observed, even when corrected for relative survivals, was between S133 and L144; at 10% relative survival, L144 yields approximately a 70-fold greater number of mutants. The possibility that fibroblast strains obtained from different biopsy sites may differ by nearly two orders of magnitude in their mutagenic sensitivity should be carefully considered when studies comparing cells from different individuals are being implemented. Rigorous control determinations must be made if any but the most obvious differences among populations are to be accurately measured and interpreted.

The validity of the analysis of differences in cytotoxic and mutagenic responses between individuals and between tissues depends upon (1) the degree of accuracy of the estimates of individual survival values and mutation frequencies and (2) the number of independent strains tested. Our goal has been to execute experiments that ensure the observation of statistically adequate numbers of colonies for each of the measurements that enter into the final estimates of relative survival and induced mutation frequencies. In addition, we are attempting to examine the mutagenic responses of an adequate sample of normal fibroblast strains. Detailed statistical analyses of the data presented here are in progress and will be published elsewhere.[20]

One goal of our mutagenesis experiments has been to continue to improve the sensitivity and accuracy of mutagenesis detection in human fibroblasts while reducing the time and effort required to do so. Knowledge gained with fibroblasts that are more easily manipulated will pave the way for mutagenesis studies with other human cell types that are more difficult to work with. In fact, Reznikoff and DeMars[36] have successfully measured the mutagenic dose–response of a diploid human endothelial cell population using MNNG and the mutagenesis protocol described here for fibroblasts. An additional major value of work with human fibroblasts *in vitro* may be the development of technical capabilities and insight for producing methods of studying mutagenic processes *in vivo* directly in the somatic or germ cells of humans.[9]

ACKNOWLEDGMENTS

We are grateful to Yvonne Horstmeyer and Jean Minter for excellent technical assistance, to Dr. Robert C. Miller for the chromosome analyses, and to Bryan Biggers for the regression analyses. The research cited in this chapter was supported by the National Cancer Institute through grant CA30450. This is paper no. 2642 from the Laboratory of Genetics, University of Wisconsin, Madison, Wisconsin.

6. References

1. M. F. Barile and R. A. Del Guidice, Isolation of mycoplasmas and their rapid identification by plate epi-immunoflourescence, in: *Pathogenic Mycoplasmas, A CIBA Foundation Symposium*, pp. 165–186, Elsevier, Amsterdam (1972).
2. P. J. Benke and N. Herrick, Azaguanine-resistance as a manifestation of a new form of metabolic overproduction of uric acid, *Am. J. Med. 52*, 547–555 (1972).
3. M. Buchwald and C. J. Ingles, Human diploid fibroblast mutants with altered RNA polymerase II, *Somat. Cell Genet. 2*, 225–233 (1976).
4. E. H. Y. Chu and H. V. Malling, Mammalian cell genetics, II. Chemical induction of specific locus mutations in Chinese hamster cells *in vitro, Proc. Natl. Acad. Sci. USA 61*, 1306–1312 (1968).
5. J. E. Cleaver, DNA repair with purines and pyrimidines in radiation- and carcinogen-damaged normal and xeroderma pigmentosum human cells, *Cancer Res. 33*, 362–369 (1973).
6. I. W. Dawes and B. L. A. Carter, Nitrosoguanidine mutagenesis during nuclear and mitochondrial gene replication, *Nature 250*, 709–712 (1974).
7. R. DeMars, Resistance of cultured human fibroblasts and other cells to purine analogues in relation to mutagenesis detection, *Mutat. Res. 24*, 335–364 (1974).
8. R. DeMars and K. R. Held, The spontaneous azaguanine-resistant mutants of diploid human fibroblasts, *Humangenetik 16*, 87–110 (1972).
9. R. DeMars and J. L. Jackson, Mutagenicity detection with human cells, *J. Environ. Pathol. Toxicol. 1*, 55–77 (1977).
10. R. DeMars, J. L. Jackson, and D. Biehrke-Nelson, Mutation rates of human somatic

cells cultivated *in vitro*, in: *Population and Biological Aspects of Human Mutation* (E. B. Hook and I. H. Porter, eds.), pp. 209–234, Academic Press, New York (1981).

11. International Union of Biochemistry, *Enzyme Nomenclature*, Elsevier North-Holland Biomedical Press, Amsterdam (1965).

12. J. S. Felix, Genetic Studies on Cultured Cells Bearing the Lesch–Nyhan Mutation (Hypoxanthine-guanine Phosphoribosyltransferase Deficiency) with Attempts to Derepress the Inactive X Chromosome in Heterozygous Female Cells, Thesis, University of Wisconsin, (1971).

13. F. D. Gillin, D. J. Roufa, A. L. Beaudet, and C. T. Caskey, 8-Azaguanine resistance in mammalian cells I. Hypoxanthine-guanine phosphoribosyltransferase, *Genetics 72*, 239–252 (1972).

14. R. G. Ham and T. T. Puck, Quantitative colonial growth of isolated mammalian cells, in: *Methods in Enzymology* (S. P. Colowick and N. O. Kaplan, eds.), Vol. 5, pp. 90–119, Academic Press, New York (1962).

15. E. Huberman and L. Sachs, Cell-mediated mutagenesis of mammalian cells with chemical carcinogens, *Int. J. Cancer 13*, 326–333 (1974).

16. L. Jacobs and C. L. Bean, Characterization of genetically-identical pairs of lung- and skin-derived human diploid fibroblast populations: Lifespans, cloning ability, growth rates, and response to hydrocortisone, in preparation (1984).

17. L. Jacobs and R. DeMars, Chemical mutagenesis with diploid human fibroblasts, in: *Handbook on Mutagenicity Test Procedures* (B. Kilbey *et al.*, eds.), pp. 193–220, Elsevier North-Holland Biomedical Press, Amsterdam (1977).

18. L. Jacobs and R. DeMars, Quantification of chemical mutagenesis in diploid human fibroblasts: Induction of azaguanine-resistant mutants by N-methyl-N'-nitro-N-nitrosoguanidine (MNNG), *Mutat. Res. 53*, 29–53 (1978).

19. L. Jacobs and R. DeMars, Chemical mutagenesis with diploid human fibroblasts, in: *Handbook on Mutagenicity Test Procedures*, 2nd ed. (B. Kilbey *et al.*, eds.), Elsevier North-Holland Biomedical Press, Amsterdam (1984).

20. L. Jacobs and J. A. Marx, Comparisons of the mutagenic responses of genetically identical pairs of lung- and skin-derived human diploid fibroblast populations, in preparation (1984).

21. L. Jacobs, C. L. Bean, and J. A. Marx, Optimal phenotypic expression times for HPRT mutants induced in foreskin-, skin-, and lung-derived human diploid fibroblasts, *Environ. Mutagen. 5*: 717–731 (1983).

22. F. T. Kao and T. T. Puck, Genetics of somatic mammalian cells IX, Quantitation of mutagenesis by physical and chemical agents, *J. Cell. Physiol. 74*, 245–258 (1969).

23. W. N. Kelley, M. L. Greene, F. M. Rosenbloom, J. E. Henderson, and J. E. Miller, Review: HG-PRT-deficiency in gout, *Ann. Int. Med. 70*, 155–206 (1969).

24. K. Kogure, H. Sasadaira, T. Kawachi, Y. Shimosato, A. Tokunaga, S. Fugimura, and T. Sugimura, Further studies on induction of stomach cancer in hamsters by N-methyl-N'-nitro-N-nitrosoguanidine, *Br. J. Cancer 29*, 132–142 (1974).

25. P. D. Lawley and C. J. Thatcher, Methylation of deoxyribonucleic acid in cultured mammalian cells by N-methyl-N'-nitro-N-nitrosoguanidine. The influence of cellular thiol concentrations on the extent of methylation and the 6-oxygen atom as a site of methylation, *Biochem. J. 116*, 693–707 (1970).

26. M. Lesch and W. Nyhan, A familial disorder of uric acid metabolism and central nervous system function, *Am. J. Med. 36*, 561–570 (1964).

27. H. V. Malling and F. J. de Serres, Mutagenicity of alkylating carcinogens, *Ann. NY Acad. Sci. 163*, 788–800 (1969).

28. H. V. Malling and F. J. de Serres, Genetic effects of N-methyl-N'-nitro-N-nitrosoguanidine in *Neurospora crassa*, *Mol. Gen. Genet. 106*, 195–207 (1970).

29. J. D. Mandell and J. Greenberg, A new chemical mutagen for bacteria, 1-methyl-3-nitro-1-nitrosoguanidine, *Biochem. Biophys. Res. Commun.* 3, 575–577 (1960).
30. R. M. Mankovitz, M. Buchwald, and R. M. Baker, Isolation of ouabain-resistant human diploid fibroblasts, *Cell 3*, 221–226 (1974).
31. G. McGarrity, Detection of mycoplasmas in cell culture, *Tissue Culture Association Manual 1*, 113–116 (1975).
32. A. J. Muller and T. Gichner, Mutagenic activity of 1-methyl-3-nitro-1-nitrosoguanidine on *Arabdopsis, Nature 201*, 1149–1150 (1964).
33. W. L. Nyhan, J. Pesek, L. Sweetman, D. G. Carpenter, and C. H. Carter, Genetics of an X-linked disorder of uric acid metabolism and cerebral function, *Pediatr. Res. 1*, 5–13 (1967).
34. S. H. Orkin and J. W. Littlefield, Nitrosoguanidine mutagenesis in synchronized hamster cells, *Exp. Cell Res. 66*, 69–74 (1971).
35. B. W. Penman and W. G. Thilly, Concentration-dependent mutation of human lymphoblasts by methylnitro-nitrosoguanidine: The importance of phenotypic lag, *Somat. Cell Genet. 2*, 325–330 (1976).
36. C. A. Reznikoff and R. DeMars, *In vitro* chemical mutagenesis and viral transformation of a human endothelial cell strain, *Cancer Res. 41*, 1114–1126 (1981).
37. T. Saito, K. Inokuchi, S. Takayama, and T. Sugimura, Sequential morphological changes in *N*-methyl-*N'*-nitro-*N*-nitrosoguanidine carcinogenesis in the glandular stomach of rats, *J. Natl. Cancer Inst. 44*, 769–783 (1970).
38. R. Schoental, Carcinogenic activity of *N*-methyl-*N*-nitroso-*N'*-nitroguanidine, *Nature 209*, 726–727 (1966).
39. M. Seabright, A rapid banding technique for human chromosomes, *Lancet 2*, 971–972 (1971).
40. Y. Shimosato, N. Tanaka, K. Kogure, S. Fugimura, T. Kauachi, and T. Sugimura, Histopathology of tumors of canine alimentary tract, with particular reference to gastric carcinomas, *J. Natl. Cancer Inst. 47*, 1053–1070 (1971).
41. C. Steglich and R. DeMars, Mutations causing deficiency of APRT in fibroblasts cultured from humans heterozygous for mutant APRT alleles, *Somat. Cell Genet. 8*, 115–141 (1982).
42. T. Sugimura and S. Fujimura, Tumour production in glandular stomach of rat by *N*-methyl-*N'*-nitro-*N*-nitrosoguanidine, *Nature 216*, 943–944 (1967).
43. W. Szybalski, E. H. Szybalska, and G. Ragni, Genetic studies in human cell lines, *Natl. Cancer Inst. Monogr. 7*, 75–89 (1962).
44. J. J. Yunis and M. E. Chandler, High resolution chromosome analysis in clinical medicine, *Prog. Clin. Pathol. 7*, 267–288 (1977).

CHAPTER 3

The L-Arabinose Resistance Test with *Salmonella typhimurium*

Carmen Pueyo and Manuel Ruiz-Rubio

1. Introduction

The evidence that most animal carcinogens are also mutagens,[20,24] strongly suggesting that DNA is the ultimate target of carcinogenic activation, has been an important support of the somatic mutation theory of the etiology of cancer. The recent demonstration that a single point mutation, a GC to TA transversion, leads to the activation of a human oncogene[30,36] is in agreement with this suggestion.

Microbial mutation systems in several bacterial and fungal species, including a variety of genetic endpoints and methodologies, are the most extensively studied of all prescreening tests for carcinogens.[16] Among them, the *Salmonella typhimurium* His reversion assay developed by Ames *et al.*[3] is the most widely used, and for this reason is the only one that meets the criteria for an established predictive test for carcinogenicity.[29]

The His reversion system uses a set of three point mutations that revert by specific mutational events.[3,17] It has been argued that no set of strains containing a small number of revertible point mutations could

Carmen Pueyo and Manuel Ruiz-Rubio • Departmento de Genética, Universidad de Extremadura, Facultad de Ciencias, Badajoz, Spain. *Present address*: Departamento de Genética, Facultad de Ciencias, Universidad de Córdoba, Córdoba, Spain.

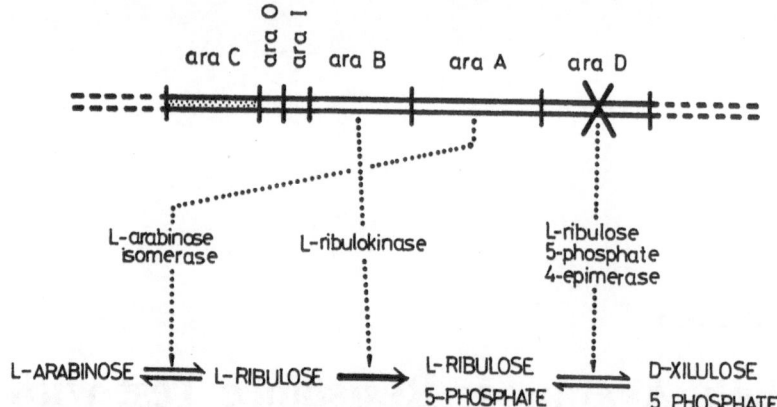

FIGURE 1. The L-arabinose gene–enzyme system of *Escherichia coli.*[9]

encompass the entire set of possible mutagenic lesions in DNA. Forward mutation systems have been offered as the genetic solution to the problem of mutagen–mutation specificity.[32,35] Several forward mutation systems have been developed in bacteria.[16] A number of mutagens undetectable by the standard His tester strains have been reported positive in some of these systems.[26,33]

The *Salmonella* L-arabinose resistance test described in this chapter is based on the fact that *araD* mutations not only block the utilization of L-arabinose as a carbon source, but also lead to the accumulation of a toxic intermediate. Bacterial growth is thus inhibited when the L-arabinose operon is expressed, even if another carbon source is available.[8,28] Our mutagen assay selects changes from an L-arabinose sensitivity to an L-arabinose resistance, a single endpoint reflecting forward mutations at several loci (Figure 1).[28]

2. Strains Sensitive to L-Arabinose

The genetic characteristics of the *Salmonella typhimurium* strains sensitive to L-arabinose are shown in Table 1.

2.1. SV Strains

Strain SV3 is derived from JL386, an auxotrophic strain for tryptophan, threonine, and uracil. Several L-arabinose-sensitive strains were obtained after *N*-methyl-*N*'-nitro-*N*-nitrosoguanidine (MNNG) and penicillin counterselection of JL386. Strain SV3 was chosen from

TABLE 1. Characteristics of the L-Arabinose-Sensitive Strains of *Salmonella typhimurium*[a]

Strain[b]	ara	his	Repair	LPS	pKM101	Ara^R/10^{8c}	Reference
SV3	araD531	+	+	+	−	74	25,32
SV19	araD531	+	−	+	−	145	25
SV20	araD531	+	+	−	−	13	25
SV21	araD531	+	−	−	−	36	25
SV50	araD531	+	−	−	+	NT	38
BA1	araD531	hisG46	+	+	−	50	31
BA3	araD531	hisG46	−	+	−	140	31
BA5	araD531	hisG46	+	−	−	70	31
BA7	araD531	hisG46	−	−	−	70	31
BA9	araD531	hisG46	−	−	+	690	31
BA2	araD531	hisD3052	+	+	−	110	31
BA4	araD531	hisD3052	−	+	−	150	31
BA6	araD531	hisD3052	+	−	−	110	31
BA8	araD531	hisD3052	−	−	−	30	31
BA10	araD531	hisD3052	−	−	+	260	31

[a] Wild-type genes and the presence of plasmid pKM101 are indicated with a plus. Excision repair deficiency (Δ*uvrB*), lipopolysaccharide barrier deficiency (*rfa*), and the absence of plasmid pKM101 are indicated with a minus.
[b] The SV strains also carry the mutations *trp-294*, *thr-115*, and *pyrB92*.
[c] The numbers are the mode of different cultures. NT, Not tested.

among them because of its high sensitivity to MNNG and a full phenotype expression of the L-arabinose-resistant mutants with no requirement of a growth period between MNNG treatment and plating in selective media.[32] Strains SV19, SV20, SV21, and SV50 are derived from SV3[25,38] and were selected as previously described.[2,21,39]

2.2. BA Strains

To compare the relative sensitivities of the Ara forward and His reversion systems, the BA set of strains were selected.[31] Strains BA1, BA2, BA3, and BA4 were obtained by phage P22 *int* transduction of the L-arabinose-sensitive mutation of strain SV3 to the auxotrophic histidine strains *hisG46*, *hisD3052*, TA1950 (*hisG46*, Δ*uvrB*), and TA1534 (*hisD3052*, Δ*uvrB*),[3] respectively. The other BA strains are derivatives selected as previously reported.[2,21,25,39]

Strain BA2 and the corresponding deep-rough derivative, strain BA6, have higher spontaneous frequencies of L-arabinose-resistant mutants than strain BA1 and its derivative BA5. Since other transduc-

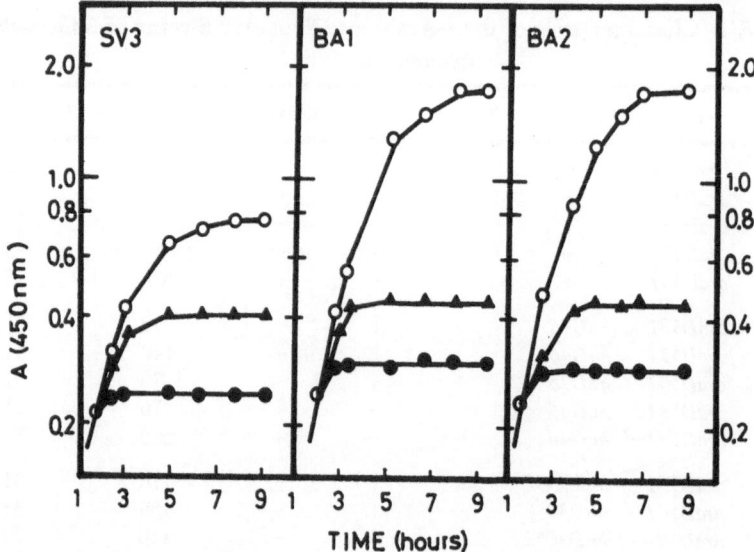

FIGURE 2. Rate of growth and inhibitory effect of L-arabinose on strains SV3, BA1, and BA2.[31] (O) Growth in minimal medium with glycerol at 2 g/liter. (●) Inhibition by L-arabinose at 2 g/liter. (▲) Release of the L-arabinose inhibition by D-glucose at 0.1 g/liter.

tants yield identical results, those higher frequencies must be due to the genetic background of strain *hisD3052* as compared with strain *hisG46*.

Strains BA1 and BA2, with the genetic backgrounds of strains *hisG46* and *hisD3052*, respectively, grow much better than strain SV3. These strains maintain, however, the same sensitivity to L-arabinose and partial inhibition release by D-glucose (Figure 2). The increased growth rate of strains BA1 and BA2 in comparison with SV3 improves the sensitivity of the Ara system toward several mutagens assayed by plate test.[31]

2.3. Properties of the Ara^S Strains

The possibility that toxic but nonmutagenic agents could yield "false positive" results was thoroughly investigated with strain SV3.[32] Ultrasonic oscillation and a combination of lysozyme exposure and osmotic shock were used as treatments that presumably lack any mutagenic effect, but produce strong physiological alterations and kill many cells. The frequency of L-arabinose-resistant mutants remains constant as the surviving fraction decreases with the treatment. This constancy is not a common property, since the auxotrophic mutations in strain SV3 and other L-arabinose-sensitive mutants show a strong dependence on lethal events.[32] With a system exhibiting such dependence, anything could be "mutagenic," even distilled water.[15] This complete lack of dependence on lethal damage is maintained in strains derived from SV3.[25]

Table 2 shows the effect of lethal events on strain BA9, which allows a direct comparison of the Ara forward and His reversion systems. The frequency of His$^+$ revertants increases markedly as the surviving fraction decreases, even if no trace of histidine is added to the selective plates. In contrast, the frequency of Ara^R mutants is practically constant in the absence of any trace of glucose, and increases slightly in its presence. The absolute number of mutants per plate does not increase in any case.

The His reversion system is assayed mostly by the standard incorporation plate test, where the absolute number of mutants is counted.[3] However, as discussed by Mattern,[19] when one wants a more quantitative statement on the mutagenic activity of a compound, the toxicity should be taken into account. This would be especially important for weak mutagens, which have to be used in toxic doses, and for comparison of the potency of mutagens differing in their mutagenicity/toxicity ratio.[19] Nevertheless, the mutagenic activity cannot be expressed with confidence as the number of induced mutants per survivor in the His reversion system, due to its strong dependence on lethal damage. In contrast, the Ara forward mutation system is exceptionally well adapted to that kind of expression of mutagenic activity.

3. Characterization of the Ara System

The L-arabinose gene–enzyme system of *Escherichia coli* is shown in Figure 1. L-arabinose is converted to D-xylulose 5-phosphate by three enzymatic reactions. Genes *araA*, *araB*, and *araD* are the structural genes for L-arabinose isomerase, L-ribulokinase, and L-ribulose 5-phosphate 4-epimerase, respectively. These genes together with their controlling sites *araI* and *araO* constitute an operon, regulated by *araC* and localized between the genetic markers threonine and leucine.[7]

3.1. Sensitivity to L-Arabinose in Strain SV3

Strain SV3 of *Salmonella typhimurium* behaves as do strains of *Escherichia coli* mutated in the structural gene (*araD*) of L-ribulose 5-phosphate 4-epimerase.[8] Neither L-arabinose isomerase nor L-ribuloki-

TABLE 2. Effect of Lethal Events on the Estimate of Ara^R and His^+ Mutant Frequencies[a]

Sonic treatment, min	Survival, %	Survivors[b] $\times 10^8$/plate	Mutants/plate				Mutants/10^8 viable[c]			
			No supplement		Glu or His		No supplement		Glu or His	
			Ara^R	His^+	Ara^R	His^+	Ara^R	His^+	Ara^R	His^+
0	100	4.8	2996	91	3702	261	624	19	771	54
1	9.3	0.45	379	50	666	170	842	111	1480	378
							(1.3)	(5.8)	(1.9)	(7)
2	1.1	0.055	53	26	89	110	964	473	1618	2000
							(1.5)	(25)	(2.1)	(37)
3	0.3	0.016	14	20	29	157	875	1250	1812	9813
							(1.4)	(66)	(2.4)	(182)

[a] An unshaken overnight nutrient broth culture of the strain BA9 resuspended in Davis-Mingioli salts[15] was treated with a sonic dismembrator for up to 3 min. At the indicated time periods, aliquots were withdrawn and dilutions were plated for survivor and mutant counts. The Ara^R and His^+ mutants were selected in plates unsupplemented and supplemented with 500 μg of glucose or with 15.5 μg of histidine.

[b] The total bacteria plated was the same in all cases.

[c] Values in parentheses are relative to the spontaneous mutant frequency.

TABLE 3. Physiological and Enzymatic Properties of the L-Arabinose-Sensitive Strain SV3 and of L-Arabinose-Resistant Mutants Selected in This Strain[a]

Strain[b]	Growth inhibition, %	Keto-sugar accumulation	Enzymatic activities units/mg protein		
			Isomerase	Kinase	Epimerase
JL386 (wild type)	0	NT	237	67	4.35
SV3 (Ara^S)	88	1.22	179	49	<0.11
SV22 (Ara$^+$)	35	NT	302	74	0.20
SV23 (Ara$^+$)	0	NT	266	50	2.20
Type I					
2 (Ara^R)	12	0.07	1.0	<0.8	NT
4 (Ara^R)	7	0.14	3.0	<0.8	NT
Type II					
1 (Ara^R)	32	1.26	67	<0.8	NT
12 (Ara^R)	34	1.87	169	<0.8	NT
Type III					
17 (Ara^R)	25	0.04	1.3	167	NT
19 (Ara^R)	16	<0.01	1.4	53	NT
Type IV					
10 (Ara^R)	11	0.11	7	7.8	NT
11 (Ara^R)	17	1.05	29	1.6	NT

[a] Data are from Pueyo and Lopez-Barea.[28] NT, Not tested.
[b] The parental strain JL386 has wild-type L-arabinose genotype. Strains SV22 and SV23 are Ara$^+$ revertants from SV3. The enumerated strains are Ara^R L-arabinose-resistant mutants.

nase activity is detected in extracts from strain SV3 uninduced by L-arabinose. After induction (Table 3) strain SV3 shows equally high isomerase and kinase activities as the parental strain JL386. However, SV3 L-ribulose 5-phosphate 4-epimerase activity is much lower. This enzymatic deficiency in the L-arabinose pathway is correlated with high sensitivity to growth inhibition by L-arabinose. The bacteriostatic effect of L-arabinose on *araD* mutants has been explained in *E. coli* as caused by the accumulation of large quantities of L-ribulose 5-phosphate, the substrate of the epimerase (Figure 1).[8]

3.2. Nature of the Mutations to L-Arabinose Resistance

3.2.1. Ara$^+$ Revertants

The L-arabinose-resistant mutants (Ara^R) selected in strain SV3 are seldom revertants (Ara$^+$) able to grow on L-arabinose as the only carbon

TABLE 4. Simultaneous Occurrence of Mutations Affecting L-Arabinose Utilization and Leucine Biosynthesis after MNNG Treatment of Strain SV3[a]

Selected phenotype	Frequency among survivors	Leu⁻ among selected mutants	Comutation, %
Ara⁺	1.5×10^{-5}	0/500	0.0
Ara⁺	9.0×10^{-6}	0/700	
Ara^R	2.5×10^{-3}	14/500	3.0
Ara^R	1.1×10^{-3}	16/500	
None	1	0/1000	

[a] Ruiz-Vazquez et al.[32]

source. Only 14 revertants were found among 500 L-arabinose-resistant mutants induced by MNNG and none among 324 spontaneous mutants.

The reversion to L-arabinose utilization results in different degrees of restoration of the epimerase activity with a concomitant decrease in the growth inhibition by L-arabinose. The revertants maintain high levels of L-arabinose isomerase and L-ribulokinase activities (Table 3).

Reversion of strain SV3 to L-arabinose utilization is most often due to unlinked suppressor mutations. Table 4 shows that MNNG-induced reversion to L-arabinose utilization does not tend to occur simultaneously with mutation to leucine auxotrophy. Simultaneous mutation ("comutation") after MNNG treatment is a property of closely linked genes.[14] In contrast, the frequency of comutation between L-arabinose resistance and leucine auxotrophy is the same as that previously reported between *ara* and *leu* mutations, indicating that the vast majority of the mutations leading to L-arabinose resistance are closely linked to *leu*. Cotransduction (Table 5) also indicates that L-arabinose-utilizing revertants of strain SV3 still have the mutation responsible for L-arabinose sensitivity, and thus must carry suppressor mutations.

3.2.2. Ara^R Mutants

Most of the L-arabinose-resistant mutants selected in strain SV3 are still unable to utilize L-arabinose as a carbon source. These mutants are classified into four groups by the criteria of L-arabinose growth inhibition, keto-sugar accumulation, and L-arabinose isomerase and L-ribulokinase activities (Table 3).

Mutants of type I (such as mutants 2 and 4) show very low levels of both isomerase and kinase activities, accumulate small quantities of

TABLE 5. Cotransduction of L-Arabinose Sensitivity and Leucine
Independence[a]

Donor		Recipient		Ara^S among Leu$^+$ transductants
Strain[b]	Phenotype	Strain[c]	Phenotype	
SV3	Leu$^+$ Ara^S	SV15	Leu$^-$ Ara$^+$	238/1000
SV16	Leu$^+$ Ara$^+$	SV15	Leu$^-$ Ara$^+$	151/700
SV17	Leu$^+$ Ara$^+$	SV15	Leu$^-$ Ara$^+$	178/700
SV18	Leu$^+$ Ara$^+$	SV15	Leu$^-$ Ara$^+$	168/600

[a] Ruiz-Vazquez *et al.*[32]
[b] SV16, SV17, and SV18 are L-arabinose-utilizing revertants obtained after MNNG treatment of SV3.
[c] SV15 is a leucine auxotroph obtained after MNNG treatment of strain JL386.

keto-sugars, and are very slightly inhibited by L-arabinose (about 10%). Most type I Ara^R mutants could carry mutations in the regulatory gene *araC* (Figure 1), turning off altogether the L-arabinose uptake and expression of the *araBAD* operon. *Escherichia coli araC* mutants are similarly deficient in both isomerase and kinase activities, resistant to L-arabinose inhibition, and fail to accumulate keto-sugars.[6]

Type II mutants (mutants 1 and 12) behave as typical *E. coli araB* mutants[6]: they show variable but high levels of isomerase activity, no detectable L-ribulokinase, accumulate large quantities of keto-sugars, and their growth inhibition by L-arabinose is the most severe, about 35%. Mutants of type III (mutants 17 and 19) behave as *E. coli araA* mutants[6]: they show very low L-arabinose isomerase activities but very high levels of L-ribulokinase, accumulate the lowest quantities of keto-sugars, and are slightly inhibited by L-arabinose, around 25%.

Finally, type IV mutants (such as mutants 10 and 11) show variable but low levels of both isomerase and kinase activities, accumulate varying quantities of keto-sugars, and are inhibited to different extents by L-arabinose. These Ara^R mutants are less easy to classify: they could be *araB* mutants with a strong polar effect on the isomerase gene, but they could also carry any *araC* or *araI* mutations leading to diminished expression of the *araBAD* operon.[9,34] Since three permeases are involved in the L-arabinose transport of *E. coli*,[7] it seems unlikely that some of the type I and type IV Ara^R mutants could be transport mutants.

From the diversity of Ara^R mutant types we have concluded that the selection of L-arabinose resistance in strain SV3 and its derivatives is an assay that mainly detects forward mutations in any of the different genes (*araA*, *araB*, *araC*) in steps preceding the production of L-ribulose 5-phosphate in the L-arabinose pathway.

4. Applications of the Ara System

4.1. Intrasanguineous Host-Mediated Assay

It is well known that mammals can convert nonmutagenic compounds to highly mutagenic metabolites; such promutagens are not mutagenic in bacteria unless mammmalian enzymes are allowed to act on the chemicals.[13]

Rat liver homogenates have been a useful supplement to *in vitro* systems for providing metabolic activation.[1,20] However, although microsomal activation is valuable in establishing whether metabolism of a chemical can give rise to a mutagen, this metabolism may not reflect the real *in vivo* situation. Thus, microsomal preparations do not usually perform conjugation reactions and may be deficient in other detoxification pathways.[5] In contrast, in the host-mediated assay the test substance is exposed to a complete intact mammalian organism, and subjected to the processes of detoxification.

The method, as initially devised by Gabridge and Legator,[13] involves the intraperitoneal injection of the indicator microorganism into the mammal, but in this way only those mutagens able to penetrate into the peritoneum can be detected. In an attempt to bring the indicator microorganisms into closer contact with mammalian target organs, modifications of the traditional host-mediated assay have been devised.[10,12,18,22,37] In the intrasanguineous host-mediated assay the indicator cells are inoculated into the blood stream and recovered from organs where they accumulate.[4,11,22,23,27]

Salmonella typhimurium strain SV3 is an excellent indicator strain to use in the intrasanguineous host-mediated assay. Bacteria are recovered from rat or mouse liver, lungs, and kidneys in quantities large enough to be screened for mutation induction.[27] In addition, the strain reveals the genetic activity of chemicals inducing different kinds of mutation,[25] which is of importance for the host-mediated assay to be a practical technique. The distribution of bacteria after intravenous injection in mice is shown in Figure 3. The number of viable bacteria recovered from the blood decreases exponentially during the first 20 min of incubation. After 30 min, when 10% or less of the injected bacteria are recovered, the clearance rate diminishes dramatically. The bacteria cleared from the blood stream are recovered primarily from the liver,

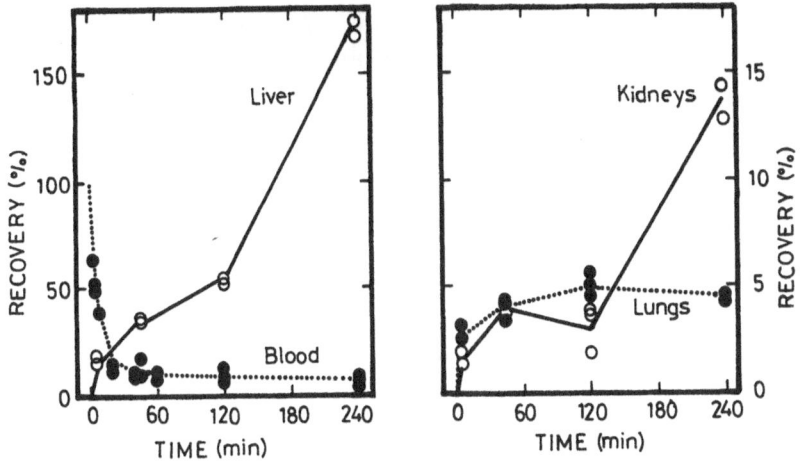

FIGURE 3. Distribution of *S. typhimurium* strain SV3 after intraveneous injection in CD₁ male mice.[27] Each symbol represents one mouse.

but also from the lungs and kidneys. The recovery of viable bacteria from these organs increases during the incubation within the animals: bacterial division must explain the sharp increase of recovery from the liver and kidneys between incubation times of 120 and 240 min, since clearance of bacteria from blood should contribute little to this increase beyond the first 20 min of incubation. Table 6 shows the high sensitivity of the intrasanguineous host-mediated assay in displaying dimethylnitrosamine (DMN) mutagenicity. The high DMN-activating potential of liver, lungs, and kidneys is clearly demonstrated by the intrasanguineous host-mediated assay. In comparison, the *in vitro* microsomal activation assay is relatively inefficient. Positive effects *in vitro* are restricted to mouse liver of kidney and rat liver microsomal preparations, and do not surpass the *in vivo* values even with a DMN concentration 65 times higher. The sensitivity of the *isolated organ* system is intermediate between the *in vivo* and the *in vitro* assays.

4.2. Plate Incorporation and Liquid Tests for Mutagenesis

The L-arabinose-resistance system of *Salmonella typhimurium* can be assayed both by plate test, where an absolute number of mutants is counted after 3 days of incubation, and by liquid test, where mutagenesis relative to survival is determined after usually 1 or 2 hr of exposure.[25–27, 31,32,38] Among the 24 mutagens assayed with the Ara system (Table 7), eight have been detected by liquid test, nine by plate test, and the remaining seven by both.

TABLE 6. Mutagenicity of Dimethylnitrosamine in Strain SV3[a]

Assay[b]	Treatment	Mutants Ara^R/10^6 viables[c]					
		Liver		Lungs		Kidneys	
		Mouse	Rat	Mouse	Rat	Mouse	Rat
In vitro	50mM, 120 min	4591	466	0	0	493	0
Isolated organs	50mM, 120 min	NT	3110	NT	39	NT	NT
Intrasanguineous host mediated assay	~0.76–0.78 mM, 90 min	5459 (±907)	3565 (±1098)	2890 (±387)	823 (±57)	4043 (±1581)	3166 (±550)

[a] Data are from Pueyo et al.[27]

[b] Microsomal assay *in vitro*, perfused *isolated organ*, and intrasanguineous host-mediated assay.

[c] Mutant frequencies after subtracting the spontaneous values. No significant increase over the spontaneous level is indicated by zero; NT, not tested. In the intrasanguineous host-mediated assay, data are the mean values of three different animals. Standard deviations are given in parentheses.

TABLE 7. Mutagenic Compounds in the Ara System[a]

Chemical	S9	Assay	Strain	Reference
Acridine orange (AO)	A	P	BA9	Unpublished
Aflatoxin B1 (AFB$_1$)	A	L,P	BA9,SV50	38, unpublished
9-Aminoacridine (9-AA)	–	L	BA9	Unpublished
2-Aminoanthracene (2-AA)	A	P	BA9,SV21,SV50	38, unpublished
Benzo(*a*)pyrene [B(*a*)P]	A	L,P	BA9,SV50	38, unpublished
Captofol	–	P	SV3	32
Crotonaldehyde (CA)	–	L	BA9	Unpublished
Dimethylnitrosamine (DMN)	U	L,I,O	BA1,BA7,BA9,SV3	25, 27, unpublished
N-Ethyl-N'-nitro-N-nitrosoguanidine (ENNG)	–	L,P	BA1,BA7	Unpublished
Ethyl methanesulfonate (EMS)	–	L,P	BA1,BA9,SV3	25, unpublished
Hycanthone methanesulfonate	–	L	SV3	25
2-Methoxy-6-chloro-9-[3-(2-chloroethyl) aminopropylamino]acridine·2HCl (ICR-191)	–	L,P	BA1,BA2,BA3,SV3,SV19, SV20,SV21,SV50	25, 38, unpublished
2-Methoxy-6-chloro-9-[3-(ethyl-2-chloroethyl) aminopropylamino]acridine·2HCl (ICR-170)	–	P	SV21,SV50	38
Methyl methanesulfonate (MMS)	–	L, P	BA1,BA9,SV3,SV50	25, 38, unpublished
N-Methyl-N'-nitro-N-nitrosoguanidine (MNNG)	–	L,P	BA1,BA9,SV3	25, unpublished
Mitomycin C	–	L	SV3	25
Natulan (procarbazine)	–	L	SV19	26
[N-(5-nitro-2-furfurylidene)-1-aminohydantoin] (NF)	–	L	BA9	Unpublished
2-Nitrofluorene (2-NF)	–	L	BA8,SV20,SV21	25, unpublished
4-Nitroquinoline-1-oxide	–	P	BA2	Unpublished
N-Nitroso carbaryl (NC)	–	P	BA1	Unpublished
Oil shale ash extract	–	P	SV21,SV50	38
β-Propiolactone	–	P	BA1,BA3,BA7,SV3	25, unpublished
Sodium azide	–	P	BA3	Unpublished

[a] Abbreviations: A, Aroclor-1254; U, uninduced; –, absence of S9; L, liquid; P, plate; I, intrasanguineous host-mediated; O, isolated organ-mediated.

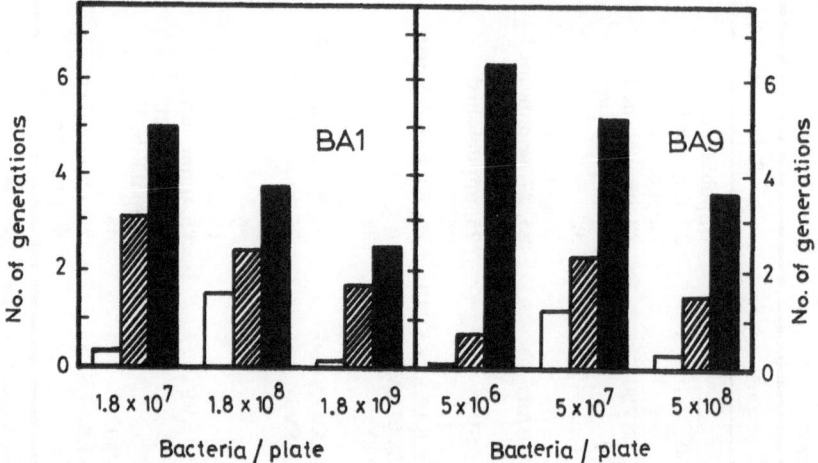

FIGURE 4. Number of generations elapsed in selective plates for Ara^R or His$^+$ mutants after 7 hr (BA1) and 9 hr (BA9) of incubation at 37° C. Some Ara^R selective plates were supplemented with 500 µg of glucose. The His$^+$ selective plates were supplemented with 15.5 µg of histidine. (□) Ara$^®$ plates not supplemental. (▨) Ara$^®$ plates + glucose. (■) His$^+$ plates + histidine.

The sensitivity of the plate incorporation assay is significatively improved by the addition of limited amounts of glucose to the selective plates.[31,32,38] Glucose partially releases the severe growth inhibition exerted by L-arabinose (Figure 2) and increases the number of bacterial generations in plates for the selection of Ara^R mutants (Figure 4). This circumstance has to be an advantage for replication-dependent mutagenesis.

In contrast to the plate test, the liquid test permits one to express the mutagenic activities of compounds as the number of induced mutants per survivor. The Ara system is especially suitable for this kind of expression, as discussed previously.

4.3. Mutagen Sensitivity of the Ara System

In order to compare the relative sensitivities of the His reversion and Ara forward mutation systems, a number of mutagens have been assayed in BA *his ara* double mutants by both liquid and plate tests (Table 8).

Since the spontaneous mutability to L-arabinose resistance is higher than that to histidine prototrophy, the method chosen to express the mutagenic response of both mutation systems to the chemicals affects

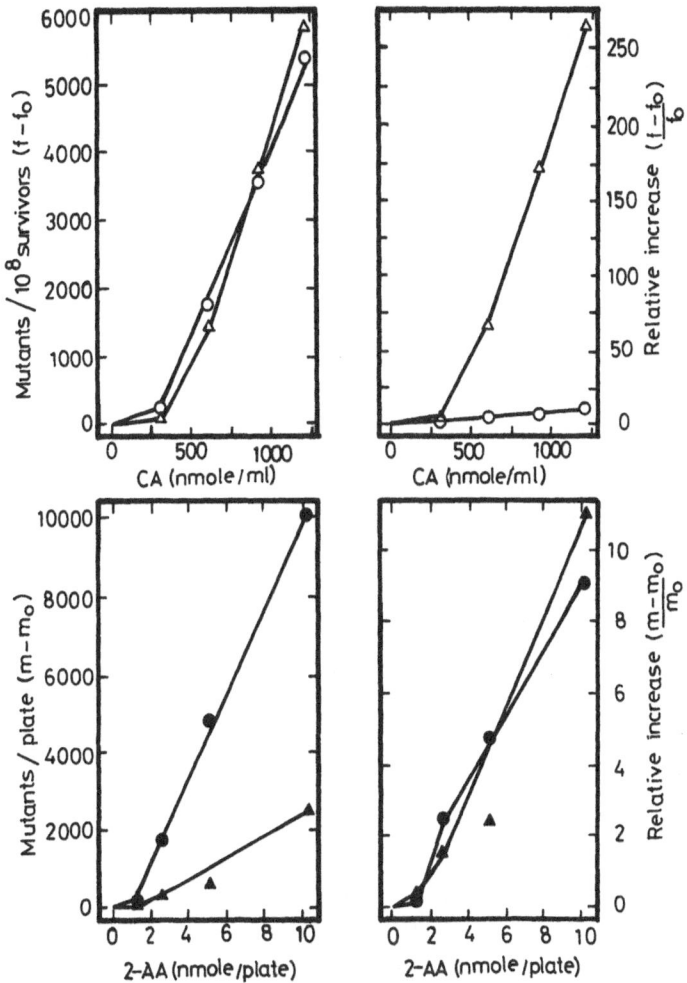

FIGURE 5. Relative sensitivities of the Ara and His systems to crotonaldehyde (CA) and 2-aminoanthracene (2-AA). The assays were performed as described in Table 8. Number of mutants in untreated plates is represented by m_0 and in treated plates by m. Mutant frequency of untreated cultures is represented by f_0 and that of treated cultures by f. ($\triangle$, $\blacktriangle$) His$^+$. ($\bigcirc$, $\bullet$) Ara$^\circledR$.

the comparison; Figure 5 illustrates this point. It is clear, when one compares the frequency of Ara^R and His$^+$ induced mutants, that both systems have the same crotonaldehyde response. On the other hand, a comparison of the increase over the spontaneous frequency shows a much higher response in the His system. Similarly, with the mutagen 2-aminoanthracene, the higher mutagenic response in the Ara system turns into an approximately equal response when, instead of the number

TABLE 8. Activity of 14 Mutagens in the Ara Forward and His Reversion Systems[a]

| Chemical[b] | Strain[c] | S9[b] | Plate test[d] | | | Liquid test[e] | | | |
| | | | Dose, μg/plate | Mutants/plate, nmole | | Survival, % | Dose, μg/ml | Mutants/10^8, nmole | |
				Ara^R	His$^+$			Ara^R	His$^+$
AO	BA9	A	20	4,352 (66)	2,842 (43)	55	30	0[f]	0
AFB$_1$	BA9	A	0.025	1,485 (18,551)	403 (5,034)	51	0.025	896 (11,193)	486 (6,071)
9-AA	BA9	–	100	T[g]	0	32	50	3,608 (18)	73
2-AA	BA9	A	2	10,066 (973)	2,492 (241)	89	2	0	0
B(a)P	BA9	A	5	4,500 (227)	334 (17)	61	5	553 (28)	47 (2.4)
CA	BA9	–	172	0	0	16	43	1,760 (2.9)	1,452 (2.4)
DMN	BA9	U	364	0	0	98	5	23,963 (355)	3,518 (52)
ENNG	BA1	–	10	1,245 (20)	4,222 (68)	4	10	52,624 (848)	20,512 (330)
	BA7	–	30	0	11,178 (60)	11	10	4,539 (73)	23,418 (377)
EMS	BA1	–	2334	578 (0.031)	211 (0.011)	76	3501	14,546 (0.52)	780 (0.028)
	BA9	–	10^4	442 (0.005)	11,931 (0.13)	35	3501	2,998 (0.11)	8,840 (0.31)

Chemical	Strain								
MMS	BA1	—	3900	207 (0.006)	302 (0.009)	10	1300	10,559 (0.89)	1,118 (0.09)
	BA9	—	858	475 (0.061)	1,453 (0.19)	11	858	8,562 (1.1)	1,722 (0.22)
MNNG	BA1	—	1.5	2,689 (264)	1,790 (176)	61	2	38,735 (2,849)	14,711 (1,082)
	BA9	—	2	6,442 (474)	10,214 (751)	41	2	11,156 (820)	10,764 (792)
NC	BA1	—	200	15,453 (18)	5,169 (6)			NT	NT
2-NF	BA8	—	30	0	2,938 (21)	59	30	120 (0.84)	603 (4.2)
NF	BA9	—	3.5	T	2,755 (187)	12	3.5	12,153 (827)	8,532 (581)

[a] For each strain and each chemical the liquid and plate tests were run in parallel, except for DMN, EMS (BA1), and MNNG (BA1). Data are from dose–response curves.

[b] For abbreviations see Table 7. Liver homogenate (S9, 9000 × g supernatant) was added at 50 µl/plate (plate test) and 333 µl/ml (liquid test).

[c] Culture conditions of the strains: BA9, unshaken overnight, nutrient broth; BA1, shaken overnight, minimal medium (ENNG, EMS, and MNNG) or logarithmic minimal medium (MMS and NC); BA7 and BA8, logarithmic minimal medium.

[d] Approximately 2 × 10^8 bacteria were plated. Mutants Ara^R were selected in plates supplemented with 500 µg of glucose and mutants His$^+$ in plates with 15.5 µg of histidine. Number of mutants per plate represents the number of mutant colonies after subtracting the spontaneous count: BA9 (1009 ± 287 Ara^R, 227 ± 122 His$^+$), BA1 (328 ± 146 Ara^R, 3 ± 2 His$^+$), BA7 (487 Ara^R, 8 His$^+$), and BA8 (200 Ara^R, 15 His$^+$). The number of mutants per nmole, calculated from the induced mutants and the molecular weight, is given in parentheses.

[e] Approximately 5 × 10^8 bacteria were exposed in minimal medium for 120 min. The exposure was stopped by centrifugation. Selective plates for Ara^R and His$^+$ mutants contained no glucose or histidine. Mutant frequency was estimated as the number of mutant colonies per 10^8 viable. The spontaneous mutant frequency has been subtracted: BA9 (772 ± 362 Ara^R, 52 ± 24 His$^+$), BA1 (136 ± 91 Ara^R, 3 ± 2 His$^+$), BA7 (208 Ara^R, 7 His$^+$), BA8 (43 Ara^R, 3 His$^+$). The number of mutants per nmole, calculated from the mutant frequency and the molecular weight, is given in parentheses.

[f] 0, no dose–response curve was obtained at the indicated concentrations.

[g] T, toxic effect.

of Ara^R and His$^+$ mutants induced per plate, the increase over the spontaneous number is compared.

Like the Ara and His systems, TA strains such as TA100 and TA1535 have significantly different spontaneous mutability. The mutagenic response of these strains has been compared as an "absolute" number of His$^+$ revertants induced per plate.[21] The conclusion of this comparison has been that the mutagen bis(2-chloroethyl)-amine is equally effective on each strain, since 50 μg of this chemical induced 2708 His$^+$ revertants in TA1535 and 2306 in TA100. On the other hand, the carcinogen benzyl chloride is considered enormously more effective in TA100, where 2000 μg induced 230 His$^+$ revertants, whereas just 12 His$^+$ revertants were induced in TA1535.[21] However, if the comparison is made in terms of relative increase over the spontaneous His$^+$ revertants, bis(2-chloroethyl)-amine should be considered much more effective in TA1535 (2708/11) than in TA100 (2306/160), and benzyl chloride equally effective in either strain (12/11 versus 230/160).

In Table 8 the Ara forward and His reversion systems are compared in the same way as for the TA strains. Thus, the mutagenic response is expressed as the number of Ara^R or His$^+$ mutants induced per plate (plate test) or per 10^8 survivors (liquid test). In the liquid test, the chemicals are at least as effective on the Ara forward as on the His reversion system, and sometimes much more effective. The two possible exceptions are ENNG, if BA7 is used as tester strain instead of BA1, and 2-NF. The comparison of the Ara and His systems has a poorer performance in the plate test, inasmuch as the bacteria seem to reproduce better in plates selective for His$^+$ revertants than in plates selective for Ara^R mutants (Figure 4). This circumstance, an advantage for replication-dependent mutagenesis, can explain the results with 2-NF and NF. These chemicals are effective on both systems in the liquid test, but in the plate test are effective only on the His system (Table 8). The greater bacterial growth in plates selective for His$^+$ revertants compared with that of plates selective for Ara^R mutants is less with the BA1 than with the BA9 strain (Figure 4). This might also explain the different relative effectiveness of EMS on the BA1 and BA9 strains. A similar explanation could be given for the ENNG results (Table 8).

5. Features of the L-Arabinose Resistance Test

To be used as a bacterial system in the screening of mutagens, the following features of the L-arabinose resistance test of *Salmonella typhimurium* must be pointed out:

1. L-Arabinose resistance is mainly the result of forward mutations at several genes in the L-arabinose operon (Table 3).
2. It can be carried out in both liquid and plate tests (Table 7).
3. It is remarkably independent of cell death and bacterial lysis (Table 2), as well as of variations in plating density.[25,32] Thus, it is exceptionally suitable for expressing the mutagenic activity of compounds as the number of mutants induced per survivor, permitting confidence in the study of weak mutagens with lethal effects, and in the comparison of mutagens differing in their mutagenicity/toxicity ratio.
4. It is very sensitive in the *in vivo* intrasanguineous host-mediated assay. In this bioassay practice, the ability to use one strain instead of several auxotrophs is an obvious advantage. In addition, metabolic activation can be performed *in vitro* or even by perfused *isolated organs* (Table 6).
5. It is sensitive to a broad spectrum of chemicals. Among them, natulan is one of the carcinogens that is negative in the His reversion system[20] (Table 7).
6. It seems to detect most mutagens with greater or equal sensitivity than the His reversion system. Among the mutagens tested to date for that comparison, only the frameshift mutagen 2-NF was clearly more effective on the His system (Table 8).

Note Added in Proof. Most of the unpublished results in Table 7 and the data summarized in Table 8 will be published in Ref. 40.

ACKNOWLEDGMENTS

Part of this work has been supported by grant 328 from the Comisión Asesora de Investigación Científica y Técnica, Spain.

The authors wish to thank Elsevier Biomedical Press B. V. for permission to reproduce data previously published in *Mutation Research*.

6. References

1. B. N. Ames, W. E. Durston, E. Yamasaki, and F. D. Lee, Carcinogens are mutagens: A simple test system combining liver homogenates for activation and bacteria for detection, *Proc. Natl. Acad. Sci. USA* 70, 2281–2285 (1973).
2. B. N. Ames, F. D. Lee, and W. E. Durston, An improved bacterial test system for the detection and classification of mutagens and carcinogens, *Proc. Natl. Acad. Sci. USA* 70, 782–786 (1973).

3. B. N. Ames, J. McCann, and E. Yamasaki, Methods for detecting carcinogens and mutagens with the Salmonella/mammalian-microsome mutagenicity test, Mutat. Res. 31, 347–364 (1975).

4. P. Arni, T. Mantel, E. Deparade, and D. Müller, Intrasanguine host-mediated assay with Salmonella typhimurium, Mutat. Res. 45, 291–307 (1977).

5. B. E. Butterworth, Recommendations for practical strategies for short-term testing for mutagens/carcinogens, in: Strategies for Short-Term Testing for Mutagens/Carcinogens (B. E. Butterworth, ed.), pp. 89–102, CRC Press, West Palm Beach (1979).

6. E. Englesberg, Enzymatic characterization of 17 L-arabinose negative mutants of Escherichia coli, J. Bacteriol. 81, 996–1006 (1961).

7. E. Englesberg, Regulation in the L-arabinose system, in: Metabolic Pathways (H. J. Vogel, ed.), Vol. 5, pp. 257–296, Academic Press, New York (1971).

8. E. Englesberg, R. L. Anderson, R. Weinberg, N. Lee, P. Hoffee, G. Huttenhauer, and H. Boyer, L-Arabinose-sensitive, L-ribulose 5-phosphate 4-epimerase-deficient mutants of Escherichia coli, J. Bacteriol. 84, 137–146 (1962).

9. E. Englesberg and G. Wilcox, Regulation: Positive control, Annu. Rev. Genet. 8, 219–242 (1974).

10. R. Fahring, Development of host-mediated mutagenicity tests, I. Differential response of yeast cells injected into testes of rats and peritoneum of mice and rats to mutagens, Mutat. Res. 26, 29–36 (1974).

11. R. Fahring, Development of host-mediated mutagenicity tests—yeast systems II. Recovery of yeast cells out of testes, liver, lungs, and peritoneum of rats, Mutat. Res. 31, 381–394 (1975).

12. G. Ficsor, R. D. Beyer, F. C. Janca, and D. M. Zimmer, An organ-specific host-mediated microbial assay for detecting chemical mutagens in vivo: Demonstration of mutagenic action in rat testes following streptozotocin treatment, Mutat. Res. 13, 283–287 (1971).

13. M. G. Gabridge and M. S. Legator, A host-mediated microbial assay for the detection of mutagenic compounds, Proc. Soc. Exp. Biol. Med. 130, 831–834 (1969).

14. N. Guerola, J. L. Ingraham, and E. Cerda-Olmedo, Induction of closely-linked multiple mutations by nitrosoguanidine, Nature 230, 122–125 (1971).

15. M. H. L. Green and W. J. Muriel, Mutagen testing using Trp⁺ reversion in Escherichia coli, Mutat. Res. 38, 3–32 (1976).

16. M. Hollstein and J. McCann, Short-term tests for carcinogens and mutagens, Mutat. Res. 65, 133–226 (1979).

17. D. E. Levin, E. Yamasaki, and B. N. Ames, A new Salmonella tester strain TA97, for the detection of frameshift mutagens. A run of cytosines as a mutational hot-spot, Mutat. Res. 94, 315–330 (1982).

18. H. V. Malling and C. N. Frantz, Metabolic activation of dimethylnitrosamine and diethylnitrosamine to mutagens, Mutat. Res. 25, 179–186 (1974).

19. I. E. Mattern, Bases of evaluation of an Ames test, in: Progress in Environmental Mutagenesis and Carcinogenesis (A. Kappas, ed.) Vol. 2, pp. 187–189, Elsevier Biomedical Press, Amsterdam (1981).

20. J. McCann, E. Choi, E. Yamasaki, and B. N. Ames, Detection of carcinogens as mutagens in the Salmonella/microsome test: Assay of 300 chemicals, Proc. Natl. Acad. Sci. USA 72, 5135–5139 (1975).

21. J. McCann, N. E. Spingarn, J. Kobori, and B. N. Ames, Detection of carcinogens as mutagens. Bacterial tester strains with R factor plasmids, Proc. Natl. Acad. Sci. USA 72, 979–983 (1975).

22. G. Mohn and J. Ellenberger, Mammalian blood-mediated mutagenicity test using a multipurpose strain of Escherichia coli K-12, Mutat. Res. 19, 257–260 (1973).

23. G. Mohn, J. Ellenberger, D. McGregor, and H. J. Merker, Mutations in plants, microorganisms and in host-mediated assay, *Mutat. Res. 29*, 221–233 (1975).

24. M. Nagao, T. Sugimura, and T. Matsushima, Environmental mutagens and carcinogens, *Annu. Rev. Genet. 12*, 117–159 (1978).

25. C. Pueyo, Forward mutations to arabinose resistance in *Salmonella typhimurium* strains. A sensitive assay for mutagenicity testing, *Mutat. Res. 54*, 311–321 (1978).

26. C. Pueyo, Natulan induces forward mutations to L-arabinose-resistance in *Salmonella typhimurium*, *Mutat Res. 67*, 189–192 (1979).

27. C. Pueyo, D. Frezza, and B. Smith, Evaluation of three metabolic activation systems by a forward mutation assay in *Salmonella*, *Mutat. Res. 64*, 183–194 (1979).

28. C. Pueyo and J. Lopez-Barea, The L-arabinose-resistance test of *Salmonella typhimurium* strain SV3 selects forward mutations at several *ara* genes, *Mutat. Res. 64*, 249–258 (1979).

29. I. F. H. Purchase, ICPEMC Working Paper 2/6. An appraisal of predictive tests for carcinogenicity, *Mutat. Res. 99*, 53–71 (1982).

30. E. P. Reddy, R. K. Reynolds, E. Santos, and M. Barbacid, A point mutation is responsible for the acquisition of transforming properties by the T24 human bladder carcinoma oncogene, *Nature 300*, 149–152 (1982).

31. M. Ruiz-Rubio and C. Pueyo, Double mutants with both His reversion and Ara forward mutation systems of *Salmonella*, *Mutat. Res. 105*, 383–386 (1982).

32. R. Ruiz-Vazquez, C. Pueyo, and E. Cerda-Olmedo, A mutagen assay detecting forward mutations in an arabinose-sensitive strain of *Salmonella typhimurium*, *Mutat. Res. 54*, 121–129 (1978).

33. L. E. Sacks and J. T. MacGregor, The *B. subtilis* multigene sporulation test for mutagens: Detection of mutagens inactive in the Salmonella *his* reversion test, *Mutat. Res. 95*, 191–202 (1982).

34. D. E. Sheppard and D. A. Walker, Polarity in gene *araB* of the L-arabinose operon in *Escherichia coli* B/r, *J. Bacteriol. 100*, 715–723 (1969).

35. T. R. Skopek, H. L. Liber, J. J. Krolewski, and W. G. Thilly, Quantitative foward mutation assay in *Salmonella typhimurium* using 8-azaguanine resistance as a genetic marker, *Proc. Natl. Acad. Sci. USA 75*, 410–414 (1978).

36. C. J. Tabin, S. M. Bradley, C. I. Bargmann, R. A. Weinberg, A. G. Papageorge, E. M. Scolnick, R. Dhar, D. R. Lowy, and E. H. Chang, Mechanism of activation of a human oncogene, *Nature 300*, 143–149 (1982).

37. L. A. Wheeler, J. H. Carter, F. B. Soderberg, and P. Goldman, Association of *Salmonella* mutants with germfree rats: Site specific model to detect carcinogens as mutagens, *Proc. Natl. Acad. Sci. USA 72*, 4607–4611 (1975).

38. W. Z. Whong, J. Stewart, and T. M. Ong, Use of the improved arabinose-resistant assay system of *Salmonella typhimurium* for mutagenesis testing, *Environ. Mutagen. 3*, 95–99 (1981).

39. R. G. Wilkinson, R. Gemski, and B. A. D. Stocker, Non-smooth mutants of *Salmonella typhimurium*. Differentiation by phage sensitivity and genetic mapping, *J. Gen. Microbiol. 70*, 527–554 (1972).

40. M. Ruiz-Rubio, C. Hera, and C. Pueyo. Comparison of a forward and a reverse mutation assay in *Salmonello typhimurium* measuring L-arabinose resistance and histidine prototrophy, *EMBO Journal*, in press (1984).

CHAPTER 4

Identification of Mutagens from the Cooking of Food

Frederick T. Hatch, James S. Felton, Daniel H. Stuermer, and Leonard F. Bjeldanes

1. Background, Focus, and General Review

1.1. Kinds of Genotoxins in Foods

The diet and nutritional status of human beings may have a variety of influences on an individual's probability and time of developing cancer. Among these may be the consumption of genotoxins or progenotoxins (agents that can be activated to form mutagens in the gut or in the body). Such mutagens may then play a role in the initiation of carcinogenesis. The overall contribution of dietary factors to the cancer death rate in the United States has a best estimate of 35%, albeit with a range of estimates from 10 to 70%.[1–6] The direct contribution owing to consumption of mutagens, carcinogens, promoters, and precursors thereof is presently unknown.

There is a wide variety of sources of genotoxins in the human diet. Natural mutagens include glycosides of flavonoids and alkylating agents,

Frederick T. Hatch, James S. Felton, and Daniel H. Stuermer • Biomedical and Environmental Research Program, Lawrence Livermore National Laboratory, University of California, Livermore, California 94550.   Leonard F. Bjeldanes • Department of Nutritional Sciences, University of California, Berkeley, California 94720.

pyrrolizidine alkaloids, hydrazines, nitrite or nitrate plus natural amines, products of oxidized fats, and a growing list of potential promoters.[7,8] Mycotoxins may contribute potent mutagens dependent on conditions of weather and crop storage. Anthropogenic sources include residues of agricultural chemicals and food additives. Finally, there are geno-toxins introduced during cooking and heat processing, which include pyrolytic and low-temperature thermic mutagens from high-protein foods, and products of browning reactions and caramelization from high-carbohydrate foods.[9-11]

1.2. Sources of Mutagens from the Cooking of Food

Although cooking of food dates to the human adoption of fire about 10,000 years ago, research on the subject of toxin formation by cooking, or at least by overcooking, began in 1964 with the demonstration by Lijinsky and Shubik of the formation of polycyclic aromatic hydrocarbons through pyrolysis of fats dripping into flames and envelopment of the food in the resulting smoke.[12]

Prior to 1977 studies of cigarette smoke mutagens revealed that the polycyclic aromatic hydrocarbons, which had been well characterized in smoke condensate, could account for only a small fraction of the total mutagenicity of the fraction.[13] Research showed that organic bases arising from the protein of tobacco were the predominant mutagens.[14,15] Sugimura *et al.*[13] reported at a 1976 Cold Spring Harbor Symposium that the mutagenic activity of both smoke condensate and the charred surface of broiled fish and beef was also much higher than could be accounted for by the benzo(*a*)pyrene present. This finding rapidly led to pyrolysis studies on proteins of food origin and on all the amino acids.[16-24] A series of polycyclic heteroaromatic amines with a broad range of potency in the Ames/*Salmonella* histidine reversion assay were discovered.[23,25] The pyrolysis studies were conducted on dry substrates at temperatures above 400°C in analogy to the cigarette burning conditions. However, they pointed to the possible involvement of the cooking of foods in mutagen formation. Additional heterocyclics (nonamine carbolines) were found with the ability to influence the mutagenicity of the amines and other mutagens.[26-28]

Commoner *et al.*[29-31] soon demonstrated that similar mutagen formation would occur at household cooking temperatures (below 200°C) and even by boiling beef extract. Further research, which is the subject of this review, followed rapidly in several laboratories in Japan and the United States.[32-44]

In 1979 and 1980 the research was extended to carbohydrate-rich foods.[9,45-49] Both nonenzymatic browning reactions and caramelization reactions appear to contribute to genotoxin formation, although the chemistry is very complex and the mutagenic or clastogenic potency of the products is generally low.

1.3. Subject of This Review: Mutagenic Organic Bases of Thermic Origin

The focus of this chapter will be confined to mutagenic organic bases of thermic or pyrolytic origin. Although the history of this work covers only the past 6 years, approximately one dozen such compounds have been discovered and a substantial literature has appeared,[50] including several review articles.[10,25,51-54] Our intent here is to render a fairly complete review of the extraction and purification methods and the chemical properties of these mutagens. Coverage of genetic toxicity and its public health implications will be more superficial.

The historical development of the subject of pyrolytic mutagens originated with studies of cigarette smoke condensate. Whereas benzo(a)pyrene and related polycyclic aromatic hydrocarbons had been isolated and quantified from the condensate previously, application of the Ames/*Salmonella* mutagenicity assay showed that only a fraction of the activity could be accounted for by the measured amounts of benzo(a)pyrene. Other inconsistencies existed. First, the bulk of the mutagenic activity was not in the neutral fraction with the PAH, but was characterized as organic bases.[13,55] Second, in pyrolyzing tobacco the optimum temperature for mutagen formation was considerably lower than that for benzo(a)pyrene formation. Finally, mutagen formation from cigarette smoking was correlated with the protein content of the tobacco.[13,15,19] These findings led directly to experimental pyrolysis of amino acids, peptides, proteins, and intact foods.[13,16,17,19] The production of mutagenic organic bases was observed in all cases. In the case of foods, mutagen formation was proportional to the protein content of the item.[56] Invariably, demonstration of mutagenic activity required the presence of a metabolic activation system (e.g., rat liver S9),[21] and was specific for frameshift-sensitive strains of *Salmonella*.[29,57] Pyrolysis of nucleic acids, fats, and carbohydrates resulted in much lower mutagen formation and a different specificity pattern for *Salmonella* strains.[57] As was true for tobacco, when proteins and foods were heated the mutagenic activity far exceeded that accountable by the content of benzo(a)pyrene. This activity was characterized as organic bases and did not have the properties of polycyclic aromatic hydrocarbons.[13]

1.4. Subjects Excluded from This Review

Because each of the following subjects has an extensive literature associated with it, readers are referred to recent reviews and these subjects will not be discussed further in this chapter: nonenzymatic browning reactions, including carbonyl-amine,[9,58–62] Maillard, and caramelization[47]; polycyclic aromatic hydrocarbons[63–67]; residues of agricultural chemicals such as herbicides, pesticides and hormones; and nitrosamines.[68–72]

1.5. Mutagen Formation during Cooking of Foods

1.5.1. Survey of Protein Foods in the American Diet

In view of the findings described in Section 1.3, laboratories in both Japan and the U. S. have emphasized studies of protein-rich foods. The major sources of cooked protein in the U. S. diet were defined from data bases obtained by the U. S. Department of Agriculture in 1964–1966 and by the U. S. Department of Health, Education and Welfare (Health and Nutrition Examination Survey) in 1971–1975.[73] In both surveys the major protein-rich foods were beef in various forms, eggs, pork, ham, bacon, and fried chicken. All of the foregoing foods plus certain species of fish contained significant amounts of mutagen (Ames/*Salmonella* assay) when fried or broiled to a well done but noncharred state.[32] Foods that are minor sources of protein, such as shellfish, other fish species, cheese, tofu, and beans, contained low or negligible amounts of mutagen unless cooked very extensively.[33]

1.5.2. Effects of Temperature, Time, Water Content, and Method of Cooking

The cooking temperature was found by Bjeldanes *et al.*,[34] Yoshida *et al.*,[22] and Pariza *et al.*[38] to be the major factor controlling mutagen formation. Ground beef fried on a stainless steel griddle at temperatures >200°C showed a rapid increase in mutagen formation until 10 min of cooking per side, followed by a plateau. Mutagenic activity formed in 6 min per side at 300°C was nearly fivefold that formed at 200°C. At reasonable cooking temperatures little mutagen formation was observed until the water content was reduced from the fresh level of ~67% to

55%. Below that point there was a nearly linear relation between water content and mutagen formation down to 35% water, the limit for edible product. In reconstitution experiments, beef with initial water content below 50% formed little mutagen when cooked at 200°C. Thus, mutagen formation appears to occur rapidly as meat passes through a narrow range of water content during the process of cooking.[34] In the next section evidence will be given that the mechanism of mutagen formation involves water-soluble precursors and reactions in aqueous solution rather than on a solid surface or in a lipid phase. Adjustment of the fat content of beef has only a minor influence on mutagen formation in the authors' laboratory,[34] but produced a greater effect in other experiments.[74]

Frying on a ceramic surface resulted in a slower rate of cooking and water loss than on stainless steel. Mutagen formation was slower on ceramic at short cooking times, but reached the same final level. However, a thin Teflon coating on steel had little influence on mutagen formation kinetics. Cooking methods involving direct contact with a hot metal surface or radiant heating (frying, broiling) result in greater mutagen formation than methods involving lower temperatures and the continued presence of water and steam (stewing, braising, simmering, and microwave cooking).[32,33] Methods involving air conduction of heat (baking, roasting) are less efficient and require substantial cooking times; intermediate levels of mutagen are produced.[34]

1.5.3. Model System for Study of Mutagen Precursors and Reaction Mechanisms

A model system based on beef has been studied extensively by Taylor *et al.* in our laboratory.[75-77] Lean round steak is homogenized, filtered, or centrifuged, and the soluble fraction boiled 30 min to remove heat-denaturable proteins. The second supernatant fraction contains only 5% of the meat total solids and less than 10% of the original protein, consisting therefore of soluble small molecules and nondenaturable peptides. Upon heating at 98–100°C (maintaining constant volume) up to 30 hr, mutagen formation occurs with exponential kinetics, and the final amount is equal to that obtained by boiling whole beef homogenate for a similar period. Thus, all components required for mutagen formation are water-soluble (<500 molecular weight) and no solid-phase reactions are involved. The separated insoluble residues yield no significant mutagen when boiled, even after treatment with various proteolytic enzymes to release free amino acids and small peptides. A battery of experiments has established that in the active soluble fraction both amino acids and unknown non-amino acid components are required for mutagen formation. A small set of mutagens is formed, which appear to differ from the major ones formed in fried beef at 200°C.

1.5.4. Genetic Toxicity of Thermic and Pyrolytic Mutagens

Approximately one dozen mutagens of the heterocyclic aromatic amine type have been identified to date.[25] Primarily these have been obtained from experiments involving dry heating of amino acids and proteins at temperatures above 300°C (pyrolysis). In a few cases some of the chemicals have been isolated from well done or more harshly cooked foods.[37,78] Invariably the detection of the mutagens has been with frameshift-sensitive strains of *Salmonella* using the His⁻ reversion assay of Ames. The thermic and pyrolytic mutagens exhibit a broad range of potency in *Salmonella*; many of them are among the most potent known mutagens in this assay, several orders of magnitude more potent than conventionally used "standard" mutagens and carcinogens.[79,80] Thus, the Ames bioassay is capable of detecting nanogram quantities of some of these agents, enabling their detection in cooked foods well below the part-per-billion level.

Application of mammalian bioassay systems to this type of mutagen has not yet been very extensive. One of the earliest isolated agents from tryptophan pyrolysis tar, Trp-P-2, gives a positive response in many mammalian cellular assays for mutagenesis and chromosome damage.[81-83] The effective dose levels are about two orders of magnitude higher for *Salmonella*. The agent is also a hepatocarcinogen for mice and rats in lifetime feeding experiments at rather high dosage.[84,85] This pattern of genetic toxicity is being repeated in a growing number of reports for several additional mutagens of this type[80]; available data are summarized for each mutagen in Section 3.

Of particular interest has been the comparison of genotoxicity between two agents of this type in our laboratory, Trp-P-2 and IQ, both of which have been identified in broiled sardines.[86,87] These agents are very potent in *Salmonella*, IQ being threefold more potent than Trp-P-2. In Chinese hamster ovary cells in culture both agents are much less potent than in *Salmonella* in assays for mutagenesis and chromosome damage. Surprisingly, in this system Trp-P-2 is at least 50-fold more potent than IQ.

1.5.5. Hazard Analysis: Implications for Public Health

Screening for the presence of toxic substances with the *Salmonella* reversion assay is widely practiced. It has played an essential role in enabling detection and purification of mutagens formed during the cooking of foods. However, the assay has proven to be extraordinarily sensitive to the heterocyclic aromatic amines, leaving open the question of whether the very small quantities present are significantly hazardous to persons consuming the foods.[79] This, of course, is the rationale for the application of a battery of mammalian assays, and for carcinogenesis feeding bioassays when sufficient quantities of the agents have been provided by organic synthesis. All of the well-recognized problems remain of extrapolation from *in vitro* and even from animal *in vivo* assays to estimate human risk.

The limited data available currently do not permit even a preliminary speculation on the question of the hazard implications of mutagen formation during the cooking of food. The screening bioassay in *Salmonella* is unusually sensitive to the particular type of mutagen involved, whereas both *in vitro* and *in vivo* mammalian assays are much less sensitive. The general result to date is that the agents are genotoxic in animal cells and carcinogenic in animals at high dose levels.[79,80] The data base of international epidemiology of cancer, particularly for the gastrointestinal tract, points strongly to dietary differences as a factor in cancer causation. However, the role of cooking practices in contributing to this factor is quite uncertain. Mutagen formation in cooked food deserves much further study in order to achieve some quantitative estimate of its contribution and significance.

2. Isolation and Identification of Mutagens in Cooked Food

2.1. Extraction of Foods

Initial extraction procedures to obtain mutagens for *Salmonella* assay or further purification have undergone considerable evolution in the 5 years of intensive research. One of the earliest studies involved collecting smoke from broiling fish and beef over charcoal on a glass fiber filter.[20] Material soluble in DMSO was assayed after removal of solids by centrifugation. The charred surface of the foods was also scraped and suspended in DMSO, centrifuged, and assayed. Early studies of amino acid pyrolysis were carried out by dissolving the product tar in aqueous HCl, extraction of nonmutagenic neutral and acidic fractions with ethyl ether, and re-extraction of the aqueous layer with ether after adjustment to alkaline pH—providing an organic base fraction.[88] A survey of 50 foods was made by Uyeta *et al.*[89] after comparison of several organic solvents as extractants. Mutagen extraction was roughly in proportion to the polarity of the solvent. *N*-Hexane extracted none; chloroform, a small yield; methanol, a moderate yield; and 1:1 chloroform–methanol, a maximal yield.

Extensive work leading to the isolation of two tryptophan pyrolysis mutagens from broiled sardines employed repeated homogenization in methanol, centrifugation, solvent evaporation, and solution of the extract in 1 N HCl.[90] Liquid–liquid extraction with diethyl ether removed acidic and neutral components. After adjustment to pH 10 with NaOH and saturation with NaCl the organic bases were extracted with diethyl ether. More than 75% of the mutagenic activity for *Salmonella* was lost between the methanol extract and the base fraction. The nature of the eliminated mutagens has never been reported. However, this type of extraction with pH control indicated that an important fraction of the mutagens in broiled food consists of organic bases.

Commoner *et al.*, continuing earlier studies of mutagens in bacterial nutrient media containing commercial beef extract,[91] reported frameshift mutagens for *Salmonella* in both commercial and home-cooked beef extract and in ground beef fried at 200°C under normal household conditions.[30] The heated samples were suspended in water, adjusted to pH 2 with HCl; and proteins were precipitated by saturation with ammonium sulfate. After filtration, acidic and neutral components were removed by extraction with methylene chloride. The aqueous phase was adjusted to pH 10 with ammonium hydroxide, and organic bases were extracted with methylene chloride. Hargraves and Pariza have extracted beef extract and fried ground beef according to this procedure, modified only by use of NaCl instead of ammonium sulfate for salting out of proteins.[41,92,93] Iwaoka *et al.* have studied the foregoing extraction conditions and have presented evidence that mutagenic artifacts may be introduced by the presence of a high concentration of ammonium ion, and also by adjustment to very low and very high pH during the solvent partitions.[94–96] Extensive control experiments in the authors' laboratory for the extraction procedures to be described below are not in agreement with this point of view. At least with fried ground beef, there appears to be no artifact from ammonium ion; but its use is unnecessary. Furthermore, adjustments to pH 2 and pH 12 during the solvent partitions are necessary to achieve an optimum yield of mutagens, presumably owing to rather extreme pK values for the important mutagenic constituents.[35] It has been reported, for example, that some aminocarbolines have a basic pK greater than 10.

In the authors' laboratory standard cooking conditions were developed for cooking well-done fried ground beef. Patties (100 g, 11 cm diameter × 1 cm thick) are cooked 6 min per side at 200°C on a stainless steel griddle.[35] Rigorous quantification of the Ames *Salmonella* assay was applied to the extract fractions.[35,97] Neutral, acid, and base fractions were obtained by the method of Commoner *et al.* described above.[30] Comparison was made with the corresponding fractions obtained with an acetone extraction procedure.[35] The cooked food was homogenized in acetone; solids were filtered and re-extracted. Protein was precipitated by storage for 18 hr at −15°C, rather than by salting out. The acetone was evaporated *in vacuo*; liquid–liquid extractions were performed with methylene chloride: first versus aqueous HCl at pH 2, and second versus aqueous NaOH at pH 12. Neutral and acid constituents were separated by back extraction. Dried fractions were dissolved in DMSO and dose–responses for mutagenicity were determined. The acetone method extracted threefold more mutagen than the salt precipitation method. Identical results were obtained with isopropanol extraction, ruling out any artifact from use of the ketonic solvent. Mutagenicity was essentially limited to the organic base fraction; acid and neutral fractions were nearly devoid of activity for any of the Ames tester strains.

Substantial further improvement in extraction yield from fried ground beef was based on the findings that the major mutagens were exclusively organic bases and that HPLC patterns indicated that some components were highly polar. Therefore, cooked meat was homogenized in aqueous HCl with careful control of pH at 2.0. Combined centrifuged supernatant solutions were neutralized to pH 7.0 with NaOH and passed through a column of XAD-2 resin.[98] All mutagenic activity was adsorbed to the resin and could be eluted with acetone; in some situations additional activity required methanol for elution. The solvent was evaporated from the eluent. The aqueous product was adjusted to pH 2, NaCl was added for emulsion control, and neutral and acid components were extracted with methylene chloride. The aqueous phase was adjusted to pH 12, and the organic bases were extracted with methylene chloride. The aqueous phase was adjusted to pH 7, evacuated to remove remaining methylene chloride, and passed over the resin column once again. The acetone eluent was combined with the organic base extract (above), concentrated *in vacuo*; and the final extract was ready for chromatographic fractionation. When carefully performed the resin method yields about fivefold more mutagen than the acetone method. This yield is believed to be well over 80% of the total mutagen present in the fried ground beef, although the latter value is difficult to determine with confidence. Use of an acidic aqueous initial extraction appears to maximize the extraction of highly polar basic organic mutagens from the relatively apolar milieu of the cooked food. The resin column has high affinity for the mutagens, and permits their isolation without the partition losses encountered in liquid–liquid extractions.

2.2. Monitoring Purification of Mutagens with a Bacterial Mutagenesis Assay

The only determining characteristic available for detection of genotoxic food constituents that may be produced during the cooking of foods is their mutagenicity in a bioassay. This fact imposes enormous logistic constraints on the pursuit of this problem. Time and labor considerations limit the choice of bioassay to a bacterial system; the histidine reversion assay of Ames in selected strains of *Salmonella* is the best known and most widely applied example.[99] Although the assay is labor intensive, a few steps can be automated and data processing can be fast and rigorous.[97] Data can be returned to the chemist approximately 48 hr after sample submission, and estimates can be given earlier when needed. The major effort, of course, is in monitoring chromatographic column fractions, which can require processing hundreds of petri plates for a single chemical step.

An important, though unexpected, feature of the thermic and pyrolytic mutagens is the extraordinary potency of many of the discovered agents for inducing mutation in bacterial systems. Certain of the aminoazaarenes are more than 1000-fold more potent than the usual mutagenic standards such as benzo(a)pyrene and the alkylating agents.[79] Except for certain dinitropyrenes, they are the most potent known mutagens in the *Salmonella* assay. This is a major advantage in the purification of the mutagens, since nanogram quantities are measurable and only small aliquots of valuable fractions need be sacrificed for assay. However, this sensitivity permits the tracking of quantities of chemicals that are insufficient for characterization and verification by most techniques of structural chemical analysis. Further, the extreme sensitivity of the assay may permit the isolation of substances that actually are present in the diet in harmless amounts.

The variety of *Salmonella* strains available from Ames allows various distinctions of mutagenic character. In the case of the thermic and pyrolytic mutagens their activity has uniformly been limited almost entirely to the frameshift-sensitive strains, and there is little difference in the presence or absence of the R factor. The *uvrB* repair deficiency mutation is required. Thus, strains TA1538 and TA98 give practically identical results for several known thermic and pyrolytic mutagens (S.

Healy and J. Felton, unpublished data). Strain TA1538 is preferred in the authors' laboratory because of favorable growth characteristics and a lower background plate count, which makes easier the detection of small increments.

All of the thermic and pyrolytic mutagens so far studied exhibit an absolute requirement for the addition of a microsomal activation system to the bacterial mutagenesis assay.[79] Thus, the agents are promutagens, requiring enzymatic conversion to active forms before detection. This is a normal characteristic of aromatic molecules that are potentially mutagenic. The activation is specific for cytochromes of the P-448 type.[100–102] Induction of the system in the animals from which livers are obtained is necessary; this can be accomplished by pretreatment with Aroclor, 3-methylcholanthrene, or β-naphthoflavone. Phenobarbital and some other inducers are ineffective. The sensitivity of the assay can be improved slightly by use of liver from Syrian hamster rather than the rat.[101,102] The metabolic activation requirement in bioassays used for chemical isolation procedures imposes the need for careful standardization of the activation technique, with cross-comparison when new batches of liver microsomes are prepared. Failure to control this aspect of the assay can lead to lack of comparability of data obtained at various stages of a purification procedure, and to loss of sensitivity of the assay.

2.3. Fractionation and Isolation of Mutagens

In early studies extracts were subjected to preliminary processing by thin-layer or liquid chromatography on silica gel. In many cases a subsequent step on a column of Sephadex LH-20 was included. More recently the availability of preparative HPLC columns has decreased the need for preliminary steps. The workhorse adsorption medium has been reverse phase silica, coated usually with C-18 hydrocarbon. In addition, C-8 coatings have also been used. Specific sequences have been worked out by individual laboratories. Generally, preparative LC or HPLC is followed by analytical HPLC on reverse phase media or on columns with amino or cyano functional groups. Solvents are appropriate to the media used. Counterions are usually employed to improve resolution, although nonvolatile counterions (e.g., sulfonic acids) used for improving resolution complicate the subsequent removal of solvent *in vacuo.*

Currently in the authors' laboratory, for purification of the major mutagens of fried ground beef, preparative or analytical scale HPLC on C-18 reverse phase columns is followed by a normal phase step on an amino column and a final return to analytical scale reverse phase. It is of interest that, despite highly complex ultraviolet absorption traces, the fractionation of mutagenic activity gives a relatively simple pattern. Three major and four to seven minor fractions appear to account for 80–90% of the recoverable mutagens.[103] Massive amounts of ultraviolet absorbance move rapidly apart from the mutagens; and after three or four steps of HPLC the mutagenic activity becomes fairly consonant with small ultraviolet peaks. Mass spectrometry reveals the presence of impurities, probably added during isolation from labware or chromatographic media. However, when at least 10 kg of well done fried beef is used as starting material, high-resolution mass spectrometry yielded definitive spectra for the major mutagenic components.

The chromatographic fractionations carried out to date have been applied to fried beef[78,92,93,103] and broiled fish.[37,104] A wider variety of foods, many of which contain substantial amounts of mutagen when cooked well done,[32,33] has not yet been studied in detail. Preliminary work on cooked egg suggests a less polar suite of mutagens, for which isolation conditions have been modified (J. Grant and L. Bjeldanes, unpublished data).

2.4. Identification Strategies

Strategies for identification of mutagens purified from cooked foods will be driven primarily by the miniscule quantities of active material. In the case of the pyrolyzates of amino acids and proteins, where large quantities of a single precursor were available, X-ray crystallography, mass spectrometry or GC/MS, and even NMR were applicable. For the identification of mutagens from cooked foods, it currently appears that mass spectrometry is the major tool.[90,105,106] It is unlikely that quantities, and perhaps purities, adequate for crystallization will be obtainable from manageable amounts of food. Even after cooking of massive quantities of food (e.g., 100 kg), for the more potent mutagens at least, the amounts obtainable will most likely be of borderline sufficiency for NMR analysis as presently available. Finally, the combination of gas chromatography/mass spectrometry (GC/MS), though very powerful, will be limited by the generally poor behavior of the aminoazaarene mutagens on GC columns (H. Kasai, personal communication; D. Stuermer, in the authors' laboratory). Their high polarity and functional groups lead to low recoveries. Limited experience with derivatization to improve GC behavior, such as trifluoroacetylation, is promising, but introduces an extra step and the necessity for obtaining high yields (D. Stuermer, D. Ng, and C. Wood, in the authors' laboratory).

Mass spectrometry has been the most frequently applied technique for the thermic and pyrolytic mutagens that have been identified. The molecules are of a favorable molecular weight, and there is a modest library of standard spectra available (see tables below). The frequent presence of one or two methyl groups linked to nitrogen or carbon gives recognizable fragment ions when they are removed. The variable number of nitrogen atoms is particularly favorable for high-resolution MS, where the separation of nitrogen from carbon mass numbers may enable refined predictions of structure.

Although some of the known mutagens of this type have infrared spectra available, the utility of infrared spectrometry and Fourier-transform infrared (FTIR) for study of food mutagens has not been thoroughly explored. With good instruments, sensitivity may not be a serious problem; however, purity of the isolated material may be essential to eliminate misleading functional group signals.

Since some of the known mutagens of this type are actually isomeric despite very different ring structures, unique structural predictions may not be possible. In particular, the position of methyl groups may be ambiguous, and methylation isomers can have markedly different mutagenic potency.[107,108] Therefore, proof of structure by identity with unambiquously synthesized candidates will probably be required as the final step in identification. This rigorous step, in addition to needing the assistance of chemists with specialized skills, can add a substantial and unpredictable period of time to the process. In our own synthetic program, the mere presence of an additional methyl group has made at least a 6-month difference in the time required to obtain two congeners. Because of the small quantities of these mutagens that will inevitably be available by isolation from cooked foods, an organic synthesis program is essential if there is any interest in going beyond the bacterial bioassay in the study of genotoxicity. Milligrams to hundreds of milligrams are required for thorough studies in mammalian cell culture systems. *In vivo* animal studies cannot be entertained without the availability of gram quantities; standard carcinogenesis bioassay by lifetime feeding of mice requires several hundred grams of agent.

2.5. Mutagen Content of Foods

There is only a small amount of published data on the specific chemical assay of the content of thermic and pyrolytic mutagens in normal foods. Our laboratory has published an extensive survey of the protein-rich foods of the American diet assayed for mutagenicity in *Salmonella* only[32,33]; no quantitative chemical analyses are yet available. Preliminary data on fried ground beef from high-resolution MS indicate concentrations of the major mutagens in the range of 0.1–1.0 ppb. Our studies of extraction methods described in an earlier section indicate that organic solvent extractions may be rather incomplete, so that even reliable chemical assays performed on solvent extracts may underestimate the amounts of mutagens present. Since most available analyses show amounts in the parts-per-billion range, within very complex organic media, there is likely to be a wide margin of error in most analyses reported to date. For example, a summary of much of the data available in 1981 was presented by Sugimura.[25] The mutagen contents of five food samples are reported. Four of the five samples were cooked over a naked flame, which is not a reproducible method and may readily lead to overcooking and charring beyond that normally practiced in the household or restaurant. Four of the five samples were extracted with methanol and carried through several steps of solvent partition and column or thin-layer chromatography before analysis by mass spectrometry. Incomplete extraction and successive losses in subsequent steps may lead to quite low yields of mutagens originally present. Several of the known mutagens were reported as "not detected" in the foods; but, in general, the lower detection limits for the analysis are not known. Therefore, the available quantitative chemical analytic data on cooked foods can at this time be regarded only as indicative of the presence of some of the recognized mutagens under certain cooking conditions.

A more promising analytical method has been developed by Yamaizumi *et al.*[106] However, it is expensive and requires refined technique. The mutagens IQ, MeIQ, and MeIQx were prepared with deuterium labeling and aliquots were added to the initial methanol extracts of broiled sardines and broiled beef. After partial purification by solvent partition and LH-20 column chromatography the samples were analyzed by GC/MS. The ratio of the molecular ion peaks containing deuterium to those with hydrogen allowed correction by isotope dilution calculations for losses occurring after the initial extraction. Such a method offers great promise for improved accuracy, but will be difficult to apply in a broad survey of foods. As reported, substantial amounts of IQ and MeIQ, but no MeIQx, were present in sardine, and a very small amount of MeIQx and a trace of IQ, but no MeIQ, were present in beef (S. Nishimura, oral presentation).

2.6. Problems in Analytical Chemistry

In addition to the "needle in the haystack" problem posed for the analysis of a mixture of several polynuclear heterocyclic compounds and possibly other unknown structures in the complex medium of a cooked food, other problems have been encountered. The limited quantities of mutagen isolated, even from multikilogram samples of prepared foods, are below 1 μg in general. This places severe restrictions on instrumental methods for compound identification; the mass spectrometer is so far the only viable choice. Coupling with gas chromatography, as in GC/MS, greatly facilitates MS identification in impure samples. However, the very high polarity and possibly other properties of the type of mutagens encountered in the analysis of cooked foods renders their behavior in GC unfavorable. Recoveries and reproducibility are poor. Derivative formation, such as trifluoroacetylation, improves the GC performance greatly, but introduces a microchemical synthesis step of uncertain yield into the sequence (D. Stuermer and C. Wood, in the authors' laboratory). Currently in our laboratory, derivatization and GC analysis with the use of bracketing standard chemicals, such as diphenylpyridine and triphenylimidazole, to normalize retention times very accurately appears to offer an excellent test for identity or nonidentity with known standard mutagens. However, the quantities of derivative exiting the GC have been generally insufficient for subsequent MS analysis. Therefore, we have found direct probe introduction to be necessary for our attempts at MS identification of unknown mutagens from cooked foods. In the case of MeIQx from fried beef and from beef extract, confirmation of the Japanese findings has been possible in our laboratory [108a] and that of Hargraves and Pariza.[92,93] Identification of other mutagens from fried beef that are not identical with the known thermic and pyrolytic mutagens is in progress.

Evidently the high polarity, and possibly other properties, of the unknown mutagens from fried ground beef results in difficulties with some of the successive steps of HPLC required for their purification. In particular, substantial losses were encountered for some fractions when subjected to normal phase HPLC on several column media. Variation of solvent mixtures and pH control are being explored to try to surmount this problem.

One possible solution to the difficulties of chemical analysis of these mutagens is, of course, to prepare much larger quantities of cooked foods. Scaleup to at least 100-kg samples will be required to allow the routine use of NMR, which would be very valuable in structure determination. However, cooking on such a scale, although feasible in commercial facilities, introduces questions about the similarity of the geometry and mechanisms of heat transfer to normal household methods. There questions must be addressed adequately in order that massive scaleup can be accomplished reliably.

2.7. Prospects for Improved Analysis

Advances in analytical chemistry are improving the general capabilities in organic analysis and structure identification. Particularly evident are the extraordinary sensitivities for analysis of trace contaminants in relatively dilute media such as air and water. Condensed and highly complex media such as foods still pose formidable problems for components at the part-per-billion level and below. Currently the best available process for identification and quantification of new mutagens isolated from cooked food appears to be the use of an extraction method of proven high recovery, a sequence of preparative and analytical HPLC steps utilizing both reverse and normal phase media, followed by probe introduction to the MS. After a molecular ion is identified, GC/MS with mass fragmentography at the specific mass number may be very useful for quantification, provided that the compound or a derivative elutes without excessive loss from the GC column and is not lost by wall adsorption during transfer to the MS. The isotope dilution method described above (Section 2.5) when coupled to mass fragmentography will improve the reliability of quantification, but requires the availability of synthetic, heavy atom-labeled mutagens.

Significant improvements over the above method may come from several directions. HPLC media with improved resolution are being developed. Linkage of HPLC to MS has long been predicted, but reliable systems are still under development. This linkage would be extremely useful in work with the mutagens from cooking because their behavior in HPLC is much better than that in GC. "Triple" MS is becoming more widely available and should prove particularly useful for samples of less than ultimate purity by allowing the study of fragments derived only from the molecular ion of interest. This would help the food analyst by possibly permitting structure determination at a stage before the amounts of mutagen have been too severely eroded by the inevitable losses of successive chromatographic steps. Proper tuning and use of the triple MS requires an experienced laboratory with advanced computer support. Stream splitting (from GC or HPLC?) with simultaneous analysis by MS and FTIR is available in some places and provides highly complementary structural information from the two types of spectra. Finally, improvements in the sensitivity of NMR to allow measurements on the very small sample sizes available in this work would be welcome. Even if sensitivity were available, the NMR analysis will probably require nearly ultimate purity of samples to avoid artifacts of interpretation.

3. Identified Thermic and Pyrolytic Mutagens

3.1. Preliminary Chemical Classification of Thermic and Pyrolytic Mutagens

Twelve compounds of this type have been identified to date, all in Japanese laboratories (Figure 1). The impetus for the work came from the characterization of certain mutagenic bases in cigarette smoke condensate, which evidently arose from nitrogenous precursors in tobacco. This was followed by many pyrolysis experiments with amino acids, singly and in mixtures, with proteins, and finally with foods (references cited in earlier sections). From a chemical viewpoint all of the reported compounds are nitrogen-containing heterocyclics with predominantly aromatic structure (aminoazaarenes, generally). There are two prototypic tricyclic nuclei, carbazole and imidazoquinoline, which allow a preliminary classification into two classes.

The carbazole nucleus identifies the more numerous class, with variations introduced by placing additional ring nitrogens, an exocyclic amino group, and one or two methyl groups at different locations. Six compounds of this class are carbolines or azacarbolines: Trp-P-1, Trp-P-2, AAC, AMAC, Glu-P-1, and Glu-P-2. One of the remaining three lacks the bridging nitrogen of carbazole and is a phenylpyridine derivative, Phe-P-1; one of them has one added aromatic ring, Orn-P-1; and one has added aromatic and saturated rings, Lys-P-1. All except Lys-P-1 have an exocyclic amino group adjacent to a ring nitrogen. For several reported members of the carbazole class this amino group is involved in the mutagenicity and is sensitive to treatment with nitrite ion at acidic pH, with marked diminution of the mutagenic response. The mutagenic potency of this class in *Salmonella* strain TA98 varies from less than 10 E2 to about 10 E5 revertants/μg. The variation appears to depend on both the location of ring nitrogen atoms and the number and location of methyl substituents. The mutagens of this class

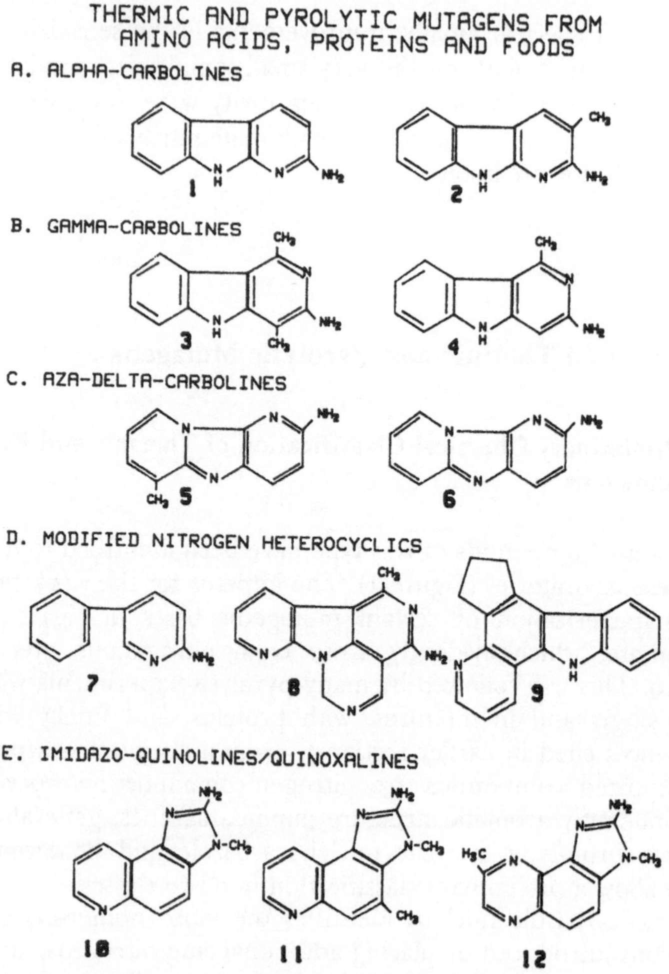

THERMIC AND PYROLYTIC MUTAGENS FROM
AMINO ACIDS, PROTEINS AND FOODS

A. ALPHA-CARBOLINES

B. GAMMA-CARBOLINES

C. AZA-DELTA-CARBOLINES

D. MODIFIED NITROGEN HETEROCYCLICS

E. IMIDAZO-QUINOLINES/QUINOXALINES

FIGURE 1. Structural formulas of identified thermic and pyrolytic mutagens. (1) AAC, (2) AMAC, (3) Trp-P-1, (4) Trp-P-2, (5) Glu-P-1, (6) Glu-P-2, (7) Phe-P-1, (8) Orn-P-1, (9) Lys-P-1, (10) IQ, (11) MeIQ, (12) MeIQx.

were all initially isolated after pyrolysis experiments with amino acids or proteins; several have since been identified in cooked foods by mass spectroscopy.

The second class of compounds is based on the imidazoquinoline nucleus and so far contains only three members: IQ, MeIQ, and MeIQx, all of which were initially isolated from cooked foods and were not reported in pyrolysis experiments. The exocyclic amino group is located on carbon between two imidazo nitrogens. It is insensitive to nitrite treatment, with little or no diminution of mutagenicity. The members of this class are some of the most potent known mutagens in *Salmonella* TA98, from 1.5 to 6.6 E5 revertants/ug, potencies exceeded only by certain dinitropyrenes.

3.2. Trp-P-1

3.2.1. Data Summary

See Table 1.

3.2.2. Isolation and Identification

Trp-P-1 was initially isolated by Sugimura *et al.* from the tar formed after pyrolysis of D,L-tryptophan over a gas flame.[88,109] The mutagenic activity for strain TA98 was extracted from an alkaline aqueous phase with ether and was chromatographed successively on silica gel, alumina, CM-Sephadex, and Sephadex LH-20. The final mutagenic yield was less than 1%. The product was crystallized as an acetate salt. Molecular weight was determined by high-resolution mass spectrometry, allowing assignment of the elemental formula. Structure was deduced from X-ray crystallography and confirmed by NMR.

3.2.3. Organic Synthesis

Trp-P-1 was synthesized by addition of nitro and carboxyl groups to 2,5-lutidine, followed by condensation with *o*-phenylenediamine.[110] Three further steps yielded the product, which was compared with that isolated from the amino acid pyrolyzate.

3.2.4. Analysis of Foods

Trp-P-1 was identified in sun-dried sardines that had been broiled over a gas flame.[90] After extraction and partial chromatographic purification, analysis was by GC/MS with mutiple-ion detection of the molecular ion and the major fragment. The estimated content of 13.3 ng/g was only a small fraction of the total mutagenicity. The major activity in sardine was due to IQ and MeIQ (which see). Subsequently Trp-P-1 was identified in very well done broiled beef (900-g sample) as the molecular ion in a mass fragmentogram after GC/MS.[111] The estimated concentration was about 50 ng/g, which was said to comprise 6% of the total mutagenic activity. The initial activity of the beef sample was reported as extremely high, and thus perhaps excessively cooked,

TABLE 1. Data Summary Table for Trp-P-1

Trivial name	Trp-P-1		
Chemical name	3-Amino-1,4-dimethyl-5H-pyrido(4,3-b)indole		
CAS reg. no.	62450-06-0		
Formula	$C_{13}H_{13}N_3$		
Molecular weight, amu	211.1079		
Mass spectrum, *m/z*	211M+, 197, 193, 169, 168, 167, 166, 142, 140, 115, 105, 85		
UV spectrum: Max, Emol			
Japan	242, 38900; 258, 41700		
SRI	210, 2200; 264, 32800; 318, 6800		
Fluorescence excitation/emission, nm			
Japan	265/400		
SRI	320/400		
IR spectrum, cm^{-1}	3360, 3100, 1655, 1405, 1340, 1265, 1238, 925, 825, 775, 754, 743		
NMR spectrum, ppm			
JPN/CH$_3$OD	1.18, 3H,s; 2.11, 3H,s; 2.68, 3H,s; 6.95–728, 3H,m; 7.75, 1H,d		
SRI/DMSO	2.22, 3H,s; 2.72, 3H,s; 7.0–7.4, 3H,m; 7.83, 1H,d; 11.0 NH,s		
Salmonella TA1538, rev/μg			
LLNL	Rat S9 1500; hamster S9 2400		
Japan			
SRI			
Salmonella TA98, rev/μg			
LLNL			
Japan	39000		
SRI	6790		
Salmonella TA100, rev/μg			
LLNL			
Japan	1700		
SRI	232		
Mammalian bioassay			
Mutation			
SCE	Positive		
Chromosome aberrations	Positive		
Rodent carcinogenesis			
Species	Rat	Hamster	Mouse
Route	Subcutaneous	Subcutaneous	Diet
Dose	1 mg/week	1 mg/week	—
Result	Positive	Positive	Positive
Organs	Subcutaneous sarcoma Liver forestomach	Subcutaneous sarcoma	Liver
Food content			
Beef	53 ng/g		
Chicken			
Sardine	13.3 ng/g		
Cuttlefish	ND[a]		

[a] ND Not detected.

by comparison with later studies. Chromatographic investigations of fried beef in other laboratories have not yielded any material positively identified as Trp-P-1.

3.2.5. Genetic Toxicology

Trp-P-1 is a potent *Salmonella* mutagen only in the frameshift sensitive strains TA98[80] and TA1538 (J. S. Felton, unpublished data). Both Aroclor-induced rat and hamster liver microsomes can activate the compound. The Japanese, SRI, and LLNL studies all found Trp-P-1 positive, but differed considerably in the number of revertants per μg.[80,112] Trp-P-1 also causes sister chromatid exchanges (SCEs) in human lymphoblastoid cells in the presence of S9[113] and chromosomal aberrations in cultured human and Chinese hamster cells.[114] *In vitro* transformation of hamster cells was seen at a Trp-P-1 concentration of 0.5 $\mu g/ml$.[115] Ishikawa *et al.* have shown that Trp-P-1 causes enzyme-altered foci in rat liver.[81] Trp-P-1 also produced tumors (subcutaneous sarcomas) at a dose of 1 mg/week in hamsters and rats following subcutaneous injections.[82] Trp-P-1 when given at a dose of 200 ppm in the diet induced hepatocellular carcinomas in mice.[85] Female mice were much more sensitive than male mice (16/26 versus 5/24). Bladder tumors were seen in 47% of the surviving mice after implantation in the bladder of a pellet containing 1 mg Trp-P-1.[116]

3.3. Trp-P-2

3.3.1. Data Summary

See Table 2.

3.3.2. Isolation and Identification

Trp-P-2 was initially isolated from the pyrolysis tar of D,L-trypto-phan as described above for Trp-P-1.[88,109] A final chromatographic step on CM-Sephadex separated the two congeners. Trp-P-2 comprised about 3% of the total mutagenic activity of the tar. Its structure was deduced from the elemental formula assigned by high-resolution MS, identity of UV spectrum with that of Trp-P-1, and NMR spectrum indicating one less methyl group than in Trp-P-1.

3.3.3. Organic Synthesis

Synthesis of Trp-P-2 was accomplished by condensation of cyano-methylindole with acetonitrile in the presence of aluminum chloride in

TABLE 2. Data Summary Table for Trp-P-2

Trivial name	Trp-P-2	
Chemical name	3-Amino-1-methyl-5H-pyrido(4,3-b)indole	
CAS reg. no.	62450-07-1	
Formula	$C_{12}H_{11}N_3$	
Molecular weight, amu	197	
Mass spectrum, *m/z*	197M+, 180, 179, 170, 169, 157, 155, 127, 98, 85	
UV spectrum: Max, Emol		
Japan	243, 25100; 263, 32400; 300–330, 4500	
SRI	210, 5600; 242, 17800; 264, 38100; 318, 5300	
Fluorescence excitation/emission, nm		
Japan	265/400	
SRI	320/410	
IR spectrum, cm^{-1}	3250, 3020, 2800–2400, 1670, 1655, 1540, 1462, 1403, 1340, 1268, 1205, 842, 734	
NMR spectrum, ppm		
JPN/CH$_3$OD	1.96, 3H,s; 2.76, 3H,s; 6.35, 1H,s; 7.0–7.4, 3H,m; 7.8, 1H,d	
SRI/DMSO	2.72, 3H,s; 6.28, 1H,s; 7.0–7.4, 3H,m; 7.84, 1H,d; 11.1,NH,s	
Salmonella TA1538, rev/μg		
LLNL	Rat S9, 65000; hamster S9, 46000	
Japan	220,000	
SRI		
Salmonella TA98, rev/μg		
LLNL		
Japan	104,000	
SRI	13600	
Salmonella TA100, rev/μg		
LLNL	1490	
Japan	1800	
SRI	270	
Mammalian bioassay		
Mutation	Positive	
SCE	Positive	
Chromosome aberrations	Positive	
Rodent carcinogenesis		
Species	Mouse	Mouse
Route	Diet	Bladder implant
Dose	0.02%	1 mg pellet
Result	Weak positive	Positive
Organs	Liver	Bladder
Food content		
Beef	ND^a	
Chicken		
Sardine	13.1 ng/g	
Cuttlefish		

^{a} ND Not detected.

a single step, followed by extensive purification.[117] Two alternative syntheses were reported; in one the cyanomethylindole was reacted with dimethylacetamide and phosphorus oxychloride.[110] Cyclization to the final product occurred after treatment with ammonia in methanol solution. The second route proceeded from α-acetamido-β-(2-indole)propionic acid, which cyclized after treatment with acidic ethanol, forming an ester in which the new ring was partially saturated. Treatment with lead acetate produced the aromatic ring, and the ester was hydrolyzed. The resulting carboxylic acid was converted to the exocyclic amino group in the final product by a Curtius rearrangement.

3.3.4. Analysis of Foods

After extraction and partial purification from broiled sardines, Trp-P-2 was identified as molecular ion and major fragment in GC/mass fragmentography.[90] The estimated content was 13.1 ng/g, which was a small fraction of the total mutagenic activity of the broiled sardines. Trp-P-2 has not been reported in fried or broiled beef.

3.3.5. Genetic Toxicology

Trp-P-2 is a potent *Salmonella* mutagen in the frameshift strains TA98 and TA1538.[87,118] Both Aroclor-induced rat and hamster liver microsomes can metabolize Trp-P-2 to active mutagens.[87,118] The base substitution sensitive strain TA100 was weakly positive. *In vitro* analysis for forward mutation (AA^r and TG^r), SCE, and chromosome aberrations using CHO cells was uniformly positive in both repair-sufficient and repair-deficient cell lines.[87] Increased SCEs were also seen in human lymphoblastoid cells activated with S9.[113] Hamster embryo cells were transformed *in vitro* at a concentration of 0.5 μg/ml.[83] Trp-P-2 also binds covalently to calf thymus DNA.[119,120]

In vivo mouse LD_{50}s by the oral gavage route were 48 mg/kg for noninduced (corn oil only) and 16 mg/kg for Aroclor-induced (500 mg/kg, 4 days prior to the Trp-P-2 dose) (J. S. Felton, unpublished results). Trp-P-2 was also tested in the mouse oocyte depletion assay[121] and gave negative results (J. S. Felton, unpublished data). Trp-P-2 is carcinogenic in mice (hepatocellular carcinomas) after feeding 200 ppm in diet.[85] Females were more susceptible than males. Liver tumors were also found in rats fed 100 ppm Trp-P-2 for 2 years.[84] Bladder tumors (transitional cell carcinomas) were seen in mice after implantation of a pellet containing 2 mg of Trp-P-2 into the bladder.[116]

3.4. IQ

3.4.1. Data Summary

See Table 3.

3.4.2. Isolation and Identification

Three kg of broiled sun-dried sardines (Japanese maruboshi) were ground and extracted with methanol (20 liters).[36,37] After solvent evaporation the residue was dissolved in 1 N HCl and extracted with ether. The ether layer (acid-neutral fraction) was nonmutagenic. The aqueous layer was reextracted with ether after adjustment to pH 10; this basic fraction contained only 20% of the mutagenic activity. The remaining aqueous layer, now called neutral fraction, contained about 40% of the original mutagenic activity. The name neutral is misleading since the principal mutagens present are highly polar aromatic bases which do not partition well into ether at pH 10. The authors' laboratory has observed better partition into methylene chloride at pH 12. Successive chromatographic fractionation was carried out on Dianon HP-20, Sephadex LH-20, silica gel, and reverse phase C-18 HPLC twice (gradient elution followed by isocratic elution). The mutagenic yield of the compound in the fraction chosen for chemical characterization was ~0.2%. The mass yield from 10 kg of sardines was 200 μg.

Structural information was obtained from low- and high-resolution MS, including confirmation of an exocyclic amino group by acetylation and trimethlysilylation. Proton NMR at 270 MHz provided good structural information, which was consistent with data from Fourier-transformed infrared spectroscopy. The structure was further confirmed by X-ray crystallography on the synthesized material.[122]

3.4.3. Organic Synthesis

The mutagen IQ was synthesized in two steps from 5,6-diaminoquinoline, which was prepared by reduction of 6-amino-5-nitroquinoline.[123] The diamine was cyclized by condensation with cyanogen bromide. An *N*-methyl group was introduced by heating with tetramethylammonium hydroxide under reduced pressure. The product was isolated by sublimation and purified on a silica gel column.

An alternative synthesis was reported by another group.[124] An *N*-methyl group was added to 6-aminoquinoline by formylation with formic acid and acetic anhydride, and reduction with lithium aluminum hydride. A nitro group was added at the 5 position with nitric–sulfuric

TABLE 3. Data Summary Table for IQ

Trivial name	IQ
Chemical name	2-Amino-3-methylimidazo(4,5-f)quinoline
CAS reg. no.	76180-96-6
Formula	$C_{11}H_{10}N_4$
Molecular weight, amu	198.0906
Mass spectrum, m/z	
Japan	198M+, 197, 183, 170, 156, 129, 99
U. Wisc.	198M+, 197, 183, 170, 155, 129
UV spectrum: Max, Emol	
Japan	213, 27000; 264, 51500; 354, 4000
SRI	
Fluorescence excitation/emission, nm	
Japan	Not significant
SRI	
IR spectrum, cm^{-1}	3400, 3350, 3250, 3090, 1670, 1550, 1375, 800
NMR spectrum, ppm	
JPN/CDCl$_3$	3.70, 3H,s; 4.58, 2H,b; 7.43, 1H,dd; 7.56, 1H,dd; 7.84, 1H,dd; 8.72, 1H,ddd; 8.84, 1H,dd
SRI/DMSO	
Salmonella TA1538, rev/μg	
LLNL	Rat S9 145,000; hamster S9 125,000
Japan	
SRI	
Salmonella TA98, rev/μg	
LLNL	
Japan	433,000
SRI	
Salmonella TA100, rev/μg	
LLNL	
Japan	7000
SRI	
Mammalian bioassay	
Mutation	Weak positive
SCE	Weak positive
Chromosome aberrations	Weak positive
Rodent carcinogenesis	
Species	Mouse
Route	Diet
Dose	
Result	Weak positive
Organs	Liver (unpublished)
Food content	
Beef	
Japan	0.24 ng/g
U. Wisc.	ND^a
LLNL	Trace at 200°C
AmHF	0.9 ng/g @ 10% fat, 20–40 ng/g @ 35% fat
Beef extract	
U. Wisc.	~50% of total
LLNL	~50% of total
Chicken	
Sardine	158 ng/g
Cuttlefish	

^{a} ND Not detected.

acids. the nitro group was reduced by hydrogenation with palladium on carbon. Closure of the imidazo ring was by condensation with cyanogen bromide in methanol. After purification the spectral properties were compared with the material isolated from broiled sardines.

3.4.4. Analysis of Foods

The IQ content of broiled sardines was reported by Yamaizumi *et al.* as 158 ng/g.[106] The content in broiled beef after extraction into a basic fraction, elution from Sephadex LH-20, and acid–base partition was very much lower, 0.24 ng/g.[105,125] These measurements were made by the isotope dilution method of mass spectroscopy described above after addition of an internal standard of the deuterated synthetic compound to the methanol extract of the cooked food. Barnes and Weisburger in a meeting report stated that beef fried on a 250°C grill (5 min per side) contained 0.5 ng/g.[74] This value was for beef containing 11% fat; fried beef with an initial fat content of 25% contained 20 ng/g of IQ. The authors' laboratory has detected IQ in fried lean beef in considerably smaller quantity (0.02 ng/g). Hargraves and Pariza[93] have fried beef at 192°C; no IQ has been detected.

3.4.5. Genetic Toxicology

IQ is a very potent *Salmonella* mutagen in the frameshift-sensitive strains (100,000–400,000 rev/μg) when activated by Aroclor-treated rat liver microsomes.[87,118] Hamster liver microsomes also activate IQ to approximately the same level as rat microsomes.[87] Japanese investigators found that the base substitution-sensitive strain TA100 gave a positive response with IQ, but the response was 60-fold less than in TA98.[80] Mammalian bioassays for mutation, SCE, and chromosome aberrations were weakly positive in CHO cells. However, these responses were less than those to Trp-P-2,[87] in contrast to the relative potency of these mutagens in *Salmonella*. Two forward mutation assays utilizing *Salmonella*, arabinose resistance and 8-azaguanine resistance, both showed very potent responses with IQ, much greater than those with Trp-P-2.[126] Unpublished reports from Japan suggest that IQ is a very weak liver carcinogen following dietary exposure (S. Takayama, personal communication).

3.5. MeIQ

3.5.1. Data Summary

See Table 4.

3.5.2. Isolation and Identification

The isolation of MeIQ from broiled sun-dried sardines was combined with that of IQ described above.[36] Eight steps were required to achieve moderate purity, and the yield of mutagen in the MeIQ fraction was less than 0.5% of the original total mutagenicity in the methanol extract. The structure was proposed from a comparison of NMR and high-resolution mass spectra with those of IQ, the only difference being an additional methyl group. This group was originally assigned at the 5 position; but this was revised to the 4 position after synthesis of both candidates and detailed comparison of all spectral data.

3.5.3. Organic Synthesis

Synthesis of MeIQ was performed from 6-amino-7-methylquinoline.[127] The acetamido derivative was nitrated to the N-nitro, followed by acid hydrolysis with migration of the nitro group to position 5. N-Methylation at position 3 followed by reduction of the nitro group and reaction with cyanogen bromide yielded ring closure and the product MeIQ. An alternative method yielded an isomeric mixture of MeIQ and the N1-methyl derivative, which was also a strong *Salmonella* mutagen. Structural congeners were synthesized for study of structure–mutagenicity relationships in *Salmonella*.[128]

3.5.4. Analysis of Foods

No MeIQ has been detected in fried or broiled beef in several laboratories. A small fraction of the total mutagen in beef extract is reported to be MeIQ.[93] In broiled sardine MeIQ is present in approximately one-half the amount of IQ.[106]

3.5.5. Genetic Toxicology

MeIQ is the most potent *Salmonella* mutagen from the entire set of pyrolytic and thermic mutagens identified to date.[36,108,118] In fact, MeIQ is one of the most potent *Salmonella* mutagens tested. Only certain dinitropyrenes give more revertants.[129] Both Aroclor-treated rat and hamster microsomes activate MeIQ to a mutagen (Felton *et al.*, manuscript in preparation). Strain TA100 also responds to MeIQ, but much less than the frameshift-sensitive strains.[118] No carcinogenicity or mammalian bioassay data are available at this time.

TABLE 4. Data Summary Table for MeIQ

Trivial name	MeIQ
Chemical name	2-Amino-3,4-dimethyl-3H-imidazo(4,5-f)quinoline
CAS reg. no.	77094-11-2
Formula	$C_{12}H_{12}N_4$
Molecular weight, amu	212.1063
Mass spectrum, *m/z*	212M+, 197
Japan	
University of Wisconsin	
UV spectrum: Max, Emol	
Japan	219, 29000; 265, 48000; 332, 4000
SRI	
Fluorescence excitation/emission, nm	
Japan	Not significant
SRI	
IR spectrum, cm^{-1}	
NMR spectrum, ppm	
JPN/CDCl$_3$	2.83, 3H,d; 3.88, 3H,s; 4.51, 2H,s; 7.36, 1H,dd;
	7.53, 1H,dd; 8.66, 1H,ddd; 8.80, 1H,dd
SRI/DMSO	
Salmonella TA1538, rev/μg	
LLNL	Rat S9 744,000; hamster S9 518,000
Japan	
SRI	
Salmonella TA98, rev/μg	
LLNL	
Japan	661,000
SRI	
Salmonella TA100, rev/μg	
LLNL	
Japan	14000 (30000)
SRI	
Mammalian bioassay	
Mutation	
SCE	
Chromosome aberrations	
Rodent carcinogenesis	
Species	
Route	
Dose	
Result	
Organs	
Food content	
Beef	
Japan	ND[a]
U. Wisc.	ND
LLNL	ND
AmHF	
Beef extract	
U. Wisc.	~13% of total
LLNL	
Chicken	
Sardine	72 ng/g
Cuttlefish	

[a] ND Not detected.

3.6. MeIQx

3.6.1. Data Summary

See Table 5.

3.6.2. Isolation and Identification

This mutagen was isolated from fried beef in the same procedure used for IQ.[78] Copurification with IQ occurred until LH-20 chromatography. The MeIQx was then purified by reverse-phase HPLC on a C18 column. The UV spectrum was distinguished from the other thermic and pyrolytic mutagens by a maximum at 274 nm. The high-resolution mass spectrum revealed the presence of five nitrogen atoms, with the NMR data suggesting a quinoxaline structure, otherwise similar to IQ. To locate the C-methyl group at either position 7 or 8, the two candidates were synthesized and spectral data compared with the data on the fraction isolated from fried beef. This comparison placed the C-methyl group at position 8, on the quinoxaline ring.

3.6.3. Organic Synthesis

Synthesis of MeIQx was similar to that used for IQ and MeIQ.[130] The starting material 6-amino-3-methylquinoxaline was prepared by condensation of 1,2,4-triaminobenzene with methylglyoxal. After separation of the desired isomeric product, tosylation, nitration, and removal of the protective tosyl group, the intermediate 6-amino-3-methyl-5-nitroquinoxaline was obtained. This compound was reductively methylated on nitrogen and reduced to the 5-amino group. Condensation with cyanogen bromide yielded MeIQx. Ultraviolet and NMR spectra were compared with the material isolated from beef for confirmation.

3.6.4. Analysis of Food

MeIQx was originally isolated from fried beef[78]; from work in several laboratories this appears to be the major one of a few mutagens that are formed during the cooking of beef under normal household conditions to a well done state. The amount present is only about 2 ng/g fresh weight of beef, as assayed by the isotope dilution mass spectrometric method that is probably the best available at this time (Nishimura, verbal presentation of Ref. 106). This mutagen has not

TABLE 5. Data Summary Table for MeIQx

Trivial name	MeIQx
Chemical name	2-Amino-3,8-dimethylimidazo(4,5-f)quinoxaline
CAS reg. no.	77500-04-0
Formula	$C_{11}H_{11}N_5$
Molecular weight, amu	213.1014
Mass spectrum, *m/z*	
Japan	213M+, 212, 197, 185, 144
U. Wisc.	213M+, 212, 198, 197, 185, 171, 159, 145, 144
UV spectrum: Max, Emol	
Japan	214, 24300; 274, 41100; 340, 3900
SRI	
Fluorescence excitation/emission, nm	
Japan	Not significant
SRI	
IR spectrum, cm^{-1}	
NMR spectrum, ppm	
JPN/CDCl$_3$	2.80, 3H,s; 3.69, 3H,s; 4.88, 2H,s; 7.51, 1H,d; 7.75, 1H,d; 8.65, 1H,s
SRI/DMSO	
Salmonella TA1538, rev/μg	
LLNL	Rat S9 58000; hamster S9 43,000
Japan	
SRI	
Salmonella TA98, rev/μg	
LLNL	
Japan	145,000
SRI	
Salmonella TA100, rev/μg	
LLNL	14400
Japan	14000
SRI	
Mammalian bioassay	
Mutation	
SCE	
Chromosome aberrations	
Rodent carcinogenesis	
Species	
Route	
Dose	
Result	
Organs	
Food content	
Beef	
Japan	2.4 ng/g, ~35% of total
U. Wisc.	26%
LLNL	27%
AmHF	
Beef extract	
U. Wisc.	21
LLNL	<10%
Chicken	
Sardine	ND[a]
Cuttlefish	

[a] ND Not detected.

been detected in broiled sardine, where IQ and MeIQ predominate. In Difco beef extract, and particularly in food-grade beef extract, MeIQx is one of the major mutagens.[92,93]

3.6.5. Genetic Toxicology

MeIQx is a somewhat weaker *Salmonella* mutagen than the imidazo quinolines IQ and MeIQ; but, like them, MeIQx is much more reactive in the frameshift-sensitive strains.[78,80] Strain TA100 was found slightly responsive in both the Japanese[80] and LLNL studies. MeIQx, like the quinolines, is also activated by both Aroclor-treated rat and hamster microsomes (J. S. Felton *et al.*, manuscript in preparation). No mammalian bioassay or carcinogenesis data are available at this time.

3.7. Glu-P-1

3.7.1. Data Summary

See Table 6.

3.7.2. Isolation and Identification

Ten kg of L-glutamic acid was pyrolyzed over a gas flame and a base fraction was obtained from the tar.[131] This fraction was distilled under reduced pressure and the mutagenic fraction was chromatographed successively on silica gel, CM-Sephadex, Sephadex LH-20, and alumina. Two bands (Glu-P-1 and Glu-P-2) were separated by TLC on silica gel, and each product was purified to 10 mg of crystalline hydrobromide on LH-20 with HBr in the solvent. Identification was by mass spectroscopy and elemental analysis supported by NMR and UV spectra. The structure of Glu-P-1 was confirmed by X-ray crystallography and by synthesis.

3.7.3. Organic Synthesis

Previously reported 3-amino-8-methylimidazo(1,2,-a)pyridine was reacted with 2-chloroacrylonitrile and aluminum chloride in nitrobenzene at 100°C for 12 hr.[132] The basic product was isolated, chromatographed on silica gel, and crystallized from methanol–ethyl acetate. The UV and NMR spectra of the hydrobromide varied with concentration and temperature.

TABLE 6. Data Summary Table for Glu-P-1

Trivial name	Glu-P-1	
Chemical name	2-Amino-6-methyldipyrido-(1,2-a:3′,2′-d)imidazole	
CAS Reg. No.	67730-11-4	
Formula	$C_{11}H_{10}N_4$	
Molecular weight, amu	198	
Mass spectrum, m/z	198M+, 180, 170, 158, 100, 93, 92	
UV spectrum: Max, Emol		
Japan	257, 21900; 263, 23400; 293, 6800; 303, 5900; 316, 5000; 360, 7900; 375, 8700; 395, 5900	
SRI	210, 12300; 260, 17900; 306, 5500; 316, 4800; 377, 12600	
Fluorescence excitation/emission, nm		
Japan	360/435	
SRI	376/450	
IR spectrum, cm^{-1}		
NMR spectrum, ppm		
JPN/CDCl$_3$	264, 3H,s; 4.61, 2H,b; 6.73, 1H,t; 6.75, 1H,d; 7.14, 1H,d; 8.05, 1H,d; 8.42, 1H,d	
SRI/DMSO		
Salmonella TA1538, rev/µg		
LLNL		
Japan		
SRI		
Salmonella TA98, rev/µg		
LLNL		
Japan	49000	
SRI	41000	
Salmonella TA100, rev/µg		
LLNL		
Japan	3200	
SRI	245	
Mammalian bioassay		
Mutation		
SCE	Positive	
Chromosome aberrations		
Rodent carcinogenesis		
Species	Mouse	Rat
Route	Diet	Diet
Dose	0.05%	0.05%
Result	Positive	Positive
Organs	Liver, subscapular hemangioendothelioma	Liver, intestine
Food content		
Beef		
Japan	ND[a]	
U. Wisc.		
LLNL		
Beef extract		
U. Wisc.		
LLNL		
Chicken		
Sardine	ND	
Cuttlefish	ND	

[a] ND Not detected.

3.7.4. Analysis of Foods

Glu-P-1 has been reported present in casein pyrolyzate, but was not found in fried or broiled beef, sardine, or squid (cuttlefish).[78,90,133]

3.7.5. Genetic Toxicology

Glu-P-1 has *Salmonella* mutagenicity in the range similar to that of MeIQx (approximately one-tenth of the potency of IQ or MeIQ).[80,131] Studies at SRI found a very weak response for TA100, in contrast to those in Japan reporting 3200 rev/μg (ten times higher).[112] Glu-P-1 is positive for SCE in human lymphoblastoid cells.[113] It also has been found to bind covalently to DNA *in vitro*[134] and rat liver DNA *in vivo*.[120] Guanine adducts at C-8 have been identified as the major adducts.[120,134–136] No evidence that these are actually the mutagenic adducts has been presented at this time. The carcinogenesis picture is rather strange. Oral feeding of mice, 0.05% in the diet, gives a large number of hepatomas and unusual hemangioendothelial sarcomas subcutaneously between the scapulae.[118]

3.8. Glu-P-2

3.8.1. Data Summary

See Table 7.

3.8.2. Isolation and Identification

Isolation of Glu-P-2 was carried out during that of Glu-P-1 from a pyrolyzate of L-glutamic acid as described in Section 3.7.2.[131] Identification was from elemental analysis, mass spectroscopy, and NMR and UV spectra, comparing these to Glu-P-1, whose structure was established by X-ray crystallography. The putative structure of Glu-P-2 was confirmed by synthesis.

3.8.3. Organic Synthesis

Glu-P-2 was prepared in 20–40% yield by condensation of 3-aminoimidazo(1,2-*a*)pyridine with 2-chloroacrylonitrile.[132] Spectroscopic confirmation was performed as for Glu-P-1.

3.8.4. Analysis of Foods

The only food in which Glu-P-2 has been reported is the squid (cuttlefish) after broiling.[137] Quantification was done in two ways: by

TABLE 7. Data Summary Table for Glu-P-2

Trivial name	Glu-P-2	
Chemical name	2-Aminodipyrido(1,2-a:3′,2′-d)imidazole	
CAS reg. no.	67730-10-3	
Formula	$C_{10}H_8N_4$	
Molecular weight, amu	184	
Mass spectrum, m/z		
UV spectrum: Max, Emol		
Japan	257, 18600; 265, 15100; 294, 5800; 303, 5200; 316, 4500; 365, 7100; 380, 7800; 402, 4900	
SRI	211, 15000; 260, 15500; 306, 5100; 317, 4700; 383, 8800	
Fluorescence excitation/emission, nm		
Japan	360/435	
SRI	382/460	
IR spectrum, cm^{-1}		
NMR spectrum, ppm		
JPN/D20	6.82, 1H,d; 7.53, 1H,t; 7.82, 1H,d; 7.88, 1H,d; 8.12, 1H,t; 8.63, 1H,d	
SRI/DMSO		
Salmonella TA1538, rev/μg		
LLNL		
Japan		
SRI		
Salmonella TA98, rev/μg		
LLNL		
Japan	1900	
SRI	245	
Salmonella TA100, rev/μg		
LLNL		
Japan	1200	
SRI	339	
Mammalian bioassay		
Mutation		
SCE		
Chromosome aberrations		
Rodent carcinogenesis		
Species	Mouse	Rat
Route	Diet	Diet
Dose	0.05%	0.05%
Result	Positive	Positive
Organs	Liver, subscapular hemangioendothelioma	Liver, intestine
Food content		
Beef		
Japan	ND[a]	
U. Wisc.		
LLNL		
Beef extract		
U. Wisc.		
LLNL		
Chicken		
Sardine		
Cuttlefish	280 ng/g	

[a] ND Not detected.

chromatographic purification from the base fraction with mass spectrometric measurement of the molecular ion; and by mass fragmentography of the crude base fraction.

3.8.5. Genetic Toxicology

Glu-P-2 is a much weaker *Salmonella* mutagen than Glu-P-1 (20–140 times) and is moderately positive in TA100.[80,131] No mammalian assay data are available. The carcinogenicity data are similar to Glu-P-1 with hepatomas and hemangioendothelial sarcomas appearing after addition of the compound to the diet.[118]

3.9. AAC

3.9.1. Data Summary

See Table 8.

3.9.2. Isolation and Identification

In connection with the isolation of new mutagens from amino acids, proteins, foods, and cigarette smoke condensate, Yoshida *et al.*[23] pyrolyzed 3 kg of soybean globulin over a gas burner, collecting the volatilized tar. After pH-adjusted liquid–liquid partition the base fraction was chromatographed on silica gel, CM-Sephadex, and Sephadex LH-20, and TLC was performed on silica. Although expected, neither Trp-P-1 nor Trp-P-2 was found. However, a highly fluorescent fraction was extracted and crystallized, and the structure was determined by X-ray crystallography. From these data and mass spectrometry two α-carbolines AAC and AMAC were identified. Subsequently these compounds were also found after pyrolysis of tryptophan,[138] other proteins[24] and tobacco.[14]

3.9.3. Organic Synthesis

Reaction of 6-bromo-2-picolinic acid with *o*-phenylenediamine was carried out in the presence of Cu and anhydrous potassium carbonate.[139] The product was treated with nitrous acid to form benzotriazolylpicolinic acid. The latter was reacted with diphenylphosphorazitate (DPPA) to substitute an amino group for the carboxylic acid via the Curtius rearrangement. After heating with polyphosphoric acid followed by alkali, AAC was obtained in 24% yield.

TABLE 8. Data Summary Table for AAC

Trivial name	AAC
Chemical name	2-Amino-9H-pyrido(2,3-b)indole
CAS reg. no.	26148-68-5
Formula	$C_{11}H_9N_3$
Molecular weight, amu	183.0797
Mass spectrum, *m/z*	
Japan	183M+, 156, 155, 129, 128
U. Wisc.	
UV spectrum: Max, Emol	
Japan	231, 41700; 262, 13800; 336, 19500
SRI	232, 20500; 264, 5600; 338, 12700
Fluorescence excitation/emission, nm	
Japan	345/380
SRI	360/402
IR spectrum, cm^{-1}	3400, 3300, 1625, 1605, 1570, 1455, 1420, 1365, 1305, 1230, 1122, 735
NMR spectrum, ppm	
JPN/DMSO	6.10, 2H,s; 6.40, 1H,d; 7.0–7.46, 3H,m; 7.82, 1H,d; 8.04, 1H,d; 11.18, 1H,s
SRI/DMSO	
Salmonella TA1538, rev/µg	
LLNL	Rat S9 460; hamster S9 670
Japan	
SRI	
Salmonella TA98, rev/µg	
LLNL	
Japan	300
SRI	206
Salmonella TA100, rev/µg	
LLNL	
Japan	20
SRI	40
Mammalian bioassay	
Mutation	
SCE	
Chromosome aberrations	
Rodent carcinogenesis	
Species	Mouse
Route	Diet
Dose	0.08%
Result	Positive
Organs	Liver, subscapular hemangioendothelioma
Food content	
Beef	
Japan	200 ng/g
U. Wisc.	
LLNL	
Beef extract	
U. Wisc.	
LLNL	
Chicken	42 ng/g
Sardine	
Cuttlefish	
Onion	0.3 ng/g
Mushroom	8 ng/g

3.9.4. Analysis of Foods

In a single report from Yoshida's laboratory beef, chicken, onion, and mushroom were grilled at high heat over a gas stove.[18] Reported weight loss during cooking suggests that the degree of cooking was extreme. The methanol extract of each food was passed through Sephadex LH-20 and into HPLC on reverse phase C-18 column, with detection by fluorescence and quantification by comparison with chromatographed synthetic AAC. Proof of identity was by comparison of UV and mass spectra. Grilled beef contained 650 ng/g (original wet weight or cooked weight not specified) of AAC. In the authors' laboratory the α-carbolines do not seem to be present in beef fried at 200°C or 250°C. Chicken contained about one-fourth as much as beef, and the vegetables considerably less. Although AAC is a much weaker mutagen in *Salmonella* than many of the other thermic mutagens, the quantity reported in cooked beef is greater, so that the contributed mutagenic activity may be nearly as great as that of some of the more potent mutagens. Preliminary animal carcinogenesis data reported by Sugimura indicate similar tumorigenic potency for thermic and pyrolytic mutagens of widely different mutagenic potency in bacteria.[79]

3.9.5. Genetic Toxicology

AAC is a moderate to weak mutagen in *Salmonella*.[23] Both hamster and rat Aroclor-induced microsomes activate AAC; and it is not significantly positive in strain TA100.[23,80,118] No mammalian assay data are available. AAC is carcinogenic and at a dose of 0.08% in the diet gives hepatomas and hemangioendothelial sarcomas, as do Glu-P-1 and Glu-P-2.[79] Interestingly, AAC is a much stronger carcinogen than would be expected from the *Salmonella* mutagenicity.

3.10. AMAC

3.10.1. Data Summary

See Table 9.

3.10.2. Isolation and Identification

The isolation of AMAC from heated soybean globulin was combined with that of AAC described above.[23] The congeners were separated on Sephadex LH-20. Identification was by high-resolution mass spectrometry and X-ray crystallography.

TABLE 9. Data Summary Table for AMAC

Trivial name	AMAC (MeAC)
Chemical name	2-Amino-3-methyl-9H-pyrido(2,3-b)indole
CAS reg. no.	68006-83-7
Formula	$C_{12}H_{11}N_3$
Molecular weight, amu	197.0912
Mass spectrum, *m/z*	
Japan	197M+, 196, 179, 169, 141, 98, 85, 84
U. Wisc.	
UV spectrum: Max, Emol	
Japan	
SRI	232, 21900; 264, 4000; 343, 13700
Fluorescence excitation/emission, nm	
Japan	348/375
SRI	364/408
IR spectrum, cm^{-1}	
NMR spectrum, ppm	
JPN/DMSO	2.21, 3H,s; 5.85, 2H,s; 7.0–7.42, 3H,m; 7.81, 1H,d; 7.90, 1H,s; 11.04, 1H,s
SRI/DMSO	
Salmonella TA1538, rev/μg	
LLNL	Rat S9 54; hamster S9 250
Japan	
SRI	
Salmonella TA98, rev/μg	
LLNL	
Japan	200
SRI	32
Salmonella TA100, rev/μg	
LLNL	
Japan	120
SRI	57
Mammalian bioassay	
Mutation	
SCE	
Chromosome aberrations	
Rodent carcinogenesis	
Species	Mouse
Route	Diet
Dose	0.08%
Result	Positive
Organs	Liver, subscapular hemangioendothelioma
Food content	
Beef	
Japan	20 ng/g
U. Wisc.	
LLNL	
Beef extract	
U. Wisc.	
LLNL	
Chicken	4 ng/g
Sardine	
Cuttlefish	
Onion	ND[a]
Mushroom	0.9 ng/g

[a] ND Not detected.

3.10.3. Organic Synthesis

One starting reagent, 2-aminoindole, was synthesized from *o*-nitrobenzylcyanide by stannous chloride reduction in acid solution.[140] The second reagent, 3-amino-2-methylacrylonitrile, was made in several steps from diethyl methylmalonate. The two reagents were condensed in the presence of sodium ethoxide to yield AMAC, which was purified on Sephadex LH-20 and HPLC on a reverse phase C-18 column. A series of alkyl group congeners was prepared for structure–mutagenic activity studies.

3.10.4. Analysis of Foods

Under the conditions given above for AAC in foods, the AMAC content of grilled beef, chicken, mushroom, and onion was measured.[18] In general, AMAC was present at about one-tenth the level of AAC, except that none was detectable in onion.

3.10.5. Genetic Toxicology

AMAC, like AAC, is a fairly weak *Salmonella* mutagen[23] compared to many of the other thermic and pyrolytic mutagens (32–250 rev/µg), but when added to the diet gives the same carcinogenic results as AAC, Glu-P-1, and Glu-P-2: hepatomas and hemangioendothelial sarcomas.[80] No mammalian assay data are available at this time.

3.11. Phe-P-1

3.11.1. Data Summary

See Table 10.

3.11.2. Isolation and Identification

In analogous experiments to those in which the Trp pyrolytic mutagens were isolated, the tar from heating of D,L-phenylalanine or L-phenylalanine was partitioned between solvents to yield a base fraction, which was chromatographed on silica gel.[109] Its structure, slightly simpler than the other pyrolytic mutagens, was deduced from elemental analysis and a suite of spectra. The substance was shown to have antifungal activity.

TABLE 10. Data Summary Table for Phe-P-1

Trivial name	Phe-P-1
Chemical name	2-Amino-5-phenylpyridine
CAS reg. no.	33421-40-8
Formula	$C_{11}H_{10}N_2$
Molecular weight, amu	170
Mass spectrum, *m/z*	
Japan	170M+, 153, 143, 115
U. Wisc.	
UV spectrum: Max, Emol	
Japan	268, 21400; 315, Sh
SRI	
Fluorescence excitation/emission, nm	
Japan	
SRI	
IR spectrum, cm^{-1}	
NMR spectrum, ppm	
JPN/CDCl$_3$	4.55, 2H,b; 6.50, 1H,d; 7.16–7.70, 6H,m; 8.25, 1H,d
SRI/DMSO	
Salmonella TA1538, rev/µg	
LLNL	
Japan	
SRI	
Salmonella TA98, rev/µg	
LLNL	
Japan	41
SRI	
Salmonella TA100, rev/µg	
LLNL	
Japan	23
SRI	
Mammalian bioassay	
Mutation	
SCE	
Chromosome aberrations	
Rodent carcinogenesis	
Species	
Route	
Dose	
Result	
Organs	
Food content	
Beef	
Japan	
U. Wisc.	
LLNL	
Beef extract	
U. Wisc.	
LLNL	
Chicken	
Sardine	8.6 ng/g
Cuttlefish	

3.11.3. Organic Synthesis

Pyrrole was condensed with dichloromethylbenzene in xylene solution in the presence of sodium ethoxide.[133] The phenylpyridine product was aminated at the 2 position with sodamide in xylene solution.

3.11.4. Analysis of Foods

Phe-P-1 has been reported only in the extract of broiled sardines, where a very small quantity was identified by mass spectrometric detection of the molecular ion and a characteristic fragment.[90] Since its mutagenic potency is very weak, Phe-P-1 can contribute only minimally to the total mutagenic activity of this food.

3.11.5. Genetic Toxicology

Phe-P-1 is a weak *Salmonella* mutagen, 41 rev/μg (TA98) and 23 rev/μg (TA100).[109] No other toxicological data are available.

3.12. Lys-P-1

3.12.1. Data Summary

See Table 11.

3.12.2. Isolation and Identification

The base fraction from the pyrolytic tar of L-lysine was prepared as for other pyrolytic mutagens.[141] The fraction was separated into several mutagenic components on Sephadex LH-20 and alumina. One nonpolar fraction was further chromatographed on CM-Sephadex and LH-20. Final purification was by TLC on silica gel, from which the mutagen was eluted and crystallized. Structure was determined by mass spectrometry and X-ray crystallography.

3.12.3. Organic Synthesis

Indane was nitrated and reduced to 4-aminoindane.[141] A Skraup reaction gave 7,8-cyclopentenoquinoline. An amino group was introduced at position 5 and a phenyl group at position 6; the amino group was diazotized and thermolyzed to cause closure of the carbazole ring structure to form the final product.

TABLE 11. Data Summary Table for Lys-P-1

Trivial name	Lys-P-1
Chemical name	3,4-Cyclopentenopyrido(3,2-a)carbazole
CAS reg. no.	
Formula	$C_{18}H_{14}N_2$
Molecular weight, amu	258
Mass spectrum, *m/z*	
Japan	258M +
U. Wisc.	
UV spectrum: Max, Emol	
Japan	
SRI	
Fluorescence excitation/emission, nm	
Japan	
SRI	
IR spectrum, cm^{-1}	
NMR spectrum, ppm	
JPN/CDCl$_3$	
SRI/DMSO	
*Salmonella*TA1538, rev/μg	
LLNL	
Japan	
SRI	
Salmonella TA98, rev/μg	
LLNL	
Japan	90
SRI	
Salmonella TA100, rev/μg	
LLNL	
Japan	100
SRI	

3.12.4. Analysis of Foods

There have been no reports of Lys-P-1 detection in foods.

3.12.5. Genetic Toxicology

Lys-P-1 is slightly more potent in *Salmonella* than Phe-P-1 (100 rev/μg)[141] but still considerably weaker than the glutamate or tryptophan pyrolosis tars, or the imidazoquinolines. No other toxicological data are available.

TABLE 12. Data Summary Table for Orn-P-1

Trivial name	Orn-P-1
Chemical name	4-Amino-6-methyl-1H-2,5,10,10b-tetraazafluoranthene
CAS reg. no.	
Formula	$C_{13}H_{11}N_5$
Molecular weight, amu	237.1014
Mass spectrum, m/z	
Japan	237M+
U. Wisc.	
UV spectrum: Max, Emol	
Japan	
SRI	
Fluorescence excitation/emission, nm	
Japan	
SRI	
IR spectrum, cm^{-1}	
NMR spectrum, ppm	
JPN/CDCl$_3$	
SRI/DMSO	
Salmonella TA1538, rev/μg	
LLNL	
Japan	
SRI	
Salmonella TA98, rev/μg	
LLNL	
Japan	57000
SRI	

3.13. Orn-P-1

3.13.1. Data Summary

See Table 12.

3.13.2. Isolation and Identification

Five kg of L-ornithine hydrochloride was pyrolyzed over a flame and 600 g of tar base fraction was obtained as in other amino acid pyrolysis experiments.[142] The base fraction was separated by countercurrent distribution and chromatography on Sephadex LH-20, alumina, and silica gel. One component was isolated at low recovery and crystallized as the hydrobromide. Its specific mutagenic activity was as high as that of Trp-P-2. Low- and high-resolution mass spectrometry and X-ray crystallography allowed deduction of the unusual tetracyclic structure with one or more nitrogen atoms in each ring.

3.13.3. Organic Synthesis

Synthesis of Orn-P-1 has not been reported.

3.13.4. Analysis of Foods

There have been no reports of detection of Orn-P-1 in foods.

3.13.5. Genetic Toxicology

Orn-P-1 is a moderately potent *Salmonella* mutagen (57,000 rev/μg) with a potency equivalent to MeIQx in strain TA98.[142] No other data are available at this time.

4. Summary

Recently concern has arisen that the cooking of foods under certain circumstances may produce genotoxic substances that could contribute to the high incidence of cancer of the stomach, large intestine, or other organs. In the past few years two groups of Japanese investigators, headed by Dr. Takashi Sugimura of the National Cancer Center Research Institute and Dr. Daisuke Yoshida of the Japan Tobacco and Salt Public Corporation's Central Research Institute, have demonstrated that when amino acids, pure proteins, or protein foods are heated to comparatively high temperatures (charring with pyrolysis) several mutagens are formed. Further study in the United States by Commoner, Pariza, and Weisburger and their co-workers and in the authors' laboratories has shown that household frying and broiling to a well done state causes the thermic formation of the same and additional mutagens. In the *Salmonella* reversion bioassay of Ames these mutagens range from weak to extremely potent.

Isolation and identification of these mutagens has been a challenging task, since they are present at the part per billion level or less. Extraction procedures giving a high yield have been important. High-performance liquid chromatography has been essential for fractionation and isolation. For those mutagens isolated from cooked foods, high-resolution mass spectrometry has been essentially the only technique of structural analysis giving sufficient sensitivity to detect the miniscule quantities of mutagen available. The mutagens identified to date, about a dozen, have introduced a new class of agents to genetic toxicology. All of them are N-heterocyclic, polycyclic aromatic amines, i.e., aminoazaarenes, or closely related structures.

This chapter reviews the formation of the pyrolytic and thermic mutagens, the isolation and identification methods that have been employed, and potential future improvements, and summarizes the available data on chemical properties, organic synthesis, food analysis, and genetic toxicity. The literature available until June 1983 is included.

ACKNOWLEDGMENTS

Research from the authors' laboratories reported here was sponsored by the National Institute of Environmental Health Sciences/National Toxicology Program through an Interagency Agreement number 222Y01-ES-10063 and was performed under the auspices of the U. S. Department of Energy by the Lawrence Livermore National Laboratory under contract number W-7405-ENG-48. We acknowledge helpful discussions with Drs. T. Sugimura, S. Nishimura, D. Yoshida, and M. Pariza. We thank Angela Riggs and Leilani Corell for preparation of the manuscript.

Reference herein to any specific commercial products, process, or service by trade name, trademark, manufacturer, or otherwise, does not necessarily constitute or imply its endorsement, recommendation, or favoring by the United States Government or the University of California.

5. References

1. R. Doll, General epidemiologic considerations in etiology of colorectal cancer, in: *Colorectal Cancer: Prevention, Epidemiology, and Screening* (S. Winawer, D. Schottenfeld, and P. Sherlock, eds.), p. 3–12, Raven Press, New York (1980).
2. J. H. Weisburger, N. E. Spingarn, Y. Y. Wang, and L. L. Vuolo, Assessment of the role of mutagens and endogenous factors in large bowel cancer, *Cancer Bull. 33*, 124–129 (1981).
3. R. Doll and R. Peto, The causes of cancer: Quantitative estimates of avoidable risks of cancer in the United States today, *J. Natl. Cancer Inst. 66*, 1191–1308 (1981).
4. J. H. Weisburger and G. M. Williams, Chemical carcinogenesis, in: *Cancer Medicine* (J. F. Holland and E. Frei III, eds.), pp. 42–95, Lea & Febiger, Philadelphia (1982).
5. J. H. Weisburger, On the etiology of gastro-intestinal tract cancers, with emphasis on dietary factors, in: *Environmental Carcinogenesis. Occurrence, Risk Evaluation and Mechanisms* (P. Emmelot and E. Kriek, eds.), pp. 215–240, Elsevier/North-Holland Biomedical Press, Amsterdam (1979).

6. E. L. Wynder and G. B. Gori, Contribution of the environment to cancer incidence: An epidemiologic exercise, *J. Natl. Cancer Inst. 58*, 825–832 (1977).

7. T. Sugimura, Tumor initiators and promoters associated with ordinary foods, in: *Molecular Interrelations of Nutrition and Cancer* (M. S. Arnott, J. van Eys, and Y. M. Wang, eds.), pp. 3–42, Raven Press, New York (1982).

8. T. Sugimura, Mutagens, carcinogens, and tumor promoters in our daily food, *Cancer 49*, 1970–1984 (1982).

9. W. D. Powrie, C. H. Wu, M. P. Rosin, and H. F. Stich, Clastogenic and mutagenic activities of Maillard reaction model systems, *J. Food Sci. 46*, 1433–1445 (1981).

10. T. Sugimura and M. Nagao, Mutagenic factors in cooked foods, *CRC Crit. Rev. Toxicol. 6*, 189–209 (1979).

11. W. D. Powrie, C. H. Wu, M. P. Rosin, and H. F. Stich, Mutagens and carcinogens in food, *Prog. Mutat. Res. 3*, 187–199 (1982).

12. W. Lijinsky and P. Shubik, Benzo(a)pyrene and other polynuclear hydrocarbons in charcoal-broiled meat, *Science 145*, 53–55 (1964).

13. T. Sugimura, M. Nagao, T. Kawachi, M. Honda, T. Yahagi, Y. Seino, S. Sato, N. Matsukura, T. Matsushima, A. Shirai, M. Sawamura, and H. Matsumoto, Mutagen-carcinogens in foods with special reference to highly mutagenic pyrolytic products in broiled foods, in: *Origins of Human Cancer* (H. H. Hiatt, J. D. Watson, and J. A. Winsten, eds.), pp. 1561–1577, Cold Spring Harbor Laboratory, Cold Spring Harbor, New York (1977).

14. D. Yoshida and T. Matsumoto, Amino-α-carbolines as mutagenic agents in cigarette smoke condensate, *Cancer Lett. 10*, 141–149 (1980).

15. S. Mizusaki, H. Okamoto, A. Akiyama, and Y. Fukuhara, Relation between chemical constituents of tobacco and mutagenic activity of cigarette smoke condensate, *Mutat. Res. 48*, 319–326 (1977).

16. T. Matsumoto, D. Yoshida, S. Mizusaki, and H. Okamoto, Mutagenic activity of amino acid pyrolysates in *Salmonella typhimurium* TA 98, *Mutat. Res. 48*, 279–286 (1977).

17. T. Matsumoto, D. Yoshida, S. Mizusaki, and H. Okamoto, Mutagenicities of the pyrolysates of peptides and proteins, *Mutat. Res. 56*, 281–288 (1978).

18. T. Matsumoto, D. Yoshida, and H. Tomita, Determination of mutagens, amino-α-carbolines in grilled foods and cigarette smoke condensate, *Cancer Lett. 12*, 105–110 (1981).

19. M. Nagao, M. Honda, Y. Seino, T. Yahagi, T. Kawachi, and T. Sugimura, Mutagenicities of protein pyrolysates, *Cancer Lett. 2*, 335–340 (1977).

20. M. Nagao, M. Honda, Y. Seino, T. Yahagi, and T. Sugimura, Mutagenicities of smoke condensates and the charred surface of fish and meat, *Cancer Lett. 2*, 221–226 (1977).

21. M. Nagao, T. Yahagi, T. Kawachi, Y. Seino, M. Honda, N. Matsukura, T. Sugimura, K. Wakabayashi, K. Tsuji, and T. Kosuge, Mutagens in foods, and especially pyrolysis products of protein, in: *Progress in Genetic Toxicology: Proceedings of the Second International Conference on Environmental Mutagens, Edinburgh (July 1977)*, (D. Scott, B. A. Bridges, and F. H. Sobels, eds.), pp. 259–264, Elsevier (1977).

22. D. Yoshida, T. Matsumoto, and H. Nishigata, Effect of heating methods on mutagenic activity and yield of mutagenic compounds in pyrolysis products of protein, *Agric. Biol. Chem. 44*, 253–255 (1980).

23. D. Yoshida, T. Matsumoto, R. Yoshimura, and T. Matsuzaki, Mutagenicity of amino-α-carbolines in pyrolysis products of soybean globulin, *Biochem. Biophys. Res. Commun. 83*, 915–920 (1978).

24. D. Yoshida, H. Nishigata, and T. Matsumoto, Pyrolytic yields of 2-amino-9H-pyrido[2,3-b]indole and 3-amino-1-methyl-5H-pyrido[4,3-b]indole as mutagens from proteins, *Agric. Biol. Chem. 43*, 1769–1770 (1979).

25. T. Sugimura, M. Nagao, and K. Wakabayashi, Mutagenic heterocyclic amines in cooked food, in: *Environmental Carcinogens, Selected Methods of Analysis* (L. Fishbein, I. K. O'Neill, M. Castegnaro, and H. Bartsch, eds.), Vol. 4, pp. 251–267, International Agency for Research on Cancer, Lyon, France (1981).

26. M. Nagao, T. Yahagi, and T. Sugimura, Differences in effects of norharman with various classes of chemical mutagens and amounts of S-9, *Biochem. Biophys. Res. Commun. 83*, 373–378 (1978).

27. M. Nagao, T. Yahagi, T. Kawachi, T. Sugimura, T. Kosuge, K. Tsuji, K. Wakabayashi, S. Mizusaki, T. Matsumoto, Comutagenic action of norharman and harman, *Proc. Jpn. Acad. B 53*(2), 95–98 (1977).

28. T. Matsumoto, D. Yoshida, and S. Mizusaki, Enhancing effect of harman on mutagenicity in *Salmonella, Mutat. Res. 56*, 85–88 (1977).

29. B. Commoner, A. Vithayathil, and P. Dolara, Mutagenic analysis as a means of detecting carcinogens in foods, *J. Food Protection 41*, 996–1003 (1978).

30. B. Commoner, A. J. Vithayathil, P. Dolara, S. Nair, P. Madyastha, and G. C. Cuca, Formation of mutagens in beef and beef extract during cooking, *Science 201*, 913–916 (1978).

31. P. Dolara, B. Commoner, A. Vithayathil, G. Cuca, E. Tuley, P. Madyastha, S. Nair, an D. Kriebel, The effect of temperature on the formation of mutagens in heated beef stock and cooked ground beef, *Mutat. Res. 60*, 231–237 (1979).

32. L. F. Bjeldanes, M. M. Morris, J. S. Felton, S. Healy, D. Stuermer, P. Berry, H. Timourian, and F. T. Hatch, Mutagens from the cooking of food II. Survey by Ames/*Salmonella* test of mutagen formation in the major protein-rich foods of the American diet, *Food Chem. Toxicol. 20*, 357–363 (1982).

33. L. F. Bjeldanes, M. M. Morris, J. S. Felton, S. Healy, D. Stuermer, P. Berry, H. Timourian, and F. T. Hatch, Mutagens from the cooking of food III. Secondary sources of cooked dietary protein, *Food Chem. Toxicol. 20*, 365–369 (1982).

34. L. F. Bjeldanes, M. M. Morris, H. Timourian, and F. T. Hatch, Effects of meat composition and cooking conditions on mutagenicity of fried ground beef, *J. Agric. Food Chem. 31*, 18–21 (1983).

35. J. Felton, S. Healy, D. Stuermer, C. Berry, H. Timourian, F. T. Hatch, M. Morris, and L. F. Bjeldanes, Mutagens from the cooking of food. (I). Improved isolation and characterization of mutagenic fractions from cooked ground beef, *Mutat. Res. 88*, 33–44 (1981).

36. H. Kasai, Z. Yamaizumi, K. Wakabayashi, M. Nagao, T. Sugimura, S. Yokoyama, T. Miyazawa, N. E. Spingarn, J. H. Weisburger, and S. Nishimura, Potent novel mutagens produced by broiling fish under normal conditions, *Proc. Jpn. Acad. B 56*(5), 278–283 (1980).

37. H. Kasai, Z. Yamaizumi, S. Nishimura, K. Wakabayashi, M. Nagao, T. Sugimura, N. E. Spingarn, J. H. Weisburger, S. Yokoyama, and T. Miyazawa, A potent mutagen in broiled fish. Part 1. 2-Amino-3-methyl-3H-imidazo[4,5-f]quinoline, *J. Chem. Soc. 1981*, 2290–2293.

38. M. W. Pariza, S. H. Ashoor, F. S. Chu, and D. B. Lund, Effects of temperature and time on mutagen formation in pan-fried hamburger, *Cancer Lett. 7*, 63–69 (1979).

39. N. E. Spingarn and J. H. Weisburger, Formation of mutagens in cooked food: I. Beef, *Cancer Lett. 7*, 259–264 (1979).

40. N. E. Spingarn, H. Kasai, L. L. Vuolo, S. Nishimura, Z. Yamaizumi, T. Sugimura, T. Matsushima, and J. H. Weisburger. Formation of mutagens in cooked foods. III. Isolation of a potent mutagen from beef, *Cancer Lett.* 9, 177–183 (1980).

41. W. A. Hargraves and M. W. Pariza, Purification and characterization of bacterial mutagens from commercial beef extracts and fried ground beef, *Proc. Am. Assoc. Cancer Res.* 23, 92 (1982).

42. M. W. Pariza, S. H. Ashoor, F. S. Chu, and D. B. Lund, Mutagens and inhibitors of mutagenesis in pan-fried hamburger, *Proc. Am. Assoc. Cancer Res.* 20, 39 (1979).

43. M. W. Pariza, Mutagens in heated foods, *Food Technol. 1982* (March), 53–56.

44. M. W. Pariza, S. H. Ashoor, D. J. Raune, R. A. Dietrich, M. B. Hugdahl, and F. S. Chu, Formation of mutagens in cooked beef and in model systems, *Proc. Am. Assoc. Cancer Res.* 21, 87 (1980).

45. R. I. M. Chan, H. F. Stich, M. P. Rosin, and W. D. Powrie, Antimutagenic activity of browning reaction products, *Cancer Lett.* 15, 27–33 (1982).

46. N. E. Spingarn, L. A. Slocum, and J. H. Weisburger, Formation of mutagens in cooked foods. II. Foods with high starch content, *Cancer Lett.* 9, 7–12 (1980).

47. H. F. Stich, W. Stich, M. P. Rosin, and W. D. Powrie, Clastogenic activity of caramel and caramelized sugars, *Mutat. Res.* 91, 129–136 (1981).

48. M. W. Pariza, S. H. Ashoor, and F. S. Chu, Mutagens in heat-processed meat, bakery and cereal products, *Food Cosmet. Toxicol.* 17, 429–430 (1979).

49. L. F. Bjeldanes and H. Chew, Mutagenicity of 1,2-dicarbonyl compounds: Maltol, kojic acid, diacetyl and related substances, *Mutat. Res.* 67, 367–371 (1979).

50. M. W. Pariza, ed., Selected abstracts on mutagens and carcinogens in cooked, smoked and charred foods, in: *Oncology Overview*, International Cancer Research Data Bank Program, National Cancer Institute, U. S. Department of Health and Human Services, National Institute of Health (1982).

51. T. Matsushima and T. Sugimura, Mutagen-carcinogens in amino acid and protein pyrolysates and in cooked food, in: *Progress in Mutation Research* (A. Kappas, ed.), Vol. 2, pp. 49–56 Elsevier/North-Holland Biomedical Press, Amsterdam (1981).

52. W. D. Powrie and S. Nakai, Processing effects on protein systems, in: *Utilization of Protein Resources* (D. W. Stanley *et al.*, eds.), Chapter 4, Food and Nutrition Press, Westport, Connecticut (1981).

53. T. Shibamoto, Heterocyclic compounds found in cooked meats, *J. Agric. Food Chem.* 28, 237–243 (1980).

54. Report of Committee on Diet, Nutrition, and Cancer, National Research Council, National Academy Press, Washington, D.C. (1982).

55. J. J. Hutton and C. Hackney, Metabolism of cigarette smoke condensates by human and rat homogenates to form mutagens detectable by *Salmonella typhimurium* TA 1538, *Cancer Res.* 35, 2461–2468 (1975).

56. T. Sugimura, T. Kawachi, M. Nagao, and T. Yahagi, Mutagens in food as causes of cancer, *Prog. Cancer Res. Ther.* 17, 59–71 (1981).

57. T. Sugimura, Naturally occurring genotoxic carcinogens, in: *Naturally Occurring Carcinogen-Mutagens and Modulators of Carcinogenesis* (E. C. Miller, J. A. Miller, and I. Hirono, eds.), pp. 241–261, University Park Press, Baltimore, Maryland (1979).

58. K. Shinohara, R.-T. Wu, N. Jahan, M. Tanaka, N. Morinaga, H. Murakami, and H. Omura, Mutagenicity of the browning mixtures of amino-carbonyl reactions on *Salmonella typhimurium* TA 100, *Agric. Biol. Chem.* 44, 671 (1980).

59. H. F. Stich, W. Stich, M. Rosin, and W. Powrie, Mutagenic activity of pyrazine derivatives: A comparative study with *Salmonella typhimurium*, *Saccharomyces cerevisiae* and Chinese hamster ovary cells, *Food Cosmet. Toxicol.* 18, 581–584 (1980).

60. T. Shibamoto, Occurrence of mutagenic products in browning model systems, *Food Technol. 1982* (March), 59–62.

61. N. E. Spingarn, C. T. Garvie-Gould, and L. A. Slocum, Formation of mutagens in sugar–amino acid model systems, *J. Agric. Food Chem. 31*, 301–304 (1983).

62. W. D. Powrie, C. H. Wu, and H. F. Stich, Browning reaction systems as sources of mutagens and modulators, in: *Carcinogens and Mutagens in the Environment. Volume I Food Products* (Hans F. Stich, ed.), pp. 121–133, CRC Press, Boca Raton, Florida (1982).

63. J. W. Howard and T. Fazio, A review of polycyclic aromatic hydrocarbons in foods, *J. Agric. Food Chem. 17*, 527–531 (1969).

64. C. Lintas, M. C. De Matthaeis, and F. Merli, Determination of benzo(a)pyrene in smoked, cooked and toasted food products, *Food Cosmet. Toxicol. 17*, 325–328 (1979).

65. T. Panalaks, Determination and identification of polycyclic aromatic hydrocarbons in smoked and charcoal-broiled food products by high pressure liquid chromatography and gas chromatography, *J. Environ. Sci. Health Bull. 1976*, 299–315.

66. L. O. MeiTein and E. Sandi, Polycyclic aromatic hydrocarbons (polynuclears) in foods, *Residue Rev. 69*, 35–86 (1978).

67. B. P. Dunn, Polycyclic aromatic hydrocarbons (PAH), in: *Carcinogens and Mutagens in the Environment. Volume I Food Products* (Hans F. Stich ed.), pp. 175–183, CRC Press, Boca Raton, Florida (1982).

68. W. Lijinsky, Carcinogenic and mutagenic N-nitroso compounds, in: *Chemical Mutagens, Principles and Methods for Their Detection* Vol. 4 (A. Hollaender, ed.), pp. 193–217, Plenum Press, New York (1976).

69. J. R. Iyengar, T. Panalaks, W. F. Miles, and N. P. Sen, A survey of fish products for volatile N-nitrosamines, in: *Oncology Overview, Selected Abstracts on Mutagens and Carcinogens in Cooked, Smoked and Charred Foods* (Michael W. Pariza, consulting reviewer), p. 8, U. S. Department of Health and Human Services, National Cancer Institute (1982).

70. T. Kawabata, H. Ohshima, T. Uibu, M. Nakamura, M. Matsui, and M. Hamano, Occurrence, formation, and precursors of N-nitroso compounds in Japanese diet, in: *Oncology Overview, Selected Abstracts on Mutagens and Carcinogens in Cooked, Smoked and Charred Foods*, p. 9 (Michael W. Pariza, consulting reviewer), U. S. Department of Health and Human Services, National Cancer Institute (1982).

71. P. E. Hartman, Nitrates and nitrites: Ingestion, pharmacodynamics, and toxicology, in: *Chemical Mutagens, Principles and Methods for Their Detection* (F. J. de Serres and A. Hollaender, eds.), Vol. 7, pp. 211–294, Plenum Press, New York (1982).

72. L. Lakritz, R. A. Gates, A. M., Gugger, and A. E. Wasserman, Nitrosamine levels in human blood, urine and gastric aspirate following ingestion of foods containing potential nitrosamine precursors or preformed nitrosamines, *Food Chem. Toxicol. 20*, 455–459 (1982).

73. C. Plumlee, L. F. Bjeldanes, and F. Hatch, Priorities assessment for studies of mutagen production in cooked foods, *J. Am. Diet. Assoc. 79*, 446–449 (1981).

74. W. S. Barnes and J. H. Weisburger, Lipid content and mutagen formation in the cooking of beef, *Proc. Am. Assoc. Cancer Res. 24*, 65 (1983).

75. R. T. Taylor, E. Fultz, and V. Shore, Mutagen formation in a model beef boiling system. I. Conditions with soluble beef derived fraction, *J. Environ. Sci. Health*, in press (1984).

76. R. T. Taylor, V. Shore, and E. Fultz, Mutagen formation in a model beef boiling system. II. Effects of proteolysis and comparison of soluble fractions from several protein sources, *J. Environ. Sci. Health*, in press (1984).

77. R. T. Taylor, V. Shore, and E. Fultz, Mutagen formation in a model beef boiling system: Stimulation studies with amino acids and other agents, *Environ. Mutagen. 4*, 368 (1982).

78. H. Kasai, Z. Yamaizumi, T. Shiomi, S. Yokoyama, T. Miyazawa, K. Wakabayashi, M. Nagao, T. Sugimura, and S. Nishimura, Structure of a potent mutagen isolated from fried beef, *Chem Lett.* 485–488 (1981).

79. T. Sugimura, Mutagens in cooked food, *Basic Life Sci. 21*, 243–269 (1982).

80. T. Sugimura and S. Sato, Mutagens-carcinogens in foods, *Cancer Res. (Suppl.) 43*, 2415s–2421s (1983).

81. T. Ishikawa, S. Takayama, T. Kitagawa, T. Kawachi, and T. Sugimura, Induction of enzyme-altered islands in rat liver by tryptophan pyrolysis products, *J. Cancer Res. Clin. Oncol. 94*, 221–224 (1979).

82. T. Ishikawa, S. Takayama, T. Kitagawa, T. Kawachi, M. Kinebuchi, N. Matsukura, E. Uchida, and T. Sugimura, *In vivo* experiments on tryptophan pyrolysis products, in: *Naturally Occurring Carcinogen-Mutagens and Modulators of Carcinogenesis* (E. C. Miller, J. A. Miller, and I. Hirono, eds.), pp. 159–167, University Park Press, Baltimore, Maryland (1979).

83. S. Takayama, T. Hirakawa, M. Tanaka, Y. Katoh, and T. Sugimura, Transformation and neoplastic development of hamster embryo cells after exposure to tryptophan pyrolysis products in tissue culture, in: *Naturally Occurring Carcinogen-Mutagens and Modulators of Carcinogenesis* (E. C. Miller, J. A. Miller, and I. Hirono, eds.), pp. 151–157, University Park Press, Baltimore, Maryland (1979).

84. S. Hosaka, T. Matsushima, I. Hirono, and T. Sugimura, Carcinogenic activity of 3-amino-1-methyl-5H-pyrido[4,3-b] indole (Trp-P-2), a pyrolysis product of tryptophan, *Cancer Lett. 13*, 23–28 (1981).

85. N. Matsukura, T. Kawachi, K. Morino, H. Ohgaki, T. Sugimura, and S. Takayama, Carcinogenicity in mice of mutagenic compounds from a tryptophan pyrolyzate, *Science 213*, 346–347 (1981).

86. F. T. Hatch, J. S. Felton, L. H. Thompson, A. V. Carrano, S. K. Healy, E. P. Salazar, and J. L. Minkler, Comparative genotoxicity of the food mutagens Trp-P-2 and IQ in *Salmonella* and CHO cells, *Environ. Mutagen. 4*, 368 (1982).

87. L. H. Thompson, A. V. Carrano, E. Salazar, J. S. Felton, and F. T. Hatch, Comparative genotoxic effects of the cooked-food-related mutagens Trp-P-2 and IQ in bacteria and cultured mammalian cells, *Mutat. Res. 117*, 243–257 (1983).

88. T. Kosuge, K. Tsuji, K. Wakabayashi, T. Okamoto, K. Shudo, Y. Iitaka, A. Itai, T. Sugimura, T. Kawachi, M. Nagao, T. Yahagi, and Y. Seino, Isolation and structure studies of mutagenic principles in amino acid pyrolysates, *Chem. Pharm. Bull. 26*, 611–619 (1978).

89. M. Uyeta, T. Kanada, M. Mazaki, S. Taue, and S. Takahashi, Assaying mutagenicity of food pyrolysis products using the Ames test, in: *Naturally Occurring Carcinogen-Mutagens and Modulators of Carcinogenesis* (E. C. Miller, J. A. Miller, and I. Hirono, eds.), pp. 169–176, University Park Press, Baltimore, Maryland (1979).

90. Z. Yamaizumi, T. Shiomi, H. Kasai, S. Nishimura, Y. Takahashi, M. Nagao, and T. Sugimura, Detection of potent mutagens, Trp-P-1 and Trp-P-2, in broiled fish, *Cancer Lett. 9*, 75–83 (1980).

91. A. J. Vithayathil, B. Commoner, S. Nair, and P. Madyastha, Isolation of mutagens from bacterial nutrients containing beef extract, *J. Toxicol. Environ. Health 4*, 189–202 (1978).

92. W. A. Hargraves, R. Dietrich, and M. W. Pariza, A new chromatographic method for separating mutagens from commercial beef extract and fried ground beef, in:

Carcinogens and Mutagens in the Environment. Volume I Food Products (Hans F. Stich, ed.), pp. 223–229, CRC Press, Boca Raton, Florida (1982).

93. W. A. Hargraves and M. W. Pariza, Purification and mass spectral characterization of bacterial mutagens from commercial beef extract, *Cancer Res. 43*, 1467–1472 (1983).

94. W. T. Iwaoka, C. A. Krone, J. J. Sullivan, and C. A. Johnson, Effect of pH and ammonium ions on mutagenic activity in cooked beef, *Cancer Lett. 12*, 335–341 (1981).

95. W. T. Iwaoka, C. A. Krone, J. J. Sullivan, E. H. Meaker, C. A. Johnson, and L. S. Miyasato, A source of error in mutagen testing of foods, *Cancer Lett. 11*, 225–230 (1981).

96. W. T. Iwaoka and C. A. Krone, Problems associated with the extraction of mutagens from foods, in: *Carcinogens and Mutagens in the Environment. Volume I Food Products* (Hans F. Stich, ed.), pp. 211–221, CRC Press, Boca Raton, Florida (1982).

97. D. Moore and J. S. Felton, A microcomputer program for analyzing Ames test data, *Mutat. Res. 119*, 95–102 (1983).

98. L. F. Bjeldanes, K. R. Grose, P. H. Davis, D. H. Stuermer, S. K. Healy, and J. S. Felton, An XAD-2 resin method for efficient extraction of mutagens from fried ground beef, *Mutat. Res. 105*, 43–49 (1982).

99. B. N. Ames, J. McCann, and E. Yamasaki, Methods for detecting carcinogens and mutagens with the *Salmonella*/mammalian-microsome mutagenicity test, *Mutat. Res. 31*, 347–364 (1975).

100. D. W. Nebert, S. W. Bigelow, A. B. Okey, T. Yahagi, Y. Mori, M. Nagao, and T. Sugimura, Pyrolysis products from amino acids and protein: Highest mutagenicity requires cytochrome P_1-450, *Proc. Natl. Acad. Sci. USA 76*, 5929–5933 (1979).

101. J. S. Felton, S. K. Healy, M. Knize, D. H. Stuermer, P. W. Berry, H. J. Timourian, F. T. Hatch, M. Morris, and L. F. Bjeldanes, *In vitro* human and rodent metabolism of mutagenic fractions from cooked ground beef, *Environ. Mutagen. 3*, 342 (1981).

102. J. S. Felton, S. K. Healy, M. Knize, D. H. Stuermer, P. W. Berry, H. Timourian, F. T. Hatch, L. F. Bjeldanes, and M. Morris, Metabolism of mutagenic fractions from cooked ground beef, *J. Supramol. Structure Cell. Biochem. (Suppl. 5) 1981*, 166.

103. M. Knize, B. Wuebbles, E. Fultz, M. Morris, R. Taylor, J. Felton, L. Bjeldanes, and F. Hatch, Characterization of the mutagens in cooked beef: Chromatographic patterns at different temperatures and cooking conditions, *Fed. Proc. 42*, 2089 (1983).

104. H. Kasai, S. Nishimura, M. Nagao, Y. Takahashi, and T. Sugimura, Fractionation of a mutagenic principle from broiled fish by high-pressure liquid chromatography, *Cancer Lett. 7*, 343–348 (1979).

105. Z. Yamaizumi, T. Shiomi, H. Kasai, K. Wakabayashi, M. Nagao, T. Sugimura, and S. Nishimura, Quantitative analysis of a novel potent mutagen, 2-amino-3-methyl-imidazo(4,5-f)quinoline, present in broiled food by GC/MS, *Koenshu-Iyo Masu Kenkyukai 5*, 245–248 (1980).

106. Z. Yamaizumi, T. Shiomi, H. Sasai, K. Wakabayashi, T. Sugimura, and S. Nishimura, Quantitative measurement of IQ, Me-IQ and Me-IQx present in broiled foods by gas chromatography/mass spectrometry (GC/MS), in: *Abstracts of the Third International Conference on Environmental Mutagens*, p. 69 (1981).

107. M. Nagao, Y. Takahashi, T. Yahagi, T. Sugimura, K. Takeda, S. Shudo, and T. Okamoto, Mutagenicities of γ-carboline derivatives related to potent mutagens found in tryptophan pyrolysates, *Carcinogenesis 1*, 451–454 (1980).

108. M. Nagao, K. Wakabayashi, H. Kasai, S. Nishimura, and T. Sugimura, Effect of methyl substitution on mutagenicities of 2-amino-3-methylimidazo[4,5-f]quinoline, isolated from broiled sardine, *Carcinogenesis, 2*, 1147–1149 (1981).

108a. J. S. Felton, M. G. Knize, C. Wood, B. J. Wuebbles, S. K. Healy, D. H. Stuermer, L. F. Bjeldanes, B. J. kimble, and F. T. Hatch, Isolation and characterization of new mutagens from fried ground beef, *Carcinogenesis 5*, 95–102 (1984).

109. T. Sugimura, T. Kawachi, M. Nagao, T. Yahagi, T. Okamoto, K. Shudo, T. Kosuge, K. Tsuji, K. Wakabayashi, Y. Iitaka, and A. Itai, Mutagenic principles in tryptophan and phenylalanine pyrolysis products, *Proc. Jpn. Acad. 53*, 58–61 (1977).

110. H. Akimoto, A. Kawai, H. Nomura, M. Nagao, T. Kawachi, and T. Sugimura, Synthesis of potent mutagens in tryptophan pyrolysates, *Chem. Lett. 1977*, 1061–1064.

111. K. Yamaguchi, K. Shudo, T. Okamoto, T. Sugimura, and T. Kosuge, Presence of 3-amino-1,4-dimethyl-5H-pyrido[4,3-b]indole in broiled beef, *Gann 71*, 745–746 (1980).

112. J. H. Peters, K. E. Mortelmans, E. J. Reist, C. C. Sigman, R. J. Spanggord, and D. W. Thomas, Synthesis, chemical characterization, and mutagenic activities of pro-mutagens produced by pyrolysis of proteinaceous substances, *Environ. Mutagen. 3*, 639–649 (1981).

113. H. Tohda, A. Oikawa, T. Kawachi, and T. Sugimura, Induction of sister-chromatid exchanges by mutagens from amino acid and protein pyrolysates, *Mutat. Res. 77*, 65–69 (1980).

114. M. Sasaki, K. Sugimura, M. Yoshida, and T. Kawachi, Chromosome aberrations and sister chromatid exchanges induced by tryptophan pyrolysates, Trp-P-1 and Trp-P-2, in cultured human and Chinese hamster cells, *Proc. Jpn. Acad. B 56*, 332–336 (1980).

115. S. Takayama, T. Hirakawa, M. Tanaka, T. Kawachi, and T. Sugimura, In vitro transformation of hamster embryo cells with a glutamic acid pyrolysis product, *Toxicol. Lett. 4*, 281–284 (1979).

116. C. Hashida, K. Nagayama, and N. Takemura, Induction of bladder cancer in mice by implanting pellets containing tryptophan pyrolysis products, *Cancer Lett. 17*, 101–105 (1982).

117. K. Takeda, T. Ohta, K. Shudo, T. Okamoto, K. Tsuji, and T. Kosuge, Synthesis of a mutagenic principle isolated from tryptophan pyrolysate, *Yakugaku Zasshi (J. Pharmaceutical Soc. Japan) 97*, 2145–2146 (1977).

118. T. Sugimura, Mutagens, carcinogens, and tumor promoters in our daily food, *Cancer 49*, 1970–1984 (1982).

119. Y. Hashimoto, K. Shudo, and T. Okamoto, Structural identification of a modified base in DNA covalently bound with mutagenic 3-amino-1-methyl-5H-pyrido[4,3-b]indole, *Chem. Pharm. Bull. 27*, 1058–1060 (1979).

120. Y. Hashimoto, K. Shudo, and T. Okamoto, Modification of nucleic acids with muta-carcinogenic heteroaromatic amines *in vivo*—identification of modified bases in DNA extracted from rats injected with 3-amino-1-methyl-5H-pyrido-[4,3-b]indole and 2-amino-6-methyldipyrido-[1,2-a:3',2'-d]-imidazole, *Mutat. Res. 105*, 9–13 (1982).

121. J. S. Felton and R. L. Dobson, The mouse oocyte toxicity assay, in: *Short-Term Bioassays in the Analysis of Complex Environmental Mixtures III* (M. Waters, S. Shandhu, J. Lewtas, L. Claxton, N. Chernoff, and S. Newsnow, eds.), pp. 245–255, Plenum Press, New York (1983).

122. S. Yokoyama, T. Miyazawa, H. Kasai, S. Nishimura, T. Sugimura, and Y. Iitaka, Crystal and molecular structures of 2-amino-3-methylimidazo-[4,5-f]quinoline, a novel potent mutagen found in broiled food, *FEBS Lett. 122*, 261–263 (1980.).

123. H. Kasai, S. Nishimura, K. Wakabayashi, M. Nagao, and T. Sugimura, Chemical synthesis of 1-amino-3-methylimidazo-[4,5-f]quinoline (IQ), a potent mutagen isolated from broiled fish, *Proc. Jpn. Acad. 56* (6), 382 (1980).

124. C.-S. Lee, Y. Hashimoto, K. Shudo, and T. Okamoto, Synthesis of mutagenic heteroaromatics: 2-Aminoimidazo[4,5-f]quinolines, *Chem. Pharm. Bull. 30*, 1857–1859 (1982).

125. Z. Yamaizumi, S. Tomoko, K. Hiroshi, K. Wakabayashi, M. Nagao, T. Sugimura, and S. Nishimura, Quantitative analysis of a novel potent mutagen 2-amino-3-methylimidazo(4,5-f)quinoline, present in broiled food by GC/MS, *Chem. Abstr. 94*, 137879d (1981).

126. J. Felton, S. Healy, L. Avalos, and B. Wuebbles, Comparison of mutagenicity of two cooking mutagens with three microbial bioassays, *Environ. Mutagen. 5*, 446 (1983).

127. H. Kasai, Z. Yamaizumi, K. Wakabayashi, M. Nagao, T. Sugimura, S. Yokoyama, T. Miyazawa, and S. Nishimura, Structure and chemical synthesis of ME-IQ, a potent mutagen isolated from broiled fish, *Chem. Lett. 1980*, 1391–1394.

128. H. Kasai and S. Nishimura, Syntheses of 2-amino-3,4-dimethyl-3H-imidazo[4,5-f]quinoline and its related compounds, *Bull. Chem. Soc. Japan 55*, 2233–2235 (1982).

129. H. S. Rosenkranz, E. D. McCoy, D. R. Sanders, M. Butler, D. K. Kiriazides, and R. Mermelstein, Nitropyrenes: Isolation, identification, and reduction of mutagenic impurities in carbon black and toners, *Science 209*, 1039–1043 (1980).

130. H. Kasai, T. Shiomi, T. Sugimura, and S. Nishimura, Synthesis of 2-amino-3,8-dimethylimidazo[4,5-f]quinoxaline (Me-IQx), a potent mutagen isolated from fried beef, *Chem. Lett.* 675–678 (1981).

131. T. Yamamoto, K. Tsuji, T. Kosuge, T. Okamoto, K. Shudo, K. Takeda, Y. Iitaka, K. Yamaguchi, Y. Seino, T. Yahagi, M. Nagao, and T. Sugimura, Isolation and structure determination of mutagenic substances in L-glutamic acid pyrolysate, *Proc. Jpn. Acad. B 54*, 248–250 (1978).

132. K. Takeda, K. Shudo, T. Okamoto, and T. Kosuge, Synthesis of mutagenic principles isolated from L-glutamic acid pyrolysate, *Chem. Pharm. Bull. 26*, 2924–2925 (1978).

133. K. Tsuji, T. Yamamoto, H. Zenda, and T. Kosuge, Studies on active principles of tar. VII. Production of biological active substances in pyrolysis of amino acids. (2) Antifungal constituents in pyrolysis products of phenylalanine, *Yakugaku Zasshi (J. Pharmaceutical Soc. Japan) 98*, 910–913 (1978).

134. Y. Hashimoto, K. Shudo, and T. Okamoto, Structure of a base in DNA modified by Glu-P-1, *Chem. Pharm. Bull. 27*, 2532–2534 (1979).

135. Y. Hashimoto, K. Shudo, and T. Okamoto, Metabolic activation of a mutagen, 2-amino-6-methyldipyrido[1,2-a:3',2'-d] imidazole. Identification of 2-hydroxyamino-6-methyldipyrido[1,2-a:3',2'-d] imidazole and its reaction with DNA, *Biochem. Biophys. Res. Commun. 92*, 971–976 (1980).

136. Y. Hashimoto, K. Shudo, and T. Okamoto, Modification of DNA with potent mutacarcinogenic 2-amino-6-methyldipyrido[1,2-a:3',2'-d] imidazole isolated from a glutamic acid pyrolysate: Structure of the modified nucleic acid base and initial chemical event caused by the mutagen, *J. Am. Chem. Soc. 104*, 7636–7640 (1982).

137. K. Yamaguchi, K. Shudo, T. Okamoto, T. Sugimura, and T. Kosuge, Presence of 3-aminodipyrido(1,2-a:3',2'-d)imidazole in broiled cuttlefish, *Gann 71*, 743–744 (1980).

138. D. Yoshida and T. Matsumoto, Isolation of 2-amino-9H-pyrido[2,3-b]indole and 2-amino-3-methyl-9H-pyrido[2,3-b]indole as mutagens from pyrolysis product of tryptophan, *Agric. Biol. Chem. 43*, 1155–1156 (1979).

139. T. Matsumoto, D. Yoshida, H. Tomita, and H. Matsushita, Synthesis of 2-amino-9H-pyrido[2,3b]indole isolated as a mutagenic principle from pyrolytic products of protein, *Agric. Biol. Chem. 43*, 675–677 (1979).

140. T. Matsumoto, D. Yoshida, and H. Tomita, Synthesis and mutagenic activity of alkyl derivatives of 2-amino-9H-pyrido[2,3-b]indole, *Agric. Biol. Chem. 45,* 2031–2035 (1981).
141. K. Wakabayashi, K. Tsuji, T. Kosuge, K. Yamaguchi, K. Shudo, K. Takeda, Y. Iitaka, T. Okamoto, T. Yahagi, M. Nagao, and T. Sugimura, Isolation and structure determination of a mutagenic substance in L-lysine pyrolysate, *Proc. Jpn. Acad. B 54,* 569–571 (1978).
142. M. Yokota, K. Narita, T. Kosuge, K. Wakabayashi, M. Nagao, T. Sugimura, K. Yamaguchi, K. Shudo, Y. Iitaka, and T. Okamoto, A potent mutagen isolated from a pyrolysate of L-ornithine, *Chem. Pharm. Bull. 29,* 1473–1475 (1981).

The *Bacillus subtilis* Multigene Sporulation Test for Detection of Environmental Mutagens

L. E. Sacks and J. T. MacGregor

1. Introduction

The value of short-term *in vitro* screening tests for environmental mutagens is well established. It is generally agreed that a small battery of tests using both prokaryotic and eukaryotic cells should be used in preliminary screening to detect potential mutagens.[7] Bacterial tests are efficient and economical methods of detecting specific locus mutations. Moreover, well-developed methods of genetic manipulation make it easier to study mechanisms in bacteria. Tests for general genetic screening should be sensitive to a wide range of genetic damage and, if possible, representative of damage to the genome as a whole.

The *Bacillus subtilis* multigene sporulation test detects forward mutations in any of ~35 scattered gene clusters controlling sporulation.[10,25] Asporogenic mutants are identified by reduction of the pigmentation normally developed after sporulation on suitable agar (Figure 1). The multigene character of the test and the ability to detect

L. E. Sacks and J. T. MacGregor • United States Department of Agriculture, Western Regional Research Center, Berkeley, California 94710.

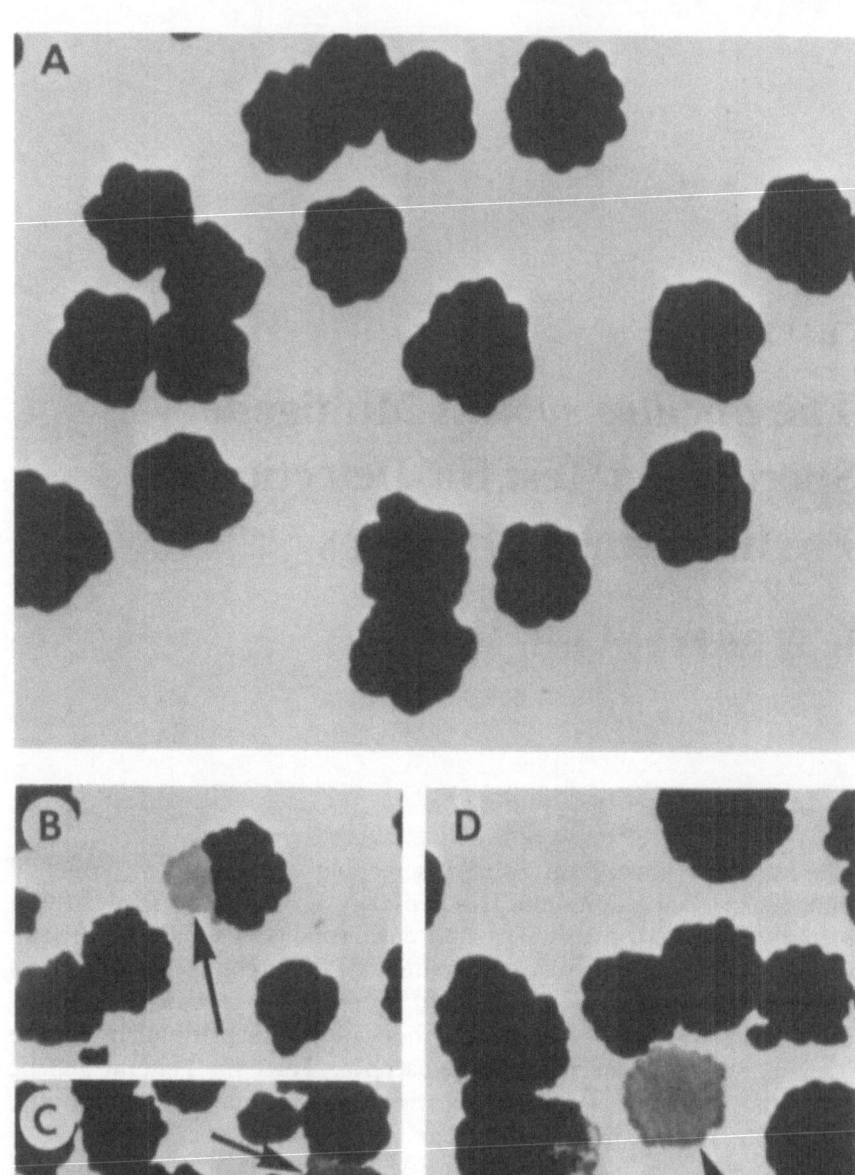

FIGURE 1. Normal and mutant colonies of *B. subtilis* 168 after 3 days at 37°C on Schaeffer's sporulation agar. (A) Normal colonies. (B–D) Arrows indicate mutant colonies of various types. (D) The minimal size sector counted. Sectors are much more common than entire mutant colonies since two DNA replications may occur before daughter cells actually separate.[31,39]

forward mutation assure that a large portion of the genome is surveyed, so mutagen-specific "hot spots" are less likely to be excluded than in systems based on a restricted region of the chromosome. It is therefore more general than tests based on reverse mutations, which are generally based on reversion of a single altered gene site, and thus may not detect certain types of inactivating mutation (e.g., deletions that span a significant fraction of one or more genes). Since the test does not require selective media for the identification of mutants, certain mutants normally selectively eliminated in minimal media form colonies and are scored.

The *B. subtilis* test has not yet been tested with an extremely large number of mutagens, but it has proven capable of detecting many major classes of mutagens, including a number not detected by standard methods of the His reversion test in *Salmonella*.[28,29] A new *Salmonella* strain, TA102, is capable of detecting some of the above.[17]

2. Techniques and Procedures

2.1. The Basic Features of the System

An aliquot of spore suspension (stable for several years at low temperature) is heat treated to destroy vegetative cells and activate the spores. After inoculation into a medium permitting germination and outgrowth, the suspension is incubated for 1 hr; the mutagen is then added and outgrowth allowed to proceed for another hour. The suspension is then diluted out and a known volume containing about 1500 cells uniformly distributed on the surface of agar plates. After incubating for 2 days at 37°C, the wild-type colonies have sporulated and formed a characteristic brown pigment. Asporogenic colonies have little or no pigment. Low-pigment colonies are counted and compared with the number of similar colonies appearing on untreated plates. The various steps of the test are illustrated diagrammatically in Figure 2. Development of the test organism at various stages of the test is indicated in Figure 3.

2.2. Bacterial Strains; Stock Spore Suspensions

Bacillus subtilis strains 168 and hcr-9 (Exc⁻)[8,24] have been found most useful in work performed to date. Spores are formed in Schaeffer's 2XSG medium,[16] incubated with shaking at 35–37°C for 42 hr. The culture is then chilled to 0–5°C and centrifuged in the cold. Sedimented

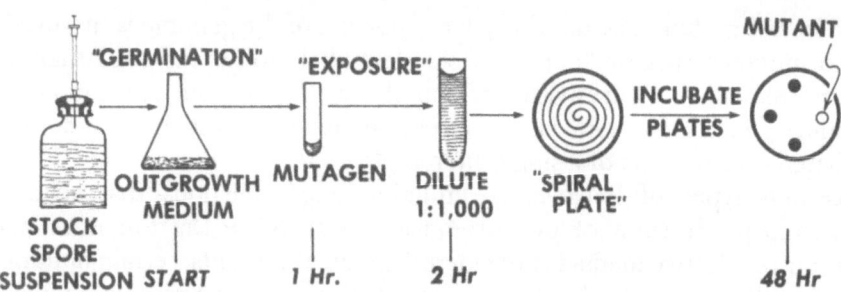

FIGURE 2. Schematic diagram of the test procedure.

cells are washed once in cold, sterile deionized water, resuspended in one-tenth volume of sterile deionized water, and stored in serum bottles at 4°C. The purpose of concentrating the cells is to make it possible to dilute out the mutagen after exposure while retaining sufficient cells to give plates of the desired colony density. The spore stocks usually contain about 10^{10} spores/ml (direct count[35]) of which over 60% form colonies on Plate Count Agar (Difco) after heat activation at 70°C for 20 min. Both AK medium[3] and Schaeffer's medium[30] are also satisfactory sporulation media.

The unusual stability of bacterial spores constitutes an advantage. A water suspension of spores of the test organism will remain viable and stable for over a year if held at 5°C. This suspension may be used as an inoculum, with no treatment other than heat activation.

2.3. Experimental Procedure

Stock spore suspension (35–70 μl) is added to 2.0 ml of sterile preheated (70°C) Spizizen medium,[36] supplemented as described in Ref. 18, in a 50-ml Erlenmeyer flask, and held at 70°C for 20 min to

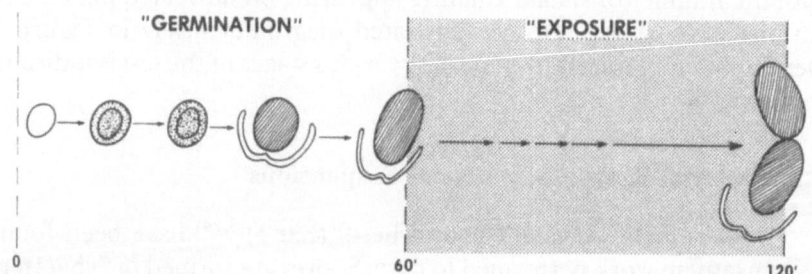

FIGURE 3. Germination and outgrowth stages of B. subtilis spores in the course of the test procedure (cf. Ref. 9).

activate spores and kill vegetative cells. The flask is then immediately transferred to a rotary shaker water bath and shaken for 60 min at 37°C ("germination"). At this time, 1 ml of the cell suspension is transferred to a sterile disposable glass tube containing 10 μl of mutagen (±S9 mix[2]). When metabolic activation is required, 50 μl of S9 mix is added to the tube containing 10 μl of mutagen and incubated at 37°C for 1 hr prior to addition of outgrowing cells. Incubation is continued ("exposure") at 37°C for another 60 min with shaking. At this time, 10 μl of the treated cell suspension is transferred to 10 ml of sterile 0.1% peptone at room temperature in a disposable glass tube (16 × 150 mm). After mixing, an aliquot (16.7 μl) is spread on the the surface of a 100-mm plate containing Schaeffer's agar, using a Spiral Plater (Spiral Systems, Bethesda, Maryland) equipped with a cam designed to provide uniform colony distribution. Plates containing Schaeffer's agar should be predried. For range-finding and preliminary test runs fresh plates need not be used. In final tests for weak mutagens, we prepare fresh plates and predry for 1 hr in a laminar flow hood. Three to ten plates of each sample with 1000–2000 colonies/plate are prepared and incubated at 37°C for 48–72 hr in an incubator not conducive to drying of plates.

The media for germination and outgrowth do not appear highly critical. Germination and outgrowth on trypticase soy broth (TSB), brain heart infusion (BHI), and supplemented Spizizen medium have all given satisfactory results, although outgrowth on BHI was characterized by morphologically unusual cells. The Mg^{2+} and phosphate content of supplemented Spizizen medium[18,36] may be undesirable when testing polyvalent substances, but this medium is better defined and less nucleophilic than TSB. The phosphate concentration of Spizizen medium can be reduced to one-fourth the prescribed phosphate with no adverse effects.

2.4. Scoring

Plates are scored at 42–72 hr, after colonies have become uniformly darkly pigmented. Spiral plating results in extremely uniform distribution of colonies, facilitating estimates of total colony counts with an electronic colony counter. Estimates of total counts may also be made from counts of at least two 1-cm^2 areas on the plate; such counts are generally slower and less accurate. Mutants are identified with a low-power (10×) dissecting microscope and transmitted illumination. Bright illumination can facilitate mutant identification; an illuminator with a blue or white frosted filter may be placed directly under the glass stage.

During scoring, the plates should be scanned in a systematic way. A series of five parallel lines on the plate bottom serves to divide the plate into six strips of equal width, with the entire width visible in the viewing field. These parallel lines may be made rapidly with a rubber inking stamp. Such a stamp is simply and cheaply constructed but may require some padding in the center to give it a slightly convex shape, which serves to compensate for the "give" of the plate during stamping. Mutant colony locations are marked with a wax pencil, and may be restreaked to confirm the mutant character. Only sectors occupying greater than one-eighth the colony area are counted. A representative number of mutant colonies should be restreaked on Schaeffer's agar. Persistence of reduced pigmentation is considered confirmation. Use of a short needle facilitates restreaking from sectors. There are varying degrees of "reduced pigmentation" and some judgment is required in making this assessment. Reduced pigmentation is usually associated with reduced sporulation levels; we prefer to count those colonies whose pigmentation level is generally associated with a sporulation level of <30%. It should be noted that various metabolic mutants will also result in oligosporogenous or asporogenous mutants.[25] These mutants will also show reduced pigmentation, as will those with altered pigment-forming enzymes. The tables of Kastenbaum and Bowman[15] may be used to evaluate the statistical significance of the results.

2.5. Purified Water, Air

We recommend that highly purified water be used for all procedures. In the work reported here, house-distilled water was filtered, deionized, passed through activated charcoal, and glass distilled. During periods of severe air pollution, it may be advisable to carry out tests under commercial "breathing air." Use of a gassing hood on the shaking water bath and a controlled atmosphere incubator with ports for removal and admission of known gas mixtures is required for such procedures.

2.6. Metabolizing System

The mammalian liver, mixed-function oxidase, metabolizing system contains 8 mM $MgCl_2$, 33 mM KCl, 100 mM sodium phosphate, pH 7.5, 5 mM glucose-6-phosphate, 4 mM NADP, and 0.3 ml liver supernatant per ml metabolizing mixture.[2] Liver supernatant is prepared using sterile technique. The experiments reported herein employed male Sprague-Dawley rat liver from animals given 500 mg/kg, i.p., Aroclor-1254 in corn oil 5 days prior to liver preparation. Liver is removed and immediately placed in ice-cold (sterile) 154 mM KCl, weighed, washed twice in KCl, minced, and homogenized in a Potter–Elvehjem homogenizer with Teflon pestle. The homogenate is made up to 25% (w/w) with 154 mM KCl and centrifuged 10 min at 9000 × *g*. The entire supernatant is mixed and aliquots (1–2 ml) are frozen in sterile ½-dram screw-top vials. Enzyme activity is retained for several months at −80°C.

2.7. Range Finding; Order of Testing

As with most mutagen assays, it is desirable that testing be carried out in a range of concentrations resulting in some killing. Thus, a preliminary test should be run to establish concentration ranges for strains 168 and hcr-9, both with and without S9 mix, resulting in survival of 20–100% of the population. Very low survival levels are undesirable because of a possible selective advantage of induced mutants and/or a small fraction of mutagen-insensitive ungerminated spores. In the absence of prior knowledge of toxicity, it is probably advisable to run a series of decimal dilutions, beginning with a concentrated solution approaching the solubility limit of the test compound. For this initial range-finding experiment it is not necessary to run a large number of plates at each concentration; two or three should suffice. To establish mutagenicity a preliminary run is made with concentrations spaced much more closely, attempting to obtain three spaced dosage points within the 20–100% survival range. The number of plates at each concentration is increased to five. Again, freshly poured plates need not be used. In the absence of any indications of mutagenicity from the range-finding experiment, and apparent lack of toxicity for one of the two strains, this test should be conducted with the sensitive strain. If both strains exhibit sensitivity to the agent, we prefer to test strain 168 first, ±S9 mix, because its colonial morphology simplifies scoring. If this proves negative, the material should be tested with strain hcr-9. For a final test ten plates should be used at each concentration, and we use only freshly poured plates.

2.8. Solvents

Water, dimethylsulfoxide, and ethanol have been used to dissolve test materials. After dilution with the outgrowth medium containing the test organism, the solvent concentration level is 1%.

2.9. Positive Controls

Furylfuramide (AF-2) serves as a highly potent positive control which does not require activation. Activity of the S9 is conveniently verified by the activation of 2-aminofluorene (2-AF) in strain hcr-9.

2.10. Contamination

Because of the low incidence of mutants it is essential that contamination be kept to an extremely low level to avoid possible confusion of mutants with contaminants. Sterile equipment must be used throughout. The possibility of contaminants in the mutagen itself may be removed by filtering the dissolved preparation. Alternatively, ten plates of Schaeffer's agar can be spiral-plated with the material to provide evidence of absence of contaminants.

2.11. Media

Schaeffer's sporulation agar.[30] In 1 liter of distilled water: 8.0 g nutrient broth (Difco), 0.25 g $MgSO_4 \cdot 7H_2O$, 1.0 g KCl; pH adjusted to 7.1–7.2, 17.0 g agar. Autoclave, then add aseptically sterile solutions of $Ca(NO_3)_2$ to 10^{-3} M, $MnCl_2$ to 10^{-5} M, $FeSO_4$ to 10^{-6} M.

Stock solutions for sporulation media: 10^{-1} M $Ca(NO_3)_2$, autoclaved; 10^{-3} M $MnCl_2$, autoclaved; 10^{-3} M $FeSO_4$ (in 0.01 M H_2SO_4), filter-sterilized; 50% glucose, filter-sterilized. Sterile solutions may be kept in serum bottles with conventional rubber sept.

Schaeffer's sporulation medium 2XSG[16]: In 1 liter distilled H_2O, 16 g nutrient broth, 0.5 g $MgSO_4 \cdot 7H_2O$, 2.0 g KCl. Add aseptically, after autoclaving: $Ca(NO_3)_2$ to 10^{-3} M, $MnCl_2$ to 10^{-4} M, $FeSO_4$ to 10^{-6} M, glucose to 0.1%.

Supplemented Spizizen medium[18,36]: 0.2% $(NH_4)_2SO_4$, 1.4% K_2HPO_4, 0.6% KH_2PO_4, 0.1% sodium citrate $(2H_2O)$, 0.02% $MgSO_4 \cdot 7H_2O$, 0.005% L-tryptophan, 0.1% peptic casein digest, 0.1% yeast extract. One volume percent of sterile 50% glucose is added after autoclaving. The phosphate concentrations may be reduced to one-fourth those listed without adverse effects.

S9 mix[2]: 80 mM (0.1ml) $MgCl_2 \cdot 6H_2O$, 330 mM (0.1 ml) KCl, 50 mM (0.1 ml) glucose-6-phosphate, 333 mM (0.3 ml sodium phosphate; pH 7.4), (0.1 ml) NADP 40 mM, liver microsome, S9 fraction[2] (0.3 ml): total 1.0 ml. The above are kept as frozen, sterile stocks. The S9 is stored at $-80°C$.

3. Classes of Mutagens Detected

3.1. Characteristic Data

A summary of mutagens showing a positive response in the *B. subtilis* multigene mutagen test is given in Table 1. The wide variety of classes of mutagens detected by the test is apparent. A more conservative scoring system was used in these tests than was used in some earlier work,[18,28] accounting for different levels of mutants at particular concentrations of mutagens. In the case of some mutagens requiring activation, the values reported here represent results obtained with S-9 preparations less active than those used previously; some variability in the activity of these preparations has been noted. The wide range of mutagens detected by the standard repair-proficient strain 168 is striking. In general, only when a known mutagen failed to provide a positive response with strain 168 was the hcr-9 strain used. Further, a number of mutagens detectable in the *Salmonella* reversion assay only after the construction of specialized tester strains are all detected in the *B. subtilis* sporulation test by a single unmodified strain. 2-Nitrosofluorene (2-NF), believed to react specifically at CGCGCGCG sequences,[12] and 9-aminoacridine, thought to react specifically with GGG sequences,[33] are both readily detected by the 168 strain. A number of mutagens that may act by inducing deletions,[28] mitomycin C, actinomycin D, bleomycin, H_2O_2, and streptonigrin, are detected by the 168 strain, but were undetectable in the *Salmonella* reversion test until very recently when the special tester strain TA102 was constructed. Caffeine and methotrexate, detected by the *B. subtilis* 168 strain, have not as yet been reported detectable by TA102. Mutagens not detected by the old standard tester strains of the classical Ames test are designated with an asterisk. Note that in most such cases, a relatively weak response was evoked. Construction of a specialized *Salmonella* tester strain, TA102,[17] has only recently made possible the detection of mitomycin C, H_2O_2, streptonigrin, and nalidixic acid.

The table also contains previously unpublished data for streptozotocin, *trans*-7,8-diol-9,10 epoxy[syn]benzo(*a*)pyrene, distamycin, streptonigrin, furylfuramide, and UV radiation. Typical dose–response curves for some of the mutagens tested are given in Figure 4. We have observed that with certain agents, such as 2-NF (Figure 4), the significant increase in mutant frequency occurs within a relatively narrow survival range. In testing materials of this type, the investigator must therefore be certain that the concentration levels are so spaced as to give three points in the 20–100% survival range.

TABLE 1. Representative Mutagens Positive in the *Bacillus subtilis* Multigene Sporulation Test

Mutagen[a]	Concentration, μg/ml	S9 Mix	Number of mutants/10^4 survivors Strain 168	Strain hcr-9	Percent survival
Control (average)	—	—	<2	5	—
Aromatic Amines, etc.					
2-Aminofluorene	50	√	—	37	97
2-Nitrosofluorene	0.5		127	—	83
ICR-191	10		19	—	100
Acriflavine	10		23	—	43
N-Acetoxy-2-acetylaminofluorene	10		—	153	63
N-Acetoxy-2-acetylaminofluorene	50		65	—	43
Nitrosamines, etc.					
N-Methyl-N'-nitro-N-nitrosoguanidine	20		121	—	52
Streptozotocin	50		73	—	100
Nitroaromatics					
4-Nitroquinoline-1-oxide	2.5		116	—	96
4-Nitro-o-phenylenediamine	200		—	44	109
Polycyclic aromatics					
Benzo(a)pyrene	400	√	—	45	87
Trans-7,8-diol-9,10 epoxy[syn]benzo(a)pyrene	1.0		—	113	85
Trans-7,8-diol-9,10 epoxy[syn]benzo(a)pyrene	1.0		11	—	84
Miscellaneous heterocyclics					
*Actinomycin D[b]	16		10	—	83
Aflatoxin B₁[e]	10	√	24	—	63
*Mitomycin C[b]	0.075		49	—	88
*Caffeine	5000		17[c]	—	42
*Nalidixic acid	200		18	—	66
Miscellaneous					
*Bleomycin[b,d]	0.003 units/ml		30	—	53
*Methotrexate	2000		8	—	90
Nitrogen mustard	250		38	—	42
*H₂O₂[b]	20		60	—	42
Distamycin	90.		14	—	64
*Streptonigrin[b]	0.6		56	—	64
Furylfuramide (AF-2)	0.2		78	—	61
Furylfuramide (AF-2)	0.05		—	457	63
UV[b]	280 μ Watt/cm² (15 sec)		10	—	97

[a] Asterisk designates mutagen negative in *Salmonella* His reversion test with tester strains TA1535, TA 1537, TA 98, TA 100.[5,10,18,25,31]
[b] Positive in newly constructed *Salmonella* TA102.
[c] Does not include unique class of phenotypic mutants.[28]
[d] Iron added to outgrowth medium.[6,27]
[e] Exposure was from 60 to 90 min.

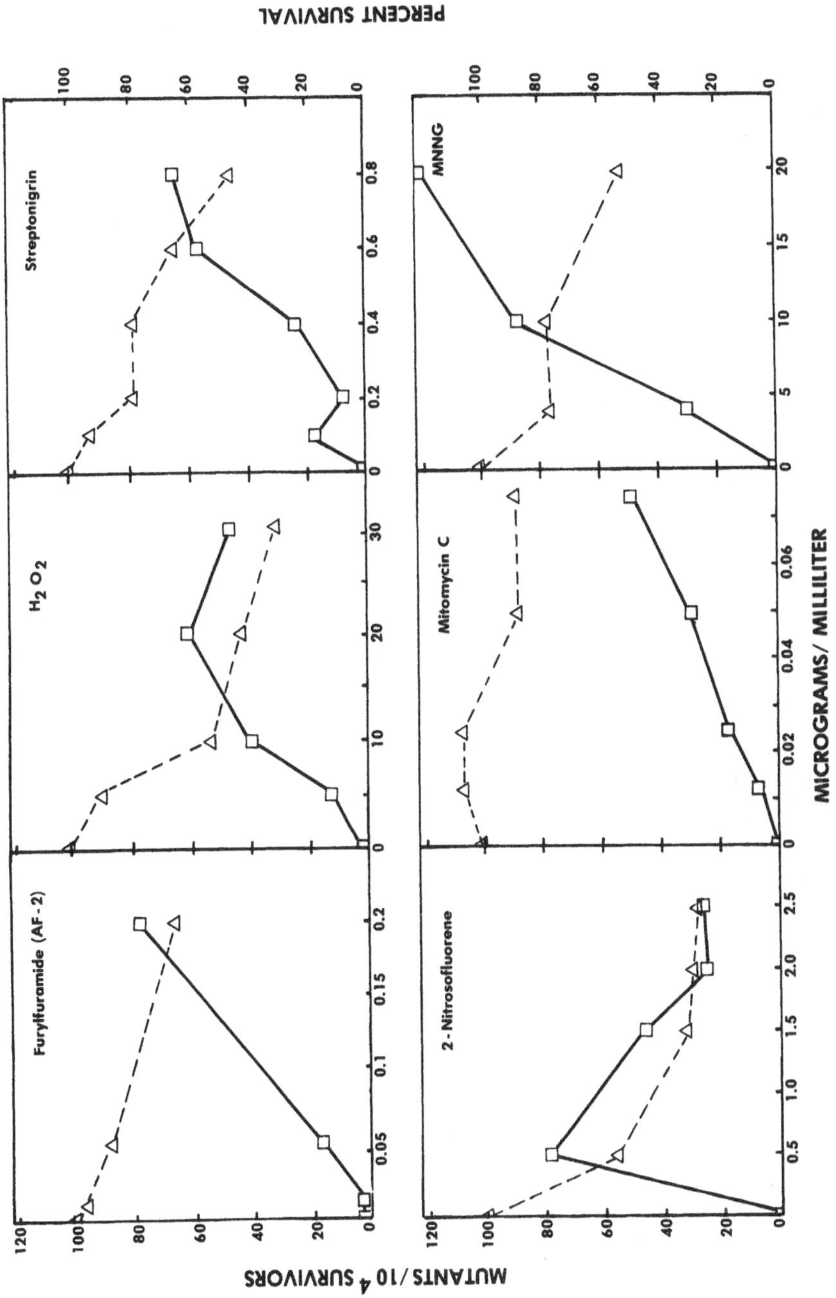

FIGURE 4. Dose–response curves for selected mutagens. (□) Mutant frequency; (△) survival.

H_2O_2 is a unique mutagen in its ability to induce a relatively high fraction of "transparent" mutant colonies.

Methylxanthines, especially caffeine, can induce large numbers of a unique phenotypic class of mutants.[29] These mutants form small, smooth, pale colonies. They appear to have some metabolic deficiency, which prevents normal sporulation.

3.2. Mutagens Not Detected

Quercetin, a moderately potent mutagen in the *Salmonella* His reversion test, has given negative responses in the *B. subtilis* multigene test.[28] It is possible that *Salmonella* metabolizes quercetin to an active form, and *B. subtilis* does not.

Benzo(*a*)pyrene responded weakly, and 2-acetylaminofluorene not at all, in the liquid culture exposure described above, although they are powerful mutagens in the Ames test. These weak responses appear to be due to inefficient activation, since the activated forms are potent mutagens (Table 1). The top agar variation of the Ames test, where S9 mix and mutagen are embedded in the agar with the test organisms, may provide a more effective exposure to the activated form.

Most known mutagens that we have tested to date have proved to be mutagenic in strain 168, which has full repair capacity and normal permeability, using the assay for sporulation mutants described above. Exceptions include the hair dye 4-nitro-*o*-phenylenediamine, which was positive only with strain hcr-9. Likewise, benzo(*a*)pyrene and aminoflu-orene derivatives have been found to give much greater mutagenic effects in strain hcr-9 than in strain 168 (Table 1; unpublished observations). These data suggest that some repair- or recombination-defi-cient mutants may be useful for certain mutagens, though we wish to emphasize the wide range of mutagens detectable by a single unmodified strain.

4. Advantages and Limitations

4.1. Advantages

The ability of the *B. subtilis* multigene test to detect a wide variety of mutagen clases[18,28] as well as a number of mutagens undetectable by standard Ames *Salmonella* tester strains[2] demonstrates the versatility of the test and suggests that it may play a useful role in preliminary screening of previously untested chemicals for mutagenic activity.

Among a number of desirable features of the test system, three are of particular importance with regard to its promise as a test system for environmental mutagens.

First, it is a *forward mutation* test. It should therefore be more general than tests based on reverse mutation, which are not expected to detect certain types of inactivating mutations (e.g., deletions that span a significant fraction of one or more genes). Other disadvantages of reverse mutation tests have been discussed by Skopek *et al.*[33,34] and Mohn *et al.*[21,22] Although forward mutation tests have only infrequently been reported to detect mutagens missed by the Ames test,[34] some instances have been published.[11,13,20]

Second, the test is a *nonselective* forward mutation test, and it is carried out in a relatively complex nutrient medium. This feature should provide greater sensitivity to certain types of mutation than other forward mutation tests, which are carried out in minimal media (e.g., extended deletions should be strongly selected against in minimal media).

Third, it is a *multigene test*. Mutations in any of a very large number of genes with distinct chromosomal separation are detected. In addition to detecting mutations of specific sporulation-associated loci, certain mutations in biochemical loci give rise to nonsporulating colonies[25] and are therefore detected, as are mutations that block pigment synthesis. Thus, mutations in any of these many loci throughout the chromosome are easily detected by observing the pigmentation of the colonies on sporulation agar. The large number of loci surveyed for mutation is a unique feature of this test system and is desirable for a number of reasons. In screening for new classes of mutagens with unknown mechanism of action, a very large fraction of the genome is surveyed for mutation, so mutagen-specific "hotspots" are less likely to be excluded than in systems based on a restricted region of the chromosome. The large target and nonselective medium give the test system high mutant recovery rates relative to other systems. These features should make the system especially sensitive to events, such as extended deletions, that give a high frequency of mutants that are incompatible with growth in minimal media. Although beyond the scope of the present discussion, these features may also be uniquely suited to addressing a number of fundamental questions regarding gene specificity and regional chromosomal differences in mutagen action.

Specificity of mutagens for certain genes or loci might be responsible for false negatives in tests based on specific loci.[18,23,33,34] We have previously discussed a number of specific lines of evidence that support the expectation of nonrandom mutagenic effects throughout the chromosome.[28]

Another advantage of the *B. subtilis* multigene sporulation test stems from the fact that Gram-positive bacteria lack the outer membrane characteristic of *Salmonella* and the other gram-negative forms. Such an outer membrane constitutes a permeability barrier and it was necessary to construct "deep-rough" mutants of *Salmonella* to obtain marked mutagenesis with compounds such as 2-NF, N-acetoxy-2-acetylaminofluorene, and ICR-191.[1] All these compounds are detectable by the 168 strain originally used in this work. Little is known about the permeability of germinating spores, but it is possible that an unusually permeable stage is passed through during this phase.

In any screening program designed to identify mutagens among groups of chemicals previously untested for genetic activity, it seems desirable to include a test that is as general as possible with regard to the type of mutational lesion detected, and which involves a large number of different chromosomal sites to minimize the chance of missing agents that exhibit gene specificity. Nonspecific tests for gene damage, such as the bacterial repair tests,[8,14,27,38] the "inductest,"[23] and mammalian cell tests for unscheduled DNA synthesis[37] or single-strand DNA breakage,[4] are often included in screening batteries to provide generality. Although these tests are simple to perform and have successfully detected mutagens scored negative in the *Salmonella* reversion test, they do not directly measure mutational events, and the inability to recover and confirm mutants is a disadvantage. We believe that the *B. subtilis* sporulation test is uniquely suited to filling this need for a simple mutation test with a representative sampling of the entire chromosome and general sensitivity to a wide variety of mutagens. We therefore suggest that this test, because of the broad spectrum of agents expected to be detected in a multilocus forward mutation test, may be better suited for "blind" screening of large numbers of chemicals than either reversion assays or forward mutation assays based on a single locus.

4.2. Limitations

A significant limitation of a multigene system based on a single endpoint is its expected insensitivity to relatively weak mutagens that are also highly selective for specific gene sequences, due to the existing background in the many genes not affected by the mutagen. This type of compound would, however, be difficult to detect with any existing system unless the specific selective site were known in advance. The chief practical limitation of the multigene sporulation test is scoring, which is exacting and tedious because of the small numbers (generally <1%) of mutants in the presence of a larger number of wild-type colonies. A test run of 60 plates can be scored in about 2 hr by an experienced technician. It seems likely that automated scoring by image analysis can significantly reduce that time.

5. References

1. B. N. Ames, F. D. Lee, and W. E. Durston, An improved bacterial test system for the detection and classification of mutagens and carcinogens, *Proc. Natl. Acad. Sci. USA* 70, 782–786 (1973).
2. B. N. Ames, J. McCann, and E. Yamasaki, Methods for detecting carcinogens and mutagens with the *Salmonella*/mammalian-microsome mutagenicity test, *Mutat. Res.* 31, 347–364 (1975).
3. B. Arret and A. Kirshbaum, A rapid disc assay method for detecting penicillin in milk, *J. Milk Food Technol.* 22, 329–331 (1959).
4. D. E. Brash and R. W. Hart, DNA damage and repair *in vivo*, *J. Environ. Pathol. Toxicol.* 2, 79–114 (1978).
5. W. R. Bruce and J. A. Heddle, The mutagenic activity of 61 agents as determined by the micronucleus, *Salmonella*, and sperm abnormality assays, *Can. J. Genet. Cytol.* 21, 319–334. (1979).
6. J. C. Dabrowiak, F. T. Greenaway, F. S. Santillo, and S. T. Crooke, The iron complexes of bleomycin and tallysomycin, *Biochem. Biophys. Res. Commun.* 91, 721–729 (1979).
7. F. J. de Serres, Mutagenicity of chemical carcinogens, *Mutat. Res.* 41, 43–50 (1976).
8. I. C. Felkner, K. M. Hoffman, and B. C. Wells, DNA-damaging and mutagenic effects of 1,2-dimethylhydrazine on *Bacillus subtilis* repair-deficient mutants, *Mutat. Res.* 68 31–40 (1979).
9. G. W. Gould, Effect of food preservatives on the growth of bacteria from spores, in: *Microbial Inhibitors in Food* (N. Molin and A. Erichsen, eds.), pp. 17–24, Almqvist and Wiksell, Stockholm (1964).
10. D. J. Henner and J. A. Hoch, The *Bacillus subtilis* chromosome, *Microbiol. Rev.* 44, 57–82 (1980).
11. M. Hollstein, J. McCann, F. A. Angelosanto, and W. W. Nichols, Short term tests for carcinogens and mutagens, *Mutat. Res.* 65, 133–226 (1979).
12. K. Isono and J. Yourno, Chemical carcinogens as frameshift mutagens: *Salmonella* DNA sequence sensitive to mutagenesis by polycyclic carcinogens, *Proc. Natl. Acad. Sci. USA* 71 1612–1617 (1974).
13. V. N. Iyer and W. Szybalski, Two simple methods for the detection of chemical mutagens, *Appl. Microbiol.* 6, 23–27 (1958).
14. T. Kada, M. Moriya, and Y. Shirasu, Screening of pesticides for DNA interactions by 'rec-assay' and mutagenesis testing, and frameshift mutagens detected, *Mutat. Res.* 26, 243–248 (1974).
15. M. A. Kastenbaum, and K. O. Bowman, Tables for determining the statistical significance of mutation frequencies, *Mutat. Res.* 9, 527–549 (1970).
16. T. J. Leighton, and R. H. Doi, The stability of messenger ribonucleic acid during sporulation in *Bacillus subtilis*, *J. Biol. Chem.* 246, 3189–3195 (1971).

17. D. E. Levin, M. Hollstein, M. F. Christman, E. A. Schwiers, and B. N. Ames, A new *Salmonella* tester strain (TA102) with A–T base pairs at the site of mutation detects oxidative mutagens, *Proc. Natl. Acad. Sci. USA 79*, 7445–7449 (1982).
18. J. T. MacGregor and L. E. Sacks, The sporulation system of *Bacillus subtilis* as the basis of a multi-gene mutagen screening test, *Mutat. Res. 38*, 271–286 (1976).
19. J. McCann, E. Choi, E. Yamasaki, and B. N. Ames, Detection of carcinogens as mutagens in the *Salmonella*/microsome test: Assay of 300 chemicals, *Proc. Natl. Acad. Sci. USA 72*, 5135–5139 (1975).
20. J. de G. Mitchell, P. A. Dixon, P. J. Gilbert, and D. J. White, Mutagenicity of antibiotics in microbial assays, Problems of evaluation, *Mutat. Res. 79*, 91–105 (1980).
21. G. Mohn, 5-Methyltryptophan resistance mutations in *Escherichia coli* K-12, Mutagenic activity of monofunctional alkylating agents including organophosphorus insecticides, *Mutat. Res. 20*, 7–15 (1973).
22. G. Mohn, J. Ellenberger, and D. McGregor, Development of mutagenicity tests using *Escherichia coli* K-12 as indicator organism, *Mutat. Res. 25*, 440–454 (1974).
23. P. Moreau, A. Bailone, and R. Devoret, Prophage induction in *Escherichia coli* K-12 *envA uvrB*: A highly sensitive test for potential carcinogens, *Proc. Natl. Acad. Sci. USA 73*, 3700–3704 (1976).
24. S. Okubo, and W. R. Romig, Impaired transformability of *Bacillus subtilis* mutant sensitive to mitomycin C and ultraviolet radiation, *J. Mol. Biol. 15*, 440–454 (1966).
25. P. J. Piggot and J. G. Coote, Genetic aspects of bacterial endospore formation, *Bacteriol. Rev. 40*, 908–962 (1976).
26. S. J. Rinkus and M. S. Legator, Chemical characterization of 465 known or suspected carcinogens and their correlation with mutagenic activity in the *Salmonella typhimurium* system, *Cancer Res. 39*, 3289–3318 (1979).
27. H. S. Rosenkranz, B. Gutter, and W. T. Speck, Mutagenicity and DNA-modifying activity: A comparison of two microbial assays, *Mutat. Res. 41*, 61–70 (1976).
28. L. E. Sacks and J. T. MacGregor, The *B. subtilis* multigene sporulation test for mutagens: Detection of mutagens inactive in the *Salmonella his* reversion test, *Mutat. Res. 95*, 191–202 (1982).
29. L. E. Sacks and K. Mihara, Induction at high frequency of a unique phenotypic class of *Bacillus subtilis* mutants by methylxanthines, *Mutat. Res. 117*, 55–65 (1983).
30. P. Schaeffer, H. Ionesco, A. Ryter, and G. Balassa, La sporulation de *Bacillus subtilis*, étude génétique et physiologique, *Coll. Int. Centre Natl. Rech. Sci. Mechanismes Regulation, Marseille 1963*, 553–563.
31. A. G. Siccardi, A. Galizzi, G. Mazza, A. Clivio, and A. M. Albertini, Synchronous germination and outgrowth of fractionated *Bacillus subtilis* spores: Tool for the analysis of differentiation and division of bacterial cells, *J. Bacteriol. 121*, 13–19 (1975).
32. V. F. Simmon, *In vitro* mutagenicity assays of chemical carcinogens and related compounds with *Salmonella typhimurium*, *J. Natl. Cancer Inst. 62*, 893–899 (1979).
33. T. R. Skopek, H. L. Liber, J. J. Krolewski, and W. G. Thilly, Quantitative forward mutation assay in *Salmonella typhimurium* using 8-azaguanine resistance as a genetic marker, *Proc. Natl. Acad. Sci. USA 75*, 410–414 (1978).
34. T. R. Skopek, H. L. Liber, D. A. Kaden, and W. G. Thilly, Relative sensitivities of forward and reverse mutation assays in *Salmonella typhimurium*, *Proc. Natl. Acad. Sci. USA 75*, 4465–4469 (1978).
35. N. Snell, Direct counts of bacterial spores on membrane filters under phase optics, *Appl. Microbiol. 16*, 436 (1968).
36. J. Spizizen, Transformation of biochemically deficient strains of *Bacillus subtilis* by deoxyribonucleate, *Proc. Natl. Acad. Sci. USA 44*, 1072–1078.(1958).

37. H. F. Stich, R. H. C. San, P. Lam, J. Koropatnick, and L. Lo, Unscheduled DNA synthesis of human cells as a short-term assay for chemical carcinogens, in: *Origins of Human Cancer* (H. H. Hiat, J. D. Watson, and J. A. Winsten, eds.), pp. 1499–1512, Cold Spring Harbor Laboratory (1977).

38. H. Tanooka, Development and applications of *Bacillus subtilis* test systems for mutagens, involving DNA-repair deficiency and suppressible auxotrophic mutations, *Mutat. Res. 42*, 19–32 (1977).

39. W. van Iterson and J. A. Aten, Nuclear and cell division in *Bacillus subtilis*: Cell development from spore germination, *J. Bacteriol. 26*, 384–399 (1976).

CHAPTER 6

The L5178Y/TK Gene Mutation Assay System

David E. Amacher

1. Introduction

The L5178Y/TK$^+$ → TK$^-$ assay is one of several recognized mammalian cell systems for estimating the genotoxic potential of environmental or commercial chemicals. In a typical test application, thymidine kinase-competent (TK$^+$) 3.7.2C L5178Y cells are exposed to five or more different concentrations of the test substance for 1–4 hr, maintained 48–72 hr in growth medium, then cloned in soft-agar medium in a single-step selection process to detect mutagen-induced thymidine kinase-deficient (TK$^-$) variants. Independent validation studies[11,15,30] have demonstrated that this assay is sensitive to more than 20 known genotoxic chemicals. This short-term mutagenesis assay is a good predictor of mammalian carcinogenicity, with few "false positive" results in those studies where noncarcinogens were tested.[8]

The purpose of this chapter is to review the development of this gene mutation assay since its introduction by Clive and co-workers 10 years ago.[27,29] Experimental procedures have been refined and a considerable number of carcinogens and several noncarcinogens have

---

David E. Amacher • Drug Safety Evaluation, Pfizer Central Research, Groton, Connecticut 06340.

been tested in the L5178Y/TK assay since the initial report[28] in Volume 3 of this series. Used strictly as a gene mutation assay with proper experimental design and carefully controlled culturing conditions, this assay can provide useful information about the genotoxic potential of chemicals.

The "wild-type" 3.7.2 cell now used in this assay originated as a spontaneous TK-competent revertant from a TK-deficient mutant and has been assigned a heterozygous TK^+/TK^- genotype based on the observation that it and eight other revertants possessed between one-third and two-thirds the TK enzyme activity of the original $TK^{+/+}$-3 cell.[29] The forward $TK^+ \rightarrow TK^-$ mutational event confers a mutagen dose-dependent increase in trifluorothymidine resistance (TFT^r) or 5-bromodeoxyuridine resistance ($BrdUrd^r$), presumably due to mutations in the structural gene for the TK enzyme.[25] However, no soluble TK enzyme was found in the original $TK^{+/+}$ cell and one revertant cell line in that study. More recent physicochemical studies have suggested that some other mutational event(s) may be responsible for the loss of TK activity following mutagen exposure, such as a mutation in an intervening sequence that prevents the processing of RNA.[75] The TFT^r and $BrdUrd^r$ mutants were first thought to be completely void of TK activity.[30] However, residual TK activity has been found in some well-characterized $TK^{-/-}$ mutants by other workers,[9] probably due to minute amounts of mitochondrial TK enzyme.[56]

In the mammalian cell, TK enzyme occurs in multiple forms, with a fetal type prevalent in the growing cultured cell.[1] TK activity is influenced by a number of factors, including the physiological state of the cell.[78] In cultured mouse fibroblasts, TK is an unstable, constitutive enzyme produced by periodic enzyme synthesis.[61] Following transport of thymidine (dThd) into the cell, phosphorylation of dThd or halogenated nucleoside analogues by TK provides an intracellular trapping mechanism.[77] Low substrate specificity is characteristic of kinases[79], and loss of TK activity indicated by BrdUrd or FdUrd resistance causes corresponding cross-resistance to either BrdUrd, IdUrd, or FdUrd.[20] When L5178Y cells are synchronized by thymidine deficiency, the addition of BrdUrd permits only one subsequent synchronous division, while upon the addition of TFT, no further cell divisions follow.[47] In human lymphoblasts, BrdUrd is cytotoxic, while TFT is cytostatic.[84] Thus, both pyrimidines select for the same TK gene locus, but by a different mechanism.

The L5178Y TK$^-$ cells derived from mutagen-treated 3.7.2C cells are: (1) resistant to 50–300 μg/ml BUdR,[67] 0.01–0.1 μg/ml FdUrd,[9] and, depending upon the medium used, either 0.05–5.0 μg/ml or 1 to

>12 μg/ml TFT[11,67]; (2) cross-resistant to BrdUrd after single-step isolation by TFT[11]; and (3) sensitive to THMG (supplemented with thymidine, hypoxanthine, methotrexate, glycine) medium containing a drug that inhibits *de novo* dTMP synthesis.[24]

5-Bromodeoxyuridine was first used in the L5178Y/TK assay as the mutant selective agent.[27,29] Toliver *et al.*[86] had demonstrated that cell division in cultured mammalian cells that were presumably TK-competent ceases after a delayed second division in the presence of 10 μg/ml BrdUrd. With 50 μg/ml BrdUrd, Clive *et al.* reported the killing of L5178Y TK$^{+/-}$ heterozygotes after 3–4 cell generations.[27] Others observed that when L5178Y cells were plated in the presence of 50 μg/ml BrdUrd, cell division continued long enough to produce a visible haze consisting of partially resistant microcolonies,[67] and have suggested that TFT be used instead. Phosphorylated TFT acts as an irreversible inhibitor of thymidylate synthetase[47] and therefore selects against the TK-competent phenotype.

2. Experimental Procedures

The basic experimental methods for this assay have been described in great detail by Clive and Spector[24] with subsequent clarifications.[30] Undoubtedly, the availability of such well-defined experimental procedures led to the early utilization of the L5178Y/TK assay for the commercial testing of chemicals. In our own laboratory, additional modifications were helpful in making the assay amenable to the rigorous demands of scheduled routine testing.[11] We found that Fischer's medium for the leukemic cells of mice,[41] used for the culturing of lymphoma cells for many years, seemed particularly prone to deterioration. Recently, Griffin *et al.*[45] studied the effects of incidental cool white fluorescent light upon this medium after noting random disturbances in the growth and cloning efficiencies of L5178Y cells. They found that exposure to light resulted in the rapid development of toxic photoproducts that are cytostatic at lower doses of irradiation and cytotoxic at higher doses. We have observed that, even when protected from light, cloning medium prepared from Fischer's medium plus 20% heat-inactivated horse serum would not support maximum colony growth when ≥300 untreated L5178Y cells were plated per 100-mm dish in the absence of any selective agent (Figure 1). Clearly a richer, more stable medium was needed.

At the suggestion of Clive in 1977, we explored the possibility of using RPMI-1640 medium, which has a lower riboflavin content,[65]

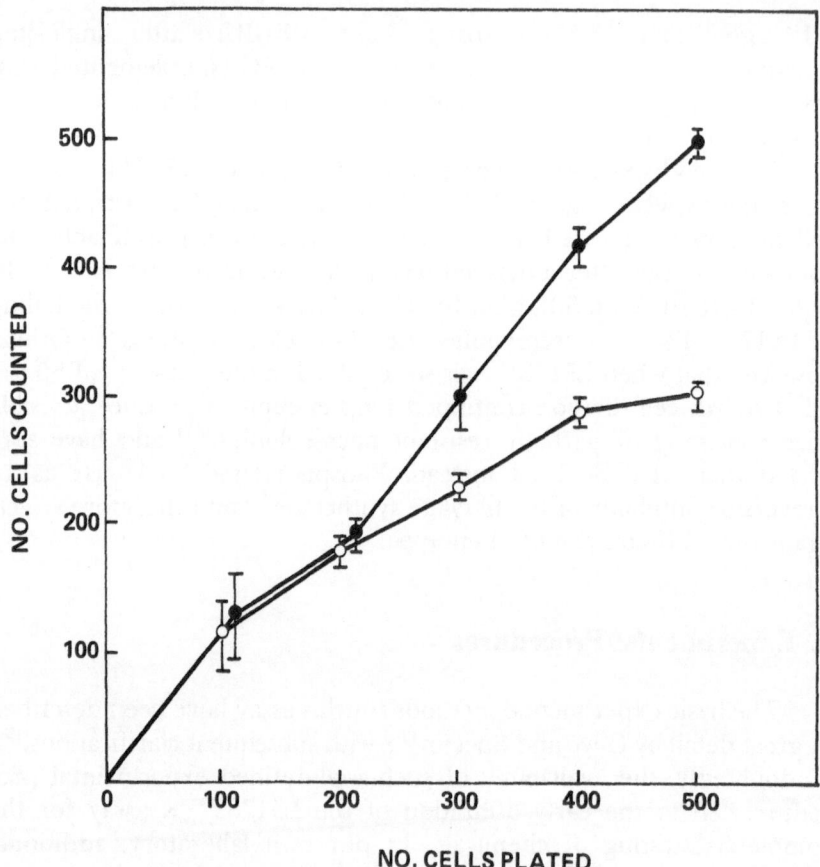

NO. CELLS PLATED

FIGURE 1. Comparison of machine versus manual colony counts in conventional soft-agar medium. Solvent-treated control L5178Y/TK$^{+/-}$ cells were plated in F_{20p} medium [24] prepared with heat-inactivated serum, incubated at 37°C in 5% CO_2–95% air for 9 days, then counted by eye and via an Artek Model 870 at maximum sensitivity. (●) Manual count. (○) Machine count.

making it less susceptible to photochemical changes. RPMI-1640 medium readily supports the growth and cloning of L5178Y cells, even at low (5%) serum concentrations.[11] Another advantage of RPMI-1640 is that, supplemented with 9 μg/ml thymidine, 15 μg/ml hypoxanthine, 0.3 μg/ml methotrexate, and 22.5 μg/ml glycine (THMG), this medium can be used routinely to suppress the accumulation of TK$^-$ mutants in TK$^{+/-}$ suspension culture.[11,15] The reported failure of THMG-supplemented Fischer's medium to eliminate spontaneously derived mutants in TK$^{+/-}$ cultures[28] may have been due to that medium's high folic acid content, which can interfere with methotrexate inhibition,[39]

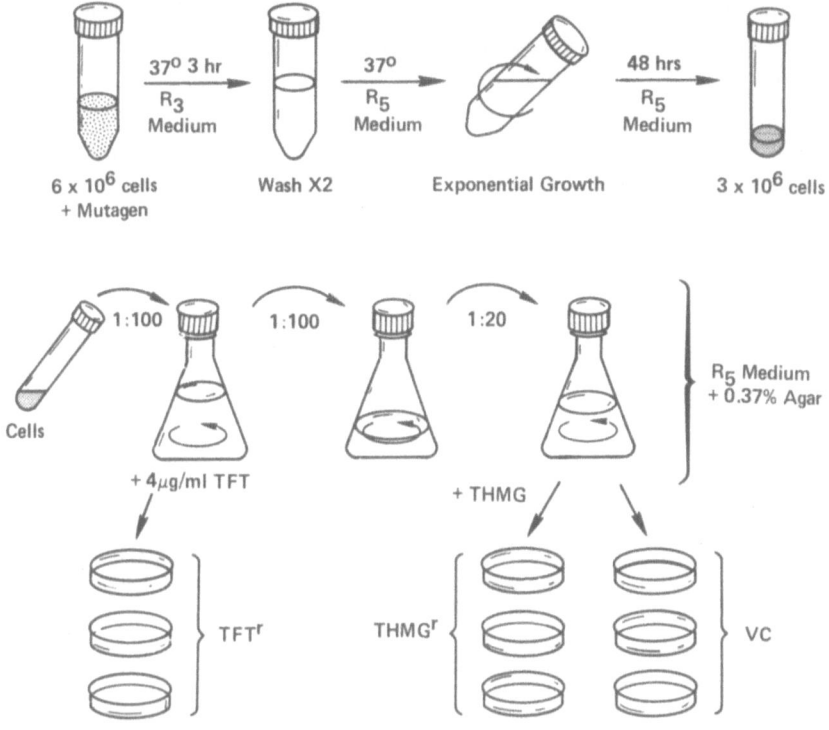

FIGURE 2. General procedure for testing presumptive mutagens in the L5178Y/TK assay. Media R_3 and R_5 are RPMI-1640 containing either 3% or 5% horse serum, respectively. The plating of cells in the presence of THMG[11] is an optional step.

coupled with early difficulties in preparing concentrated methotrexate solutions. Clive and co-workers later reversed their opinion on the utility of THMG-supplemented Fischer's medium.[24,30] The proper selection of serum to be used in this assay is another important consideration that will be discussed elsewhere in this chapter.

Our general procedures for mutagenesis testing in the L5178Y/TK assay are illustrated in Figure 2. This protocol is similar to that described by Clive et al.[30] except for (1) the duration of treatment (3 versus 4 hr), (2) the use of plastic Erlenmeyer flasks during cloning, (3) the use of a pipetting machine rather than graduated cylinders for medium transfer, (4) lower serum concentrations (3–5% versus 3–20%), (5) the exclusion of sodium pyruvate or conditioned medium during mutagen exposure from the protocol, and (6) a final concentration of 4 µg/ml instead of 1 µg/ml TFT. The rationale for the first four modifications is the conservation of serum, time, and labor.

At least for some mutagens, the kinds of DNA lesions that inhibit DNA synthesis and are associated with cytotoxicity are not necessarily associated with gene mutations.[76] In V79 Chinese hamster cells, mutagenic and cytotoxic responses to alkylating agents were described as being independent, although both increase with increasing dose levels.[82] The independence of the two responses may indicate that different lesions or different repair processes or both are involved,[63] a situation that may also be true for the L5178Y cell. For many mutagens, cytotoxicity is usually accompanied by mutagenicity in the *in vitro* test system, but the converse is not necessarily true, as demonstrated for ultraviolet light,[53] ethyl methanesulfonate (EMS),[15,82] 3-methylcholanthrene (3MCA),[13] N-methyl-N'-nitro-N-nitrosoguanidine (MNNG), and dipropylnitrosamine.[12] The minimum treatment time in this assay system is that which permits sufficient cytotoxicity to assure that any associated mutagenic event will also have occurred. Many important mutagens have limited stability in tissue culture medium[52] or are rapidly converted to mutagenic metabolites under appropriate test conditions.[87] Successful treatment periods of less than 1 hr have been reported for this system.[44] On the other hand, a maximum treatment or dosing period of 4 hr has been described[30] based upon S9 toxicity, which increases with exposure time. A maximum treatment period without S9 can be considered as a point of diminishing return and prolonged (≥ 24 hr) exposure to a test chemical increases the possibility of a selective rather than a mutagenic response. A typical strategy is to use a fixed treatment period of 3 or 4 hr and a range of test chemical concentrations.

Plastic Erlenmeyer flasks (250 ml, Corning #25600) are extremely useful during the soft-agar cloning step. They are easily washed, sterilized, and reused without breakage. These flasks are not suitable for long-term culture (>2 weeks) due to a direct association between contact time and an increased incidence of spontaneous TFT^r cells in untreated cell stock (Table 1). These results suggest caution, should the investigator wish to use plasticware for routine cell culture of L5178Y cells.

3. Serum Quality

3.1. The Importance of Serum Quality

Serum quality is a critical factor in this assay. A minimum concentration of 4% horse serum is required to sustain the exponential growth

TABLE 1. The Mutagenic Effects of Long-Term Culture Vessels on L5178Y Cells

Culture conditions prior to soft-agar cloning[a]	Soft-agar cloning results[b]		
	Percent absolute cloning efficiency	TFT Colony count per 10^6 cells	Mutant frequency per 10^6 survivors
Grown in glass flasks only	80, 81	10, 10	12, 12
Grown in both plastic and glass flasks	74, 79	24, 33	32, 42
Grown only in plastic flasks	72, 81	69, 83	96, 102

[a] Untreated L5178Y/TK$^{+/-}$ stock cells were grown under three conditions: (1) 2 weeks in glass flasks with weekly THMG "cleansing"; (2) 2 weeks in plastic flasks, transferred to glass flasks for 2 weeks with weekly THMG "cleansing"; or (3) 4 weeks in plastic flasks with weekly THMG "cleansing" during the last 2 weeks.
[b] Results are shown for duplicate cultures for each condition.

of L5178Y cells in RPMI-1640,[11] compared to 10% serum in Fischer's medium[30]; however, some commercial serum preparations can interfere with mutant selection by trifluorothymidine, perhaps due to the presence of alternate pyrimidine substrates[70] or stable catabolic enzymes, and can even inhibit cell growth at higher concentrations.[3] Table 2 illustrates the moderate but steady increase in background TFT^r frequency as serum content is increased in the soft-agar cloning

TABLE 2. Mutant Frequencies per 10^6 Survivors for Untreated L5178Y/TK$^{+/-}$ Cells in Selective Medium Containing Varied Serum Content

Test conditions[a]	Percent serum in selective medium[b]			
	5%	10%	15%	20%
Cells not cleansed				
Heat-inactivated serum	—[c]	88	92	108
Normal serum	68	79	95	117
Cells cleansed				
Heat-inactivated serum	33	40	53	78
Normal serum	31	52	63	54

[a] Cleansing refers to 24 hr treatment in THMG medium (see Ref. 11), which selects against preexisting TFT^r variants.
[b] Cells had been grown in Fischer's medium (F_{10p})[24] and were cloned in Fischer's soft-agar medium containing 4 µg/ml TFT and 5, 10, 15, or 20% serum. Values are averages for three plates.
[c] Colonies too small to count.

TABLE 3. Mutant Frequencies per 10^6 Survivors for
Untreated L5178Y/TK$^{+/-}$in Two Types of Selective
Soft-Agar Cloning Medium Containing 4 μg/ml TFT
and the Indicated Serum Content

	Serum content in the selective medium	
Test medium[a]	5%	20%
RPMI-1640	113	200
Fischer's medium	149	188

[a] Cells had been grown continuously in Fischer's medium (F_{10p})[24] for 132 days without THMG treatment. All sera were heat inactivated (56°C for 30 min).

medium. This effect is independent of the medium used and is observed even though the serum is heat-inactivated (Table 3). Similarly, a serum factor that interferes with the selection of azaguanine-resistant somatic cells by breaking down the selective agent has been described by Van Zeeland and Simons.[88]

We routinely screen for two characteristics in reserved lots of serum prior to purchase: (1) the ability to sustain optimal cell growth and (2) the potential to interfere with TFT selection against TK-competent cells. By determining serum quality before purchase, we eliminate the inconvenience of artificially high spontaneous background mutant frequencies in L5178Y cell stock (see Section 3.2).

3.2. Serum Selection Procedure

Many of the published studies for the L5178Y/TK assay include solvent control data where the spontaneous mutant frequencies are quite elevated, for example, $(100-297) \times 10^{-6}$ mutants per survivor.[30] Some have argued that the high background mutant frequencies in those studies were linked to mycoplasma contamination of the cells.[22] It is known that phosphorylases resulting from low-level mycoplasma infections may cause the breakdown or alteration of thymidine[37] and by inference, BrdUrd or TFT.

On the other hand, it has been emphasized that TK$^{-/-}$ mutants (BrdUrd-resistant colonies) cannot be recovered from mycoplasma-contaminated L5178Y cell cultures either spontaneously or following mutagen treatment.[24,28] Also, there was no evidence of mycoplasma contamination of L5178Y cells in other studies where high background mutant frequencies were occasionally observed.[54] We have found that the greatest single cause of elevated mutant frequencies without prior mutagen treatment in the L5178Y/TK assay is attributable to the serum used in the culture medium during the TFT (or BrdUrd) selection step.

Following the example of Fischer,[40] we prescreen horse serum prior to purchase according to three criteria: (1) that it permits the rapid growth of L5178Y cells in soft-agar cloning medium containing 5% serum (R_5) to machine-countable size within 7–8 days at ~100% cloning efficiency; (2) that the spontaneous mutant frequency in the presence of 4 μg/ml TFT in R_5 medium is acceptably low [$(10–40) \times 10^{-6}$, for example]); and (3) that there is no demonstrable increase in the spontaneous mutant frequency of solvent control cells when the serum content in the cloning medium is increased from 5% to 10%.

Experimental data from serum screening trials are shown in Figure 3. Serum sample lots 08 and 53 did not meet criterion 1 above; serum sample 33 did not meet criterion 2; and serum sample 70 obviously did not meet criterion 3. Although not shown here, further testing of sample 70 serum at up to 1 ml/plate in Ames *Salmonella typhimurium* tester strains TA1537, TA98, and TA100 produced no significant increase in the histidine reversion frequency over background. In the L5178Y/TK assay, both overnight dialysis and heat-inactivation of sample 70 at 56°C for 30 min failed to eliminate this "mutagenic" effect, suggesting that the serum may have introduced degradative enzymes to the extracellular environment, which then prevented effective levels of TFT from reaching the intracellular compartment. In support of this, cloning with 36 μg/ml TFT diminished this high background mutant frequency (not shown).

4. The Metabolic Activation of Mutagens

There are many potential mutagens or "promutagens" among the various hydrocarbons, nitrosamines, aromatic amines, fungal toxins, and other environmental chemicals that require enzymatic conversion (activation) to be detected by gene mutational events in the target cell. Microsomal monooxygenase enzymes play a key role in the initial activation step, particularly for the polycyclic hydrocarbon series. Animal strain or species differences in monooxygenases can result in very dramatic differences in mutagen activation.[73] Cultured cells vary in hydrocarbon metabolizing activity, which is high in some low-passage rodent cells compared to intermediate- or high-passage cells.[57] In

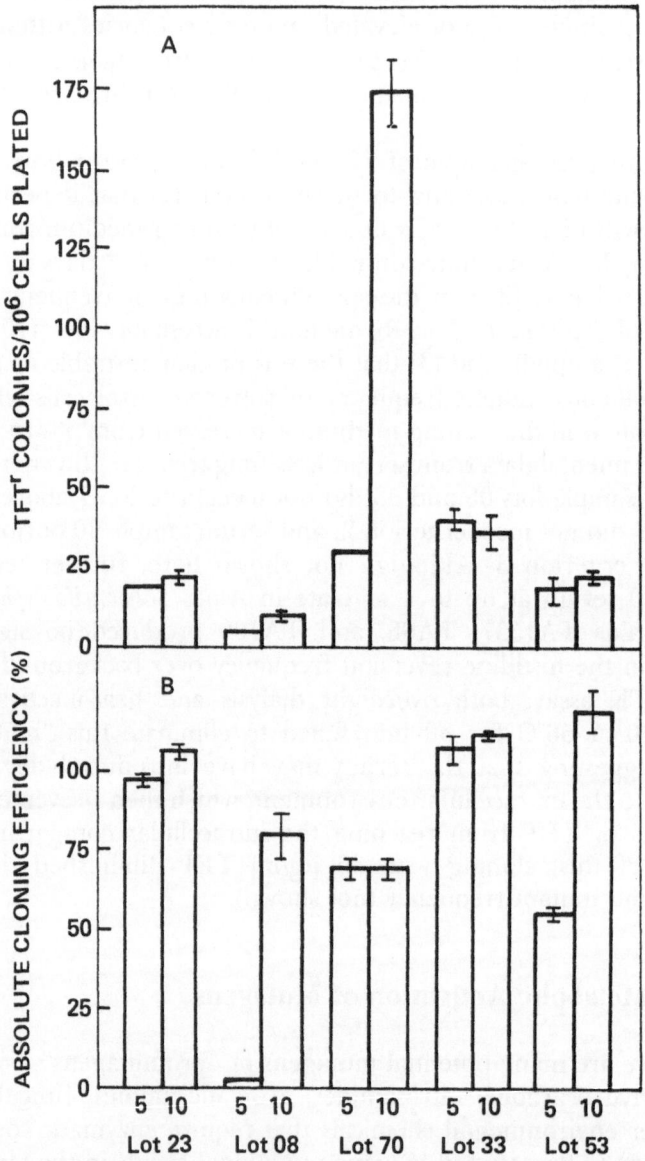

FIGURE 3. The influence of serum concentration upon the cloning efficiencies of untreated L5178Y/TK$^{+/-}$ cells in the presence and absence of trifluorothymidine. Five different serum lots were compared at a final concentration of either 5% or 10% in RPMI-1640 soft-agar medium. (A) Incidence of TFT-resistant colonies in soft-agar medium containing 4 μg/ml TFT (1×10^6 cells seeded/plate); (B) the corresponding absolute cloning efficiencies (500 cells seeded/plate).

cultured embryo cells, enzyme activity has been observed to drop in each succeeding passage after primary culture.[48] Monooxygenase activity is generally too low in many established cell lines for those cells to be useful in drug metabolism studies.[89]

While generalizations can be made, the enzyme specificity and kinetics for the biotransformation processes leading to an active species can differ considerably among promutagens. Chemicals such as 4-nitroquinoline-1-oxide are activated to mutagens through metabolic activation systems common to many microbial and mammalian cells.[69] Others, such as N-acetylaminofluorene (AAF) and benzo(a)pyrene [B(a)P], are activated through mammalian microsomal enzymes that may be tissue or organ specific.[50,60] Still other chemical mutagens, for example, dimethylnitrosamine (DMN), may be activated or inactivated by complex mechanisms.[46,59] On the other hand, N-methyl-N'-nitro-N-nitrosoguanidine and mitomycin C are direct-acting mutagens that are inactivated by microsomal enzymes.[30]

The L5178Y cell does have the capability to directly detect the mutagens lucanthone and cyclophosphamide, which require the presence of liver microsomal fraction for activation and detection in *Salmonella typhimurium*,[14] and procarbazine, which is negative in *S. typhimurium*. The L5178Y cell can also directly detect mutagenic substances in the β-glucuronidase-treated urine of mice dosed with 2-aminofluorene, cyclosphosphamide, safrole, or lucanthone.[9,14] Nevertheless, the inclusion of S9 liver microsomal preparation (9,000 × g supernatant)[16] along with the test chemical extends the utilization of the L5178Y assay for the detection of a wide variety of promutagens.

4.1. Rodent Liver S9 Activation

Mouse liver and kidney S9 were first used by Matheson and Creasy[62] in the L5178Y/TK assay to activate AAF, 2,4-diaminotoluene, DMN, diethylnitrosamine, B(a)P, dimethylbenzanthracene, and m-phenylenediamine. During that same year Frantz and Malling[44] reported the activation of DMN by calcium-precipitated mouse microsomes in the L5178Y cell. Over the next 4 years, rat liver S9 was used to activate benzanthracene, 3MCA, acridine orange, aniline, B(a)P, AAF, diethylnitrosamine, DMN, and cyclophosphamide to mutagens in the L5178Y/TK assay.[15,30]

In contrast to some reports,[31] AAF is weakly mutagenic when tested directly in the L5178Y/TK assay,[10] but B(a)P is negative when tested in the absence of exogenous metabolic activation,[11] as is 3MCA. These results are consistent with the observation that the L5178Y

TK$^{+/-}$ 3.7.2C cell possesses measurable AAF-N-hydroxylase activity, but negligible benzo(a)pyrene hydroxylase or N-demethylase activities.[10] The selection of 3MCA or B(a)P instead of AAF as positive controls when S9 activation is used is thus supported by established experimental data.

Following the example of Ames and co-workers,[16,17] many of the first mutagenicity experiments with promutagens in the L5178Y/TK assay used liver S9 from Aroclor-treated Sprague–Dawley (CD) rats at a final concentration of 10% (v/v) (2.5% when corrected for dilution with 0.15 M KCl). combined with arbitrarily determined concentrations of NADP and isocitric acid.[15,30] While these earlier studies were qualitatively successful, the quantitative aspects of S9-mediated activation were not explored. In 1979, Thornton and Hite[85] reported that increasing S9 concentration caused a significant decrease in mutant frequencies for five polycyclic aromatic hydrocarbons in the L5178Y/TK assay, presumably due to the progressive metabolism of these promutagens to nonmutagenic metabolites. This concentration effect for S9 from Aroclor-induced rats has been confirmed for B(a)P along with the report of a similar effect (decreasing mutagenicity with increasing S9 concentration) for L5178Y/TK$^+$ cells exposed to AAF, 2-aminoanthracene, 1,2-benzanthracene (BA), or 3MCA.[7] However, DMN produced a moderate S9 concentration-dependent increase in mutagenicity, in agreement with the earlier observation of Kuroki et al.,[58] who used 15,000 × g liver supernatant and V79 Chinese hamster cells. Moreover, the deactivation of many promutagens other than DMN by high concentrations of liver S9 obtained from Aroclor-dosed rats may be a general phenomenon in mammalian cell systems, a complicating factor that must be considered when S9 from Aroclor-dosed animals is used. The previously recommended final concentration of liver S9[30] from Aroclor-pretreated rats might be excessive for the optimal detection of some promutagens.

Another consideration is whether liver S9 prepared from animals pretreated with polychlorinated biphenyls (Aroclors) should be used at all. In a preliminary report where 11 chemicals were tested for mutagenicity in the L5178Y/TK assay within the same laboratory with both "induced" (from Aroclor-dosed rats) and "noninduced" S9, the results were qualitatively (positive or negative) the same.[72] Some mutagens, such as B(a)P and DMN, reportedly induce higher mutant frequencies at either lower toxicity or lower doses, i.e., are more easily detected, in the presence of noninduced S9 than with "induced" S9.[30] When compared by using although not necessarily optimal concentrations of S9, Aroclor pretreatment of Sprague–Dawley (CD) male rats did not substantially increase the mutagenicity of BA, B(a)P, or 3MCA.[6] In that same study, liver S9 prepared from at least eight different untreated rodent strains could be used in the L5178Y/TK assay, although some differences were observed in the relative ability to activate three polycyclic hydrocarbons versus one representative aromatic amine. Most studies use S9 prepared from male rodents, but we found that liver S9 samples prepared from male or female Syrian hamsters are equivalent for the activation of 3MCA.[9] When ten known carcinogens were tested in the presence of liver S9 prepared from untreated Sprague–Dawley (CD) rats, eight produced mutant yields that were significantly greater than the solvent control mutant yields in the L5178Y/TK assay.[8] Included was AAF, which reportedly could be activated only with S9 from Aroclor-treated rats.[30] Ethyl carbamate and thioacetamide, both negative in the latter study, were still negative in the presence of liver S9 prepared from Aroclor-treated rats. In the absence of evidence to the contrary, there is little to recommend the pretreatment of rodents with polychlorinated biphenyls when preparing S9 for this assay.

4.2. Cell-Mediated Activation

Because they can be grown only in suspension culture, L5178Y cells are ideally suited for cocultivation experiments where the mouse lymphoma target cell is temporarily combined with a monolayer culture of metabolically competent cells in the same stationary culture vessel. Cocultivation conditions are designed to optimize the uptake and biotransformation of the test chemical by the activating cell and the intercellular transfer of any metabolites to the target cell. Test conditions selected to maximize total cell densities (activating cells + target cells) or duration of exposure are generally not the culture conditions most conducive to the general maintenance of exponential growth of either cell type, a conflict that may limit the usefulness of this type of activation for some mutagens. Nevertheless, cell-mediated activation is an attractive concept. The resulting intercellular transfer of active metabolites to the L5178Y target may be far more realistic than the more commonly used broken cell S9 preparation.

Syrian hamster embryo cells were first used to activate 3MCA and B(a)P to ultimate mutagenic forms producing increased TFT resistance in L5178Y/TK$^{+/-}$ cells.[13] This was followed by the use of both hamster and rat hepatocytes to activate B(a)P, 3MCA, diethylnitrosamine, DMN, dipropylnitrosamine, 2-aminoanthracene, and AAF in the L5178Y/TK gene mutation assay.[4,5] One other promutagen, 1,2-benzanthracene, was activated by hamster but not rat hepatocytes. These results were qualitatively the same as those obtained with the corresponding S9 except for 1,2-benzanthracene, a mouse carcinogen,[81] which was negative with hamster liver S9 but a positive mutagen in the presence of rat liver S9 in the L5178Y/TK assay. It remains to be seen if species specificity will be observed if mouse and rat hepatocytes, for example, are compared in a cell-mediated L5178Y/TK assay.

In summary, initial testing of a suspected promutagen is easily accomplished in the L5178Y/TK assay using conventional liver S9 fraction prepared from untreated rats. Where further information is desired regarding organ or tissue biotransformation specificity or the effect of membrane barriers upon the transport of potentially short-lived metabolites, embryonic fibroblasts, hepatocytes, and perhaps other cell types can be used in cocultivation experiments with the L5178Y/TK target cell.

5. The Small Mutant Colony Phenomenon

The discovery that TFT^r L5178Y cell colonies formed in soft-agar medium following mutagen exposure were typically bimodal in size distribution[30] has led to extensive studies to determine what if any genetic basis is responsible. All previous studies had presumably dealt with only or predominantly the large, so-called lambda TFT^r colony population. Through a series of elaborate and painstaking experiments, Clive, Moore, and co-workers[23,26] made the following conclusions about the small TFT^r (sigma) colony component. Sigma colonies: (1) are heritably slow-growing with a heritable reversion rate to rapid growth, (2) are mutagen specific in the sense that some mutagens produce only or predominantly small colonies, (3) possess minimal TK enzyme activity, (4) are in most cases attributable to detectable chromosomal aberrations, particularly involving chromosome 11, the locus for TK.

In addition to the cited work by Clive and co-workers, some other observations on L5178Y cell colony size have appeared in the literature. Frantz and Malling[44] first observed a marked increase in the number of very small BrdUrd^r colonies in treated versus control cultures after TK$^{+/-}$-3.2.1 L5178Y cells had been exposed for 15 min to DMN in the presence of calcium-precipitated mouse liver microsomes. Both large and small TFT^r colonies were observed by Meltz and MacGregor[64] in L5178Y cells that had been treated with quercetin for 4 hr. We have found that virtually all TFT^r colony size distributions are bimodal, including spontaneously arising mutants from solvent control cultures,[11,15] which vary from approximately 30–60% of the total mutant fraction, and that small colonies can arise after treatment with the sigma colony mutant methyl methanesulfonate in soft agar-medium containing no TFT,[3] the first indication that the slow-growth condition itself could be a transient manifestation of cytotoxicity unrelated to TFT selection, i.e., independent of the TK locus. As further evidence that the small colony phenomenon can be completely independent of the TK locus, we have recently observed (D. Amacher and E. Dunn, unpublished observations) that mutagens such as DMN, EMS, methyl methanesulfonate, B(a)P, N-methyl-N'-nitrosoguanidine, iodomethane, and epichlorohydria all produce both large and small colonies at the oubain-resistance locus of $3.7.2C$ TK$^{+/-}$ cells.

A mutagen-induced lag in cell division rates has been observed in other mammalian cell gene mutation systems, which in some cases was associated with small colony formation. With dThd as a selective agent, Cole and Arlett[33] noted that methyl methanesulfonate produced a dose-dependent division delay in L5178Y cells and that many of the induced resistant colonies were small. A division delay phenomenon might explain the observation of Clive et al.[30] that methyl methane-sulfonate (MMS) is a sigma-colony-former and ethyl methanesulfonate (EMS) is a lambda-colony-former, for L5178Y cells slowly recover from MMS-induced cytotoxicity[33] but rapidly recover from EMS toxicity.[32] Other investigators have noted that hydrazine and two hydrazine derivatives differ in the proportion of small to large dThd-resistant colonies induced in L5178Y cells.[80] These large and small clones had substantially different doubling times in nonselective medium; however, the small clones increased in size when returned to selective medium after being maintained for 7 weeks in nonselective medium, an observation noted elsewhere for some MMS-derived small clones.[3] In contrast, Clive et al.[30] reported that cultures derived from small colonies produce only small colonies at greatly reduced cloning efficiencies. The phenomenon of mutagen-induced division delay is not restricted to the L5178Y cell. For instance, while attempting to explain the decay in mutation frequency at the hypoxanthine-guanine phosphoribosyl trans-ferase locus in V79 cells with increasing expression time, Diamond et al.[38] noted that each of three B(a)P-induced 6-thioguanine-resistant colonies had a 30% longer lag time before reaching logarithmic growth compared to parent V79 cells. Interestingly, B(a)P has been described as a small-colony mutagen in the L5178Y/TK assay.[30]

It has been demonstrated that many but not all sigma-colony TFT^r L5178Y cells carry characteristic structural chromosomal anomalies and that mutagen-derived sigma colonies are generally authentic TK$^{-/-}$

variants.[49] However, a cause and effect relationship between "chromosomal mutations"[26] and genetically determined slow growth (sigma colonies) is difficult to establish at this point. The frequency of small colonies in a L5178Y cell population cloned in the absence of selection (<5%[30] or 5–10%[26]) indicates the presence of a stable subpopulation that greatly exceeds the frequency of both large and small TFT^r colonies typically observed after any mutagen treatment. In one study, cells derived from this small colony subpopulation were not BrdUrd resistant, but some (3/5) did retain altered growth characteristics when recloned.[3]

In an impressive study, Moore-Brown et al.[67] studied the phenotypic stability of 194 large- and small-colony-derived TFT^r variants of both spontaneous and mutagen-induced origin. With four exceptions, all variants had retained the TFT^r phenotype after 1 week (7–17 cell doublings) in nonselective Fischer's medium. Based upon a medium-dependent difference in growth rate,[68] this same time interval would be equivalent to a greater number of doublings in RPMI-1640 medium. However, this brief interval in nonselective Fischer's medium may not have been adequate for the small-colony stock, in view of the gradual loss of phenotypic stability clearly demonstrated for one hycanthone-induced small-colony TK$^{-/-}$ cell line, C.210B, approximately 50 days subsequent to isolation and transfer to nonselective conditions.[23] Long-term stability studies of the sigma TFT^r phenotype are needed to rule out the possibility of a temporarily increased resistance to the selective agent of an non-mutational origin.[43]

A possible explanation for the small-colony phenomenon in L5178Y cells is that small colonies are a manifestation of mitotic delay caused by transient gene inactivation or altered gene expression in a portion of a chromosome carrying alleles necessary for robust growth in soft agar. A suppression of gene activity which can extend for an appreciable distance along the chromosome from a break site has been described in Drosophila and in the mouse.[18] For some chemicals such as MMS, hycanthone, and pyrimethamine[49] this may coincide with those mutagen concentrations that produce considerable clastogenic activity (rearrangements involving chromosome 11). The division delay effects of MMS are well documented[42] and it has been reported that hycanthone produced maximum mutagenicity over a concentration range producing minimal growth.[21]

Percent total relative growth,[24] commonly used as an estimator of cell survival in the L5178Y/TK assay, combines both relative growth in suspension culture and soft-agar plating efficiency of treated versus solvent control cultures and thus can be influenced by both cytostatic and cytocidal events. We have previously demonstrated that mutant

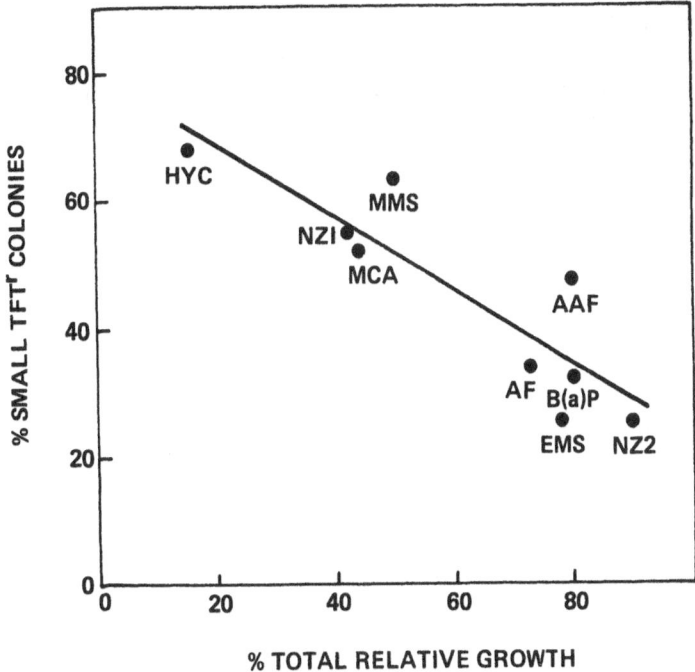

FIGURE 4. Incidence of small TFT[r] colonies in the total mutant population versus total relative growth. Test conditions (3-hr exposure) were: HYC: 10 μg/ml hycanthone methanesulfonate; NZI and NZ2: 372 and 209 μg/ml niradazole, respectively; MCA: 5.37 μg/ml 3-methylcholanthrene + rat S9; MMS: μg/ml methyl methanesulfonate; AF: 30 μg/ml 2-aminofluorene + rat S9; EMS: 310 μg/ml ethyl methanesulfonate; B(a)P: 5 μg/ml benzo(a)pyrene + rat S9; and AAF: 150 μg/ml acetylaminofluorene + rat S9. Small colonies were those <0.3 mm 8 days after cloning.

colony size in methyl methanesulfonate-treated L5178Y cells is associated with chemically induced cytotoxicity as measured by percent total relative growth.[3] A general correlation between the incidence of sigma TFT[r] colonies in soft-agar medium and percent total relative growth is illustrated in Figure 4, where sizing was performed by microscopic examination as previously described.[3] The relative proportion of sigma TFT[r] colonies for mutagenically active concentrations of eight mutagens is highly correlated ($r = -0.87$) with cytotoxicity as estimated by percent total relative growth. These results, and the data for ICR-191,[9] methyl iodide,[30] and perhaps epichlorohydrin,[66] all suggest that the relative proportion of sigma colonies in a TFT[r] cell population increases with cytotoxicity, at least to a point. For ethyl methanesulfonate, however, this relationship is apparently not maintained at very high cytotoxicity

(12% survival).[66] For many of these chemicals, the data of Hozier *et al.*[49] would predict an increased incidence of structural chromosomal damage in addition to the explicit loss of TK activity and temporarily delayed cell cycle. Although the rates of increase differ, we have observed that an increase in the lambda TFT^r component over background is always accompanied by an increase in the sigma TFT^r component over background and have yet to observe a mutagen that produced only sigma TFT^r colonies. If the distinction between single-gene versus chromosomal mutations on the basis of colony size[23] is correct, the two events do not appear to be mutually exclusive.

One aspect of the sigma mutant phenomenon that apparently has not been explored is a possible solvent comutagen effect. The mutagens AAF, B(*a*)P, and 3-methylcholanthrene among others are typically dissolved in dimethylsulfoxide (DMSO) and administered to the L5178Y cultures in a final DMSO concentration of 1% (~11 mg/ml) for 3–4 hr, exceeding the dosage of the mutagen itself. DMSO has been shown to significantly elevate rat chromosomal aberrations *in vivo*[55] and Chinese hamster cell chromosomal aberrations in the presence of S9 *in vitro*[83] and has mutagenic activity itself at high concentrations in the assay.[15] Interestingly, DMSO (0.1%) was found to enhance both the growth-inhibitory and mutagenic effects of hycanthone in the L5178Y/TK assay.[21] The possibility of an additive or synergistic effect between a high concentration of DMSO and the dissolved test chemical resulting in an enhanced sigma TFT^r colony yield requires additional study.

In summary, recognition that TFT^r colonies form a bimodal colony size distribution in soft-agar medium,[30] and the subsequent cytogenetic characterization[49] of representative small TFT^r L5178Y cell colonies were notable accomplishments. There is considerable supporting evidence that sigma colonies from mutagen-treated cells remain TK-deficient after a minimal growth period (7–17 doublings) in nonselective medium and that the majority of those examined to date possess visible chromosome anomalies. What is less certain is the genetic basis for heritable slow growth and a cause and effect relationship between detectable chromosomal damage and subsequent loss of TK activity. Additional research is likely to resolve these last two questions and determine what effect if any the co-administration of the clastogen DMSO with small colony mutagens might have on the incidence of detectable chromosomal damage. Because most if not all chemical mutagens produce both large and small colonies at the TK locus as do at least seven mutagens at the ouabain-resistance locus, the physical sizing of TFT^r L5178Y cell colonies has limited usefulness.

6. Long-Term Stability of the Trifluorothymidine-Resistant Phenotype

The most important criterion for establishing the mutational origin of a resistant cell line is the demonstration of long-term stability of the variant phenotype in the absence of selective pressure.[43] Frantz and Malling[44] reported that ten BrdUrd-resistant colonies derived from DMN-treated L5178Y cells and grown for 10 days in nonselective medium grew well when cloned again in the presence of BrdUrd. Similarly, Jacobson et al.[51] demonstrated that 11 BrdUrd-resistant isolates that had been maintained in nonselective culture for 8 days were still BrdUrd resistant, yet THMG sensitive. The authors of those studies did not indicate colony size when describing their results. In the course of studies in our laboratory, we have isolated and characterized 45 colonies that originated as large TFT[r] colonies in soft-agar medium after mutagen treatment and 48 hr expression of L5178Y cells. Following growth of the isolates in nonselective medium for times varying from 7 days to 17.5 weeks, 45/45 were BrdUrd resistant, 10/10 were TFT resistant, 35/35 were THMG sensitive, and 10/10 demonstrated negligible dThd uptake.[2,3,10,11] Unquestionably, large-colony TFT[r] colonies represent cells with a heritable TK[−] phenotype, presumably due to single-gene mutations involving the TK locus.[30]

The authenticity of spontaneously arising TFT[r] colonies is not as certain. We have shown that out of six large spontaneous TFT[r] colonies, five retained TFT[r] after 7 days growth in nonselective medium.[11] However, out of 11 small spontaneous TFT[r] colony isolates grown under nonselective conditions for at least 7 days, 3/6 had lost most or all TFT resistance (1 μg/ml TFT was used in the first selection) and 2/5 (4 μg/ml TFT was used in the original isolation) were no longer resistant to BrdUrd but were resistant to THMG medium.[3,10] Some phenotypically unstable sigma TK[−/−] clones of spontaneous origin have also been described by Clive and Moore-Brown.[23]

It is of interest that Moore-Brown and Howard[68] found that when 29 spontaneous sigma TFT[r] colonies were isolated from RPMI-1640 cloning medium containing either 1 μg/ml (14 individual mutants) or 4 μg/ml TFT (15 individual mutants), grown in nonselective medium for approximately 1 week, then tested in suspension culture for the TFT[r] phenotype, only 43% (6/14) were still TFT[r] when 1 μg/ml TFT had been used for selection, but 73% (11/15) were still TFT[r] when 4 μg/ml TFT had been used for selection. This compared to 93% (52/56) retention of TFT[r] phenotype when 56 spontaneously arising small

TFT^r colonies isolated from Fischer's medium were retested under presumably similar circumstances. These data combined with the previously cited (Section 1) higher optimal range of TFT necessary for mutant selection when RPMI-1640 versus Fischer's medium is used in the soft-agar cloning medium suggest that spontaneous sigma TFT^r colonies might represent leaky mutants, i.e., those variants whose gene products have some but not all the functional features of the wild-type (TK$^+$) protein. Leaky mutants would presumably be selected against only at the higher end of the optimal TFT concentration range for a given medium. A second possibility is that each original spontaneous small colony was formed from a subpopulation of TK$^+$ cells with temporarily altered cell cycle characteristics that rendered them partially refractory to TFT and that this condition is more easily overcome in a richer culture medium. A consistent portion (1–4%) of the normal, untreated TK$^+$ cell population forms small colonies in soft-agar medium in the absence of TFT, a phenomenon that has not been given much attention.

7. Data Interpretation Criteria

Using a hypothetical example, Brusick[19] has demonstrated a potential selection error in the derivation of mutation dose–response data due to conditions that permit a faster rate of killing for nonmutant cells. Under actual testing conditions, elevated mutant frequencies are occasionally encountered that are not meaningful due to extreme toxicity or other mitigating factors. For example, equivocal or "equivocal questionable" results and a "variable active" mutagen have been described in the L5178Y/TK assay.[31] Using empirical data, we have developed the following acceptibility and response criteria for the interpretation of all L5178Y/TK results generated subsequent to the formulation period. These criteria were designed to assure that (1) the experimental data are of high quality, and (2) enough data are available to allow an informed judgement. They are included here as representative examples, not as recommended procedures. These criteria are as follows.

I. *General Considerations*
    A. A complete routine L5178Y/TK assay consists of two individual components or trials. In one trial, serial dilutions of the test article plus appropriate controls are tested in the presence of 5% (v/v) liver S9 fraction prepared from normal (i.e., untreated)

CD male rats (200–260 g). In the second trial, serial dilutions of the test article plus appropriate controls are tested in the absence of S9. Dose levels of the test article are based upon preliminary cytotoxicity testing results.

B. A chemical unknown (test article) should be tested to toxic levels or to the limits of solubility in a suitable solvent vehicle (usually Pierce silylation grade DMSO), whichever can be obtained first.

C. A special test is defined as the testing of a chemical unknown under some other exogenous metabolic activation condition (other S9 source, concentration, etc.) or under conditions not described in parts IA and IB, as, for example, where the test article is added directly to the test medium at its desired final concentration.

D. Based on the quality and interpretation of experimental results, a completed trial will yield one of five types of data: (1) unacceptable data, (2) acceptable, conclusive, positive data, (3) acceptable, conclusive, negative data, (4) acceptable, conditionally negative data, or (5) acceptable, inconclusive data (no conclusion reached). These terms are defined in parts II–VI, which follow.

II. *Acceptability Criteria*

An individual trial would be acceptable for further analysis if both of the following criteria are satisfied. Acceptable means characterized by sufficient quality so as to convey useful information concerning *in vitro* mutagenic potential. Data are conclusive if either the criteria for a positive mutagenic response (part III) or criteria for a negative mutagenic response (part IV) are fulfilled for an individual trial.

A. A positive control was included in the trial and produced a positive mutagenic response. No limitations of cytotoxicity are placed on the positive control; the concentrations specified are considered generally but not excessively toxic and always mutagenic based on historical evidence. The specified positive controls are either EMS (ethyl methanesulfonate) (in the absence of S9) or 3MCA (3-methylcholanthrene) (in the presence of S9). For clarification reasons or test improvement purposes, other positive controls or other concentrations of these positive controls might also be used. Expected or "normal" mutagenic responses for the positive controls are to be empirically defined.

B. Two concomitant solvent controls were included in the assay and produced acceptable results. As used here and subsequently, concomitant means prepared at the same time and with the same activation conditions as the test article of interest. Complete results from at least two solvent controls are necessary in order to estimate experimental variation within the biological system in the absence of a mutagenic stimulus. Empirically derived acceptable responses for solvent controls in our laboratory are mutant frequencies of $(12-53) \times 10^{-6}$ survivors[9] and absolute cloning efficiencies of >70% but <120%.

III. *Criteria for a Positive Mutagenic Response*

A positive mutagenic response in the L5178Y/TK assay can be defined as follows: A positive mutagenic response is indicated when three or more data points (three concentrations of the test article that ultimately produced mutagenicity and cytotoxicity data) are obtained with all of the following characteristics:

A. Total relative growth, a measure of cell survival previously defined, between 20% and 80%.

B. Calculated mutant frequencies greater than the mutant frequencies for concomitant solvent controls. The observation of a positive correlation between test article concentration and increasing mutant frequencies over some interval within a range of test article concentrations producing 20–80% total relative growth should be possible.

C. An absolute mean mutant yield (average number of TFT^r colonies/plate) for all dose levels of the test article meeting criterion IIIA significantly ($p < 0.05$) greater than the corresponding mean mutant yield for concomitant solvent controls.

D. Calculated mutant frequencies for experimental data that meet criterion IIIA must also exceed the range of acceptable responses for solvent control mutant frequencies [$(12-53) \times 10^{-6}$ survivors] defined in IIB.

IV. *Criteria for a Negative Mutagenic Response*

A. A negative response for a test article is indicated by the following criteria:

1. The absence of a positive mutagenic response as defined in part III, provided that condition IIIA is met.

2. If a chemical unknown is tested to the limit of solubility (part IB) or to an arbitrary upper limit (VA) and condition IIIA cannot be met, yet some evidence of cytotoxicity was observed, the chemical should be considered a nonmutagen under routine test conditions.

V. *Criteria for a Conditional Negative Response*
    A. An assay should be considered conditionally negative under the following condition: The testing of a test article to a predetermined, fixed upper limit without obtaining evidence of appreciable cytotoxicity or compound insolubility.
    B. A conditional negative result cannot be further resolved in this system under routine conditions. The declaration of a conditional negative result means that the test article produced no evidence of untoward biological activity as measured by toxicity or mutagenicity within these defined test limits and that further testing appears pointless.

VI. *Inconclusive Data*
When data meet the acceptability criteria II but not response criteria III–V, they are inconclusive, i.e., testing is incomplete and should be repeated, perhaps with different test article concentrations.

In summary, aberrant or equivocal results are occasionally encountered in any assay system. Because the independent verification of test results is rarely possible in a typical test application, careful experimental design and adherence to a rigid set of acceptability and response criteria can improve the quality and meaning of test-derived conclusions. The specific content of these criteria may vary among testing labs, but their purpose is universal.

8. The Testing of Chemicals of Commercial Interest

The L5178Y/TK gene mutation assay has been used to test a number of drugs or therapeutic agents, including cyclophosphamide, hycanthone methanesulfonate, lucanthone hydrochloride, methotrexate, mitomycin C, Myleran, and procarbazine.[14,30] This assay is extremely useful for the testing of certain antibiotics or other antimicrobial agents but is not useful for the testing of halogenated pyrimidine antitumor or cytotoxic agents, i.e., substances that may act as a substrate for thymidine kinase.

A representative application of the L5178Y/TK assay for the mutagenic classification of two drugs is found in Figure 5. In a preliminary report, streptomycin sulfate was recently found weakly positive in *S. typhimurium*.[71] In sensitive bacteria, streptomycin combines with ribosome 30S subunits[34] and causes a misreading of ribosomes, interfering with the accurate translation of the RNA code into pro-

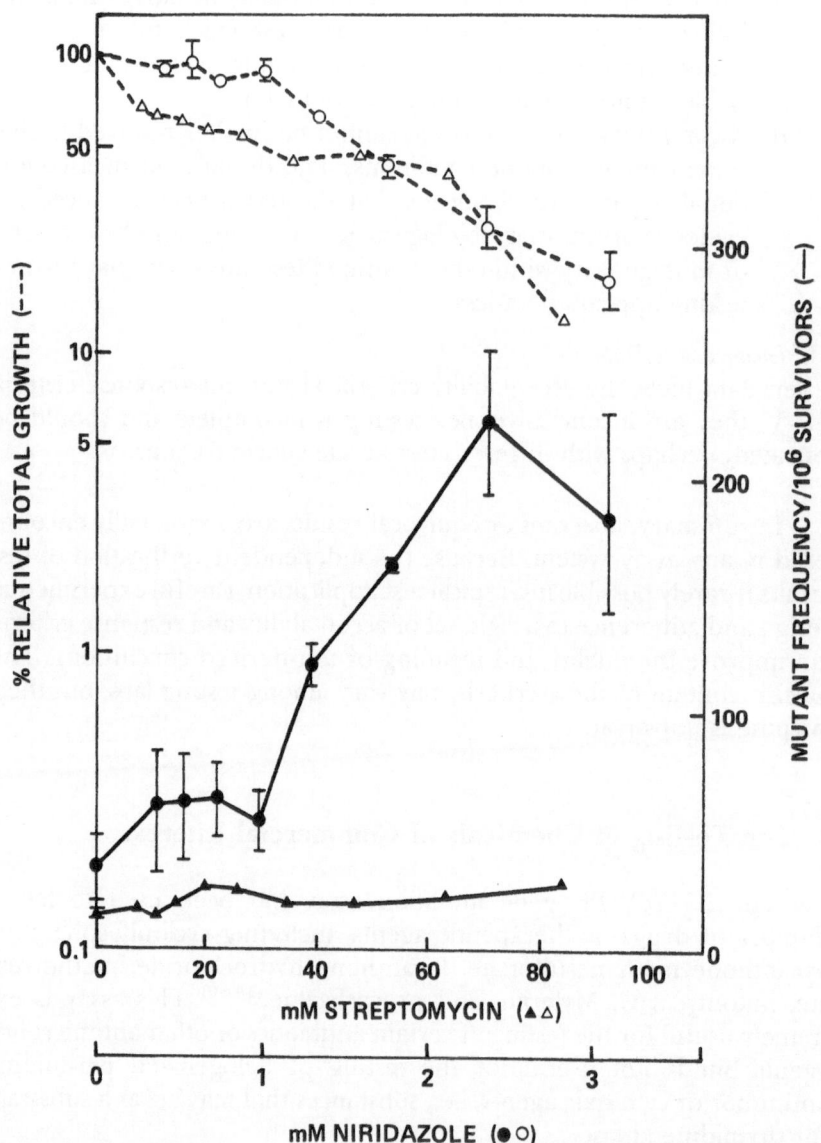

FIGURE 5. Cytotoxicity and mutagenicity of two drugs in the L5178Y/TK assay. Test conditions were: 3-hr treatment, 48-hr expression. Relative total growth and large-colony TFT^r mutant frequencies are shown. The combined results from two independent trials performed 4 years apart are shown for niridazole.

tein,[35] perhaps correcting the *his*⁻ mutation. Niridazole, a mouse carcinogen,[81] is a strong mutagen in *S. typhimurium* strain TA 100 but was negative at the ouabain and 6-thioguanine loci of Chinese hamster V79 cells due to differences in nitroreductase activity between the two cell types.[36] Niridazole produces base-pair substitution mutations but not multilocus deletion mutants in *Neurospora*.[74] In this single mammalian cell assay, our results in Figure 5 indicate a negative mutagenic response for streptomycin sulfate and a significant and highly reproducible positive mutagenic response to niridazole when both were tested in the absence of any S9 fraction at equitoxic doses.

9. Concluding Remarks

The L5178Y/TK gene mutation assay is a powerful tool for investigating the mutagenic potential of chemicals. Careful attention to serum and culture medium quality and general test conditions is essential for obtaining a low background mutant count and meaningful data. A variety of exogenous metabolic activation procedures, including liver microsomal enzyme preparations from untreated mice, rats, or hamsters and cocultivation with primary rat or hamster cells or exposure of L5178Y cells to the urine from carcinogen-dosed mice, can extend the application of the assay to promutagens. Much effort has been directed toward determining whether mutant colony size is genetically determined or has a bearing upon the mutagenic classification of the test agent. It is yet to be determined what role, if any, transient gene inactivation or altered gene expression may have in small colony etiology. There is little doubt that large trifluorothymidine-resistant clones arising after mutagen treatment of L5178Y cells represent an authentic TK⁻ phenotype, but spontaneously arising variants, particularly small ones, are another matter and require further study. Acceptability and response criteria have been developed in our laboratory and elsewhere to aid in test interpretation and lessen dependence upon subjective judgements. This assay is well suited for the screening of chemicals for mutagenic potential without the relevance problems inherent in microbial models.

ACKNOWLEDGMENTS

I wish to thank Gail Turner and Simone Paillet for their technical assistance and Janet O'Connor for her help in preparing this manuscript.

10. References

1. R. Adler and B. R. McAuslan, Expression of thymidine kinase variants is a function of the replicative state of cells, *Cell 2*, 113–117 (1974).
2. D. E. Amacher and S. C. Paillet, Induction of trifluorothymidine-resistant mutants by metal ions in L5178Y/TK$^{+/-}$ cells, *Mutat. Res. 78*, 279–288 (1980).
3. D. E. Amacher and S. C. Paillet, Trifluorothymidine-resistance and colony size in L5178Y/TK$^{+/-}$ cells treated with methyl methanesulfonate, *J. Cell Physiol. 106*, 349–360 (1981).
4. D. E. Amacher and S. C. Paillet, Hamster hepatocyte-mediated activation of procarcinogens to mutagens in the L5178Y/TK mutation assay, *Mutat. Res. 106*, 305–316 (1982).
5. D. E. Amacher and S. C. Paillet, The activation of procarcinogens to mutagens by cultured rat hepatocytes in the L5178Y/TK mutation assay, *Mutat. Res. 113*, 77–78 (1983).
6. D. E. Amacher and G. N. Turner, Promutagen activation by rodent-liver postmitochondrial fractions in the L5178Y/TK cell mutation assay, *Mutat. Res. 74*, 485–501 (1980).
7. D. E. Amacher and G. N. Turner, The effect of liver postmitochondrial fraction concentration from Aroclor 1254-treated rats on promutagen activation in L5178Y cells, *Mutat. Res. 97*, 131–137 (1982).
8. D. E. Amacher and G. N. Turner, Mutagenic evaluation of carcinogens and noncarcinogens in the L5178Y/TK assay utilizing postmitochondrial fractions (S9) from normal rat liver, *Mutat. Res. 97*, 49–65 (1982).
9. D. E. Amacher, and G. N. Turner, The L5178Y/TK gene mutation assay for the detection of chemical mutagens, *Ann. N.Y. Acad. Sci., 407*, 239–252 (1983).
10. D. E. Amacher, S. C. Paillet, and J. A. Elliott, The metabolism of N-acetyl-2-aminofluorene to a mutagen in L5178Y/TK$^{+/-}$ mouse lymphoma cells, *Mutat. Res. 89*, 311–320 (1981).
11. D. E. Amacher, S. Paillet, and V. A. Ray, Point mutations at the thymidine kinase locus in L5178Y mouse lymphoma cells. I. Application to genetic toxicological testing, *Mutat. Res. 64*, 391–406 (1979).
12. D. E. Amacher, S. C. Paillet, and G. N. Turner, Utility of the mouse lymphoma L5178Y/TK assay for the detection of chemical mutagens, in: *Banbury Report 2, Mammalian Cell Mutagenesis: The Maturation of Test Systems* (A. W. Hsie *et al.*, eds), pp. 277–293, Cold Spring Harbor Laboratory (1979).
13. D. E. Amacher, S. C. Paillet, and I. Zelljadt, Metabolic activation of 3-methylcholanthrene and benzo(a)pyrene to mutagens in the L5178Y/TK assay by cultured embryonic rodent cells, *Environ. Mutagen. 4*, 109–119 (1982).
14. D. E. Amacher, G. N. Turner, and J. H. Ellis, Detection of mammalian cell mutagens in urine from carcinogen-dosed mice, *Mutat. Res. 90*, 79–90 (1981).
15. D. E. Amacher, S. C. Paillet, G. N. Turner, V. A. Ray, and D. S. Salsburg, Point mutations at the thymidine kinase locus in L5178Y mouse lymphoma cells. II. Test validation and interpretation, *Mutat. Res. 72*, 447–474 (1980).
16. B. N. Ames, W. E. Durston, E. Yamaski, and F. D. Lee, Carcinogens are mutagens: A simple test system combining liver homogenates for activation and bacteria for detection, *Proc. Natl. Acad. Sci. USA 70*, 2281–2285 (1973).
17. B. N. Ames, J. McCann, and E. Yamaski, Methods for detecting carcinogens and mutagens with *Salmonella*/mammalian-microsome mutagenicity test, *Mutat. Res. 31*, 347–364 (1975).

18. W. K. Baker, Position-effect variegation, *Adv. Gen. 14*, 133–169 (1968).

19. D. Brusick, *Principles of Genetic Toxicology*, Plenum Press, New York (1980).

20. M. Caboche, Comparison of the frequencies of spontaneous and chemically-induced 5-bromodeoxy-uridine-resistance mutations in wild-type and revertant BHK-21/13 cells, *Genetics 77*, 309–322 (1974).

21. D. Clive, Mutagenicity of thioxanthenes (hycanthone, lucanthone and four indazole derivatives) at the TK locus in cultured mammalian cells, *Mutat. Res. 26*, 307–318 (1974).

22. D. Clive and G. Hajian, Letter to the editor, *Mutat. Res. 89*, 250–253 (1981).

23. D. Clive and M. M. Moore-Brown, The L5178Y/TK mutagen assay system: mutant analysis, in: *Banbury Report 2, Mammalian Cell Mutagenesis: The Maturation of Test Systems* (A. W. Hsie *et al.*, eds.), pp. 421–429, Cold Spring Harbor Laboratory (1979).

24. D. Clive and J. F. S. Spector, Laboratory procedure for assessing specific locus mutations at the TK locus in cultured L5178Y mouse lymphoma cells, *Mutat. Res. 31*, 17–29 (1975).

25. D. Clive and P. Voytek, Evidence for chemically-induced structural gene mutations at the thymidine kinase locus in cultured L5178Y mouse lymphoma cells, *Mutat. Res. 44*, 269–278 (1977).

26. D. Clive, A. G. Batson, and N. T. Turner, The ability of L5178Y/TK$^+$ mouse lymphoma cells to detect single gene and viable chromosome mutations: Evaluation and relevance to mutagen and carcinogen screening, in: *The Predictive Value of Short-Term Screening Tests in Carcinogenicity* (G. M. Williams *et al.*, eds.), pp. 103–123, Elsevier/North-Holland, New York (1980).

27. D. Clive, W. G. Flamm, and M. R. Machesko, Mutagenicity of hycanthone in mammalian cells, *Mutat. Res. 14*, 262–264 (1972).

28. D. Clive, W. G. Flamm, and J. B. Patterson, Specific-locus mutational assay systems for mouse lymphoma cells, in: *Chemical Mutagens, Principles and Methods for Their Detection*, Vol. 3 (A. Hollander, ed.), pp. 79–103, Plenum Press, New York (1973).

29. D. Clive, W. G. Flamm, M. R. Machesko, and N. J. Bernheim, A mutational assay system using the thymidine kinase locus in mouse lymphoma cells, *Mutat. Res. 16*, 77–87 (1972).

30. D. Clive, K. O. Johnson, J. F. S. Spector, A. G. Batson, and M. M. M. Brown, Validation and characterization of the L5178Y/TK$^{+/-}$ mouse lymphoma mutagen assay system, *Mutat. Res. 59*, 61–108 (1979).

31. D. Clive, R. McCuen, J. F. S. Spector, C. Piper, and K. H. Mavournin, Specific gene mutations in L5178Y cells in culture: A report of the U. S. EPA Gene-Tox Program, *Mutat. Res., 115*, 225–251 (1983).

32. J. Cole and C. F. Arlett, Ethyl methanesulphonate mutagenesis with L5178Y mouse lymphoma cells: A comparison of ouabain, thioguanine and excess thymidine resistance, *Mutat. Res. 34*, 507–526 (1976).

33. J. Cole and C. F. Arlett, Methyl methanesulphonate mutagenesis in L5178Y mouse lymphoma cells, *Mutat. Res. 50*, 111–120 (1978).

34. E. C. Cox, J. R. White, and J. G. Flaks, Streptomycin action and the ribosome, *Proc. Natl. Acad. Sci. USA 51*, 703–785 (1964).

35. J. Davies, L. Gorini, and B. D. Davis, Misreading of RNA codewords induced by aminoglycoside antibiotics, *Mol. Pharmacol. 1*, 93–106 (1965).

36. J. Dayan, M. C. Crajer, and S. Deguingand, Mutagenic activity of 4 active-principle forms of pharmaceutical drugs. Comparative study of the *Salmonella typhimurium* microsome test, and the HGPRT and Na$^+$/K$^+$ ATPase systems in cultured mammalian cells, *Mutat. Res. 102*, 1–12 (1982).

37. P. B. Dent, S. K. Lia, G. Ettin, and G. B. Cleland, Characterization of an inhibitor of thymidine uptake produced by cultured human melanoma cells, *Oncology 35*, 235–241 (1978).

38. L. Diamond, F. Kruszewski, and W. Baird, Expression time for benzo(a)pyrene-induced 6-thioguanine-resistant mutations in V79 Chinese hamster cells, *Mutat. Res. 95*, 353–362 (1982).

39. G. A. Fischer, Studies on the culture of leukemic cells *in vitro*, *Ann. N. Y. Acad. Sci. 76*, 673–680 (1958).

40. G. A. Fischer, The host-mediated mammalian cell assay, *Agents Actions 3/2*, 93–98 (1973).

41. G. A. Fischer and A. S. Sartorelli, I. Development, maintenance of assay of drug resistance, *Meth. Med. Res. 10*, 247–262 (1964).

42. M. Fox and B. W. Fox, Effects of methyl methanesulfonate on the growth of P-388 cells *in vitro* and on their rate of progress through the cell cycle, *Cancer Res. 27*, 1805–1812 (1967).

43. M. Fox and M. Radacic, Adaptational origin of some purine-analogue resistant phenotypes in cultured mammalian cells, *Mutat. Res. 49*, 275–296 (1978).

44. C. N. Frantz and H. V. Malling, Bromodeoxyuridine resistance induced in mouse lymphoma cells by microsomal activation of dimethylnitrosamine, *J. Toxicol. Environ. Health 2*, 179–187 (1976).

45. F. M. Griffin, G. Ashland, and R. L. Capizzi, Kinetics of phototoxicity of Fischer's medium for L5178Y leukemic cells, *Cancer Res. 41*, 2241–2248 (1981).

46. S. M. Haag and I. G. Sipes, Differential effects of acetone or Aroclor 1254 pretreatment on the microsomal activation of dimethylnitrosamine to a mutagen, *Mutat. Res. 74*, 431–438 (1980).

47. C. Heidelberger, Chemical carcinogenesis, chemotherapy: Cancer's continuing core challenges—G. H. A. Clower memorial lecture, *Cancer Res. 30*, 1549–1569 (1970).

48. T. Hirakawa, N. Nemoto, M. Yamada, and S. Takayama, Metabolism of benzo(a)pyrene and the related enzyme activities in hamster embryo cells, *Chem. Biol. Interact. 25*, 189–195 (1979).

49. J. Hozier, J. Sawyer, M. Moore, B. Howard, and D. Clive, Cytogenetic analysis of the L5178Y/TK$^{+/-}$ → TK$^{-/-}$ mouse lymphoma mutagenesis assay system, *Mutat. Res. 84*, 169–181 (1981).

50. E. Huberman and C. A. Jones, The use of liver cell cultures in mutagenesis studies, *Ann. N. Y. Acad. Sci. 349*, 264–272 (1980).

51. E. D. Jacobson, K. Krell, M. J. Dempsey, M. H. Lugo, O. Ellingson, and C. W. Hench II, Toxicity and mutagenicity of radiation from fluorescent lamps and a sunlamp in L5178Y mouse lymphoma cells, *Mutat. Res. 51*, 61–75 (1978).

52. E. M. Jensen, R. J. LaPolla, P. E. Kirby, and S. R. Haworth, *In vitro* studies of chemical mutagens and carcinogens. Stability studies in cell culture medium, *J. Natl. Cancer Inst. 59*, 941–944 (1977).

53. D. Henssen and C. Ramel, Relationship between chemical damage of DNA and mutations in mammalian cells. I. Dose–response curves for the induction of 6-thioguanine-resistant mutants by low doses of monofunctional alkylating agents, x-rays and UV radiation in V79 Chinese hamster cells, *Mutat. Res. 73*, 339–347 (1980).

54. M. M. Jotz and A. D. Mitchell, Effects of 20 coded chemicals on the forward mutation frequency at the thymidine kinase locus in L5178Y mouse lymphoma cells, in: *Progress in Mutation Research, Evaluation of Short-Term Tests for Carcinogenesis*, (F. J. deSerres and J. A. Ashby, eds.) Vol. 1 pp. 580–593, Elsevier/North-Holland, New York (1981).

55. R. W. Kapp, Jr. and B. E. Eventoff, Mutagenicity of dimethyl sulfoxide (DMSO): *In vivo* cytogenetics study in the rat, *Teratogen. Carcinogen. Mutagen. 1*, 141–145 (1980).

56. S. Kit, Thymidine kinase, DNA synthesis and cancer, *Mol. Cell. Biochem. 11*, 161–182 (1976).
57. R. E. Kouri, R. Kiefer, and E. M. Zimmerman, Hydrocarbon-metabolizing activity of various mammalian cells in culture, *In Vitro 10*, 18–25 (1974).
58. T. Kuroki, C. Malaveille, C. Drevon, C. Piccoli, M. Macleod, and J. K. Selkirk, Critical importance of microsome concentration in mutagensis assay with V79 Chinese hamster cells, *Mutat. Res. 63*, 259–272 (1979).
59. D. Y. Lai, S. C. Myers, Y. Woo, E. J. Greene, M. A. Friedman, M. F. Argus, and J. C. Areos, Role of dimethylnitrosamine-demethylase in the metabolic activation of dimethylnitrosamine, *Chem. Biol. Interact. 28*, 107–126 (1979).
60. R. Langenbach, S. Nesrow, A. Tompa, R. Gingell, and C. Kuszynski, Lung and liver cell-mediated mutagenesis systems: Specificities in the activation of chemical carcinogens, *Carcinogenesis 2*, 851–858 (1981).
61. J. W. Littlefield, The periodic synthesis of thymidine kinase in mouse fibroblasts, *Biochim. Biophys. Acta 114*, 398–403 (1966).
62. D. Matheson and B. Creasy, Use of the L5178Y ($TK^{+/-}$) mouse lymphoma cell line coupled with an *in vitro* microsomal enzyme activation system to study chemical promutagens, *Mutat. Res. 38*, 400–401 (1976).
63. S. McMillan and M. Fox, Failure of caffeine to influence induced mutation frequencies and the independence of cell killing and mutation induction in V79 Chinese hamster cells, *Mutat. Res. 60*, 91–107 (1979).
64. M. L. Meltz and J. T. MacGregor, Activity of the plant flavanol quercetin in the mouse lymphoma L5178Y $TK^{+/-}$ mutation, DNA single-strand break, and Balb/c 3T3 chemical transformation, *Mutat. Res. 88*, 317–324 (1981).
65. G. E. Moore, R. E. Gerner, and H. A. Franklin, Culture of normal human leukocytes, *J. Am. Med. Assoc. 199*, 519–524 (1967).
66. M. M. Moore-Brown and D. Clive, The L5178Y/$TK^{+/-}$ mutagen assay system: *In situ* results, in: *Banbury Report 2, Mammalian Cell Mutagenesis: The Maturation of Test Systems* (A. W. Hsie, *et al.*, eds.), pp. 71–88, Cold Spring Harbor Laboratory (1979).
67. M. M. Moore-Brown, D. Clive, B. E. Howard, A. G. Batson, and K. O. Johnson, The utilization of trifluorothymidine (TFT) to select for thymidine kinase-deficient ($TK^{-/-}$) mutants from L5178Y/$TK^{+/-}$ mouse lymphoma cells, *Mutat. Res. 85*, 363–378 (1981).
68. M. M. Moore-Brown and B. E. Howard, Quantitation of small colony trifluorothymidine-resistant mutants of L5178Y/$TK^{+/-}$ mouse lymphoma cells in RPMI-1640 medium, *Mutat. Res. 104*, 287–294 (1982).
69. M. Nagao and T. Sugimura, Molecular biology of the carcinogen, 4-nitroquinoline-1-oxide, in: *Advances in Cancer Research* (G. Klein *et al.*, eds.), Vol. 23, pp. 131–169, Academic Press, New York (1976).
70. H. Nottebrock and R. Then, Thymidine concentrations in serum and urine of different animal species and man, *Biochem. Pharmacol. 26*, 2175–2179 (1977).
71. NTP Technical Bulletin #6, p. 6, National Institute of Environmental Health Sciences (January 1982).
72. NTP Technical Bulletin #8, pp. 6–7, National Institute of Environmental Health Sciences (July 1982).
73. F. Oesch, D. Raphael, H. Schwind, and H. R. Glatt, Species differences in activating and inactivating enzymes related to the control of mutagenic metabolites, *Arch. Toxicol. 39*, 97–108 (1977).
74. T. Ong, B. Slade, and F. J. de Serres, Mutagenicity and mutagenic specificity of metronidazole and niridazole in *Neurospora crassa*, *J. Environ. Pathol. Toxicol. 2*, 1109–1118 (1979).

75. K. A. Palmer, Characteristics of thymidine kinse in L5178Y mouse lymphoma cell lines TK$^{+/+}$ P4, TK$^{+/-}$ p4.3.4, TK$^{+/-}$ 3.7.2, and TK$^{-/-}$ p4.3. *Diss. Abstr. Int. B 41*, 4382-B (1981).

76. A. R. Peterson, DNA synthesis, mutagenesis, DNA damage and cytotoxicity in cultured mammalian cells treated with alkylating agents, *Cancer Res. 40*, 682–683 (1980).

77. P. G. W. Plagemann and D. P. Richey, Transport of nucleosides, nucleic acid bases, choline and glucose by animal cells in culture, *Biochim. Biophys. Acta 344*, 263–305 (1974).

78. E. H. Postel and A. J. Levine, Studies on the regulation of deoxypyrimidine kinases in normal, SV40-transformed and SV40- and adenovirus-infected mouse cells in culture, *Virology 63*, 404–420 (1975).

79. P. Reyes and C. Heidelberger, Fluorinated pyrimidines XXV. The inhibition of thymidylate synthetase from Ehrlich ascites carcinoma cells by pyrimidine analogs, *Biochem. Biophys. Acta 103*, 177–179 (1965).

80. A. M. Rogers and K. C. Back, Comparative mutagenicity of hydrazine and 3 methylated derivatives in L5178Y mouse lymphoma cells, *Mutat. Res. 89*, 321–328 (1981).

81. J. V. Soderman, *CRC Handbook of Identified Carcinogens and Noncarcinogens:Carcinogenicity-Mutagenicity Database*, CRC Press, Boca Raton, Florida (1982).

82. W. Suter, J. Brennand, S. McMillan, and M. Fox, Relative mutagenicity of antineoplastic drugs an other alkylating agents in V79 Chinese hamster cells, independence of cytotoxic and mutagenic responses, *Mutat. Res. 73*, 171–181 (1980).

83. A. D. Tates and E. Kriek, Induction of chromosomal aberrations and sister-chromatid exchanges in chinese hamster cells *in vitro* by some proximate and ultimate carcinogenic arylamide derivatives, *Mutat. Res. 88*, 397–410 (1981).

84. W. G. Thilly, J. G. DeLuca, E. E. Furth, H. Hoppe, D. A. Kaden, J. J. Krolewski, H. L. Liber, T. R. Skopek, S. A. Slapikoff, R. J. Tizard, and B. W. Penman, Gene-locus mutation assays in diploid human lymphoblast lines. in: *Chemical Mutagens, Principles and Methods for Their Detection* Vol. 6 (F. J. de Serres and A. Hollaender, eds.), pp. 331–364, Plenum, Press, New York (1980).

85. S. Thornton and M. Hite, Inhibitory effect of high S-9 concentrations on activation of hydrocarbons in the L5178Y mouse lymphoma assay, 10th Annual Meeting of the Environmental Mutagen Society, New Orleans, Louisiana (1979).

86. A. Toliver, E. H. Simon, and P. T. Gilham, On the mechanism of 5-bromouracil inhibition of DNA synthesis and cell division, *Exp. Cell Res. 53*, 506–518 (1968).

87. J. Unger and J. B. Guttenplan, Kinetics of benzo(a)pyrene-induced mutagensis in a highly sensitive *Salmonella*/microsome assay, *Mutat. Res. 77*, 221–228 (1980).

88. A. A. Van Zeeland and J. W. I. M. Simons, The effect of calf serum on the toxicity of 8-azaguanine, *Mutat. Res. 27*, 135–138 (1975).

89. F. J. Wiebel, S. Brown, H. L. Waters, and J. K. Selkirk, Activation of xenobiotics by monooxygenases: Cultures of mammalian cells as an analytical tool, *Arch. Toxicol. 39*, 133–148 (1977).

CHAPTER 7

Induction of Bacteriophage Lambda by DNA-Interacting Chemicals

Rosalie K. Elespuru

1. Mechanism of Prophage Induction

Induction of resident prophages in coliform bacteria occurs as an indirect consequence of damage to the DNA of the host bacteria. The change in bacteriophage gene expression from the repressed to the active state (induction) is one of the manifestations of the "SOS response," among such others as mutagenesis, increased DNA repair activity, filamentation, and suppression of respiration.[33,50,53,64,71] The SOS functions have in common the property of being controlled at the level of transcription by repressors that bind to DNA and prevent its expression. Prophages are regulated by specific repressors encoded by phage genes, while many other SOS functions are repressed by a single repressor protein called LexA. The operator regions adjacent to the genes encoding these SOS functions contain sequences to which LexA protein,[10,58] or phage repressors such as lambda, bind.

DNA damage or other conditions (e.g., thymine starvation) inhibiting the movement of the replication fork are postulated to generate an "inducing signal," leading to the synthesis and activation of a protein,

Rosalie K. Elespuru • Fermentation Program NCI-Frederick Cancer Research Facility, Frederick, Maryland 21701. *Present address:* LBI-Basic Research Program, NCI-Frederick Cancer Research Facility, Frederick, Maryland 21701.

RecA, which mediates the expression of the SOS response by cleaving the repressor proteins.[37,50,53–55,69,72] The RecA protein possesses several distinct biochemical activities, including proteolytic and DNA-binding activities.[12,37,54,55,69,70] Binding of RecA protein to single-stranded regions of DNA in the presence of ATP is required for the acquisition of proteolytic activity.[13] Single-stranded regions and gaps in DNA are generated as a result of DNA repair activity on damaged templates, or may ensue spontaneously following some types of DNA-chemical interactions (e.g., from compounds such as bleomycin).[38,64] Certain types of bacteriophages, then, have the capability of responding to the cascade of events following potentially lethal insults to the host genome. The resultant expression of phage genes leads to the production of multiple copies of the phage, lysis of the host cell, and distribution of phage progeny to the environment for infection of new hosts. A summary of the evidence supporting the current view of the induction process has been published recently by Roberts and Devoret.[53]

2. Historical Development of Prophage Induction Assays

Prophage induction assays have been popular intermittently since the suggestion in the 1950s by Lwoff[39] that they would be suitable vehicles for detection of carcinogenic and carcinostatic compounds. The early data on prophage-inducing activity of compounds classified as mutagens, carcinogens, and teratogens were summarized by Heinemann.[28] Although little serious use of them was made for the detection of chemical carcinogens,[22,41] prophage induction assays were used routinely in many parts of the world for antitumor agent screening.[21,23,28] Following the development of improved screening methodologies for chemical carcinogens, namely, the addition of a metabolic activating mixture and the use of sensitive bacterial mutants, prophage induction assays have witnessed a renaissance. A number of lambda prophage induction assays have been introduced, many of which have utilized the principles of the Ames *Salmonella* mutagenesis assay,[2] along with specific contributions from lambda phage and *Escherichia coli* molecular genetics. The prototype assay of this genre was the Inductest designed in the laboratory of Raymond Devoret. For this assay, a lambda lysogen (*E. coli* with an integrated lambda prophage) containing mutations affecting bacterial permeability (*envA*) and DNA repair (*uvrB*) was utilized. Prophage induction was measured by a classical plaque assay (zones of lysis on a sensitive indicator lawn of bacteria representing individual inducing events).[44] Chemical carcinogens requiring metabolic activation, such as aflatoxin B_1 and benzo(a)pyrene, were detected for the first time as phage inducers in the presence of rat liver S9 metabolizing mixtures.[26,44]

Because prophage induction is a physiological response to DNA damage occurring in a majority of the cell population, it may be measured biochemically. This is in contrast to mutagenesis assays, in which events occurring at a particular gene are measured in a small minority of the cells. Many of the advances in prophage induction assays involve the development of rapid, simple biochemical assays. Ho and Ho measured an intermediate protein product, the endolysin responsible for lysis of the host, as the endpoint in a prophage induction assay. They utilized a lambda lysogen carrying a mutant repressor (cI_{857}) of lambda, which changes conformation as a function of temperature, and is more easily cleaved at 37°C than is the wild-type lambda repressor. These workers detected much lower concentrations of chemical carcinogens in bacteria containing the mutant repressor.[31]

Other assays of prophage induction developed more recently have taken advantage of genetic constructions in which lambda DNA is fused to genes coding for easily measurable products. In these cases, the derepression of phage genes results in the transcription of the adjacent DNA as well.

In 1960 Yarmolinsky and Wiesmeyer[73] reported a small level of "escape synthesis" of RNA transcribed from the *gal* operon immediately adjacent to lambda, following derepression of lambda genes. Several workers have constructed strains in which adjacent *gal* genes are regulated efficiently by a lambda promoter.

Radman and co-workers devised one such lambda lysogen for use in an assay called the "Zorotest."[15] In the strain developed for this assay,[67] the autoregulatory circuit of lambda that results in new synthesis of repressor and reestablishment of the repressed state was eliminated. The repressor-negative state is expressed stably as a gal^+ phenotype, detected as colored colonies on appropriate indicator plates. Because of the heritable nature of single inducing events, the prediction was made that this assay would have great sensitivity.[67] At the same time, however, the traditional advantage of speed of prophage induction plaque assays over mutant colony assays was lost.

Smith and Oishi[60,61] utilized a phage–*trp* gene fusion in which the kinetics of derepression of phi 80 (a close relative of lambda) was monitored biochemically via the appearance of anthranilate synthase, encoded by the cotranscribed *trp* DNA. Fusion phages such as this generally cannot integrate into host DNA to form lysogens; hence, they must be infected into an appropriate lysogenic host for each assay. This problem was overcome by Levine *et al.*,[35] who designed a biochemical assay for lambda induction utilizing a product encoded by adjacent *gal* genes. The product, galactokinase, was measured following addition of a radioactive substrate.[35] The lysogen constructed for this assay contains a large deletion in lambda DNA, eliminating several functions that terminate induction, including one for cell lysis and one regulating the transcription of the region containing the *gal* genes. The anthranilate synthase assay and the galactokinase assay of prophage induction were not practical for routine screening, for two reasons: the lack of mutations for sensitivity in the parent lysogens, and the requirement for uncommon or expensive substrates.

Elespuru and Yarmolinsky[20] utilized a phage constructed by Reznikoff and co-workers[4] in which lambda was fused to a portion of the *trp–lac* genes of *E. coli*. Derepression of this prophage led to the synthesis of the product of the *lacZ* gene, β-galactosidase, an enzyme measured easily and inexpensively using a variety of substrates. The use of *lacZ* fusions has now become commonplace for study of gene products without easily measurable phenotypes. Several genetic manipulations were required for the generation of a screening strain utilizing the Reznikoff phage, including its integration into *E. coli* to form a stable lysogen. A mutation in lambda (P_rt11) eliminated the expression of genes encoding lysis and autoregulatory functions. Mutations in the host (*envA*, $\Delta uvrB$) increased the strain's sensitivity to chemicals.[20]

Another assay, recently described by Hofnung and co-workers,[49] employs a *lacZ–sfiA* fusion, taking advantage of an SOS function, that of filamentation, repressed by LexA protein. The reported increase in sensitivity achieved with this strain is attributed to the greater rate of cleavage of the *lexA* repressor relative to that of the lambda repressor by RecA protease. This strain also has mutations affecting permeability and repair of DNA damage similar to those of the Ames *Salmonella* strains.

Other LexA-repressed genes have been fused to *lacZ*, including *recA* by George Weinstock (unpublished data) and the *din* (damage-inducible) genes by Kenyon and Walker.[33] The *din* genes were detected in bacteria able to synthesize β-galactosidase in response to an SOS-inducing dose of ultraviolet light.[33] The latter strains were constructed using the Mu d(*lac*) phage of Casadaban and Cohen,[11] which inserts the *lacZ* gene at random in the *E. coli* chromosome. Although these strains may offer potential for screening, they have not undergone further development for this purpose.

In addition to the assays with new bacterial strains that take advantage of current knowledge and genetic techniques, some of the classical lysogens yielding viable phage particles are being handled in new ways. One example includes the work of Rossman et al.,[57] using a lambda lysogen of E. coli WP2. The bacteria are incubated with chemicals at low cell density in microtiter plates for extended incubation periods. The bacteria are thus exposed to chemicals over many cell generations prior to sampling for phage counts. They report greatly enhanced sensitivity using this method for carcinogens and cocarcinogens not usually detected in bacterial assays, including metals.[57]

The interesting work of Wheeler et al.[71] has dealt with the inducible prophages in the widely used histidine-requiring Salmonella strains of Ames. The induction of a prophage in a Salmonella bacterium precludes its detection as a mutant colony, because of phage-mediated cell lysis. Information on the induction properties of these phages is therefore of interest to workers performing mutagenesis assays, as well as to those interested in prophage induction. As a screening assay, however, the extent of induction of these prophages is probably not sufficient to make it competitive with the other assays described.

3. Chemical Inducers of Prophage

Little is known about the specific types of chemical–DNA interactions leading to induction of the "SOS response." Categories of chemical inducers and a rough classification by mode of action are depicted in Figure 1. As is shown, a diverse set of compounds acting in quite different ways are known as inducers of bacteriophage lambda. The common feature has been considered to be an effect on DNA replication, which accounts for induction by compounds that do not interact directly with DNA. Many chemical carcinogens (such as the aromatic amines, the polycyclic aromatic hydrocarbons, and fungal metabolites such as aflatoxin B_1) add the whole molecule to DNA, generally at the C_8 or N_2 position of guanine.[29] This results in the presence of a "bulky lesion," which could presumably interfere with the passage of the replication fork. The work of Grunberger and Weinstein[27] has indicated that the major addition products of acetoxy acetylaminofluorene (acetoxy-AAF) and polycyclic hydrocarbons do result in conformational changes in DNA structure. In the case of acetoxy-AAF, the addition product to guanine completely disrupts the coding capacity of this and surrounding nucleotides, resulting in RNA transcripts with premature terminations. However, less is known about the consequences of adduct formation on DNA synthesis. Benzo(a)pyrene diol epoxides, residing in the minor groove, have a completely differrent effect on DNA

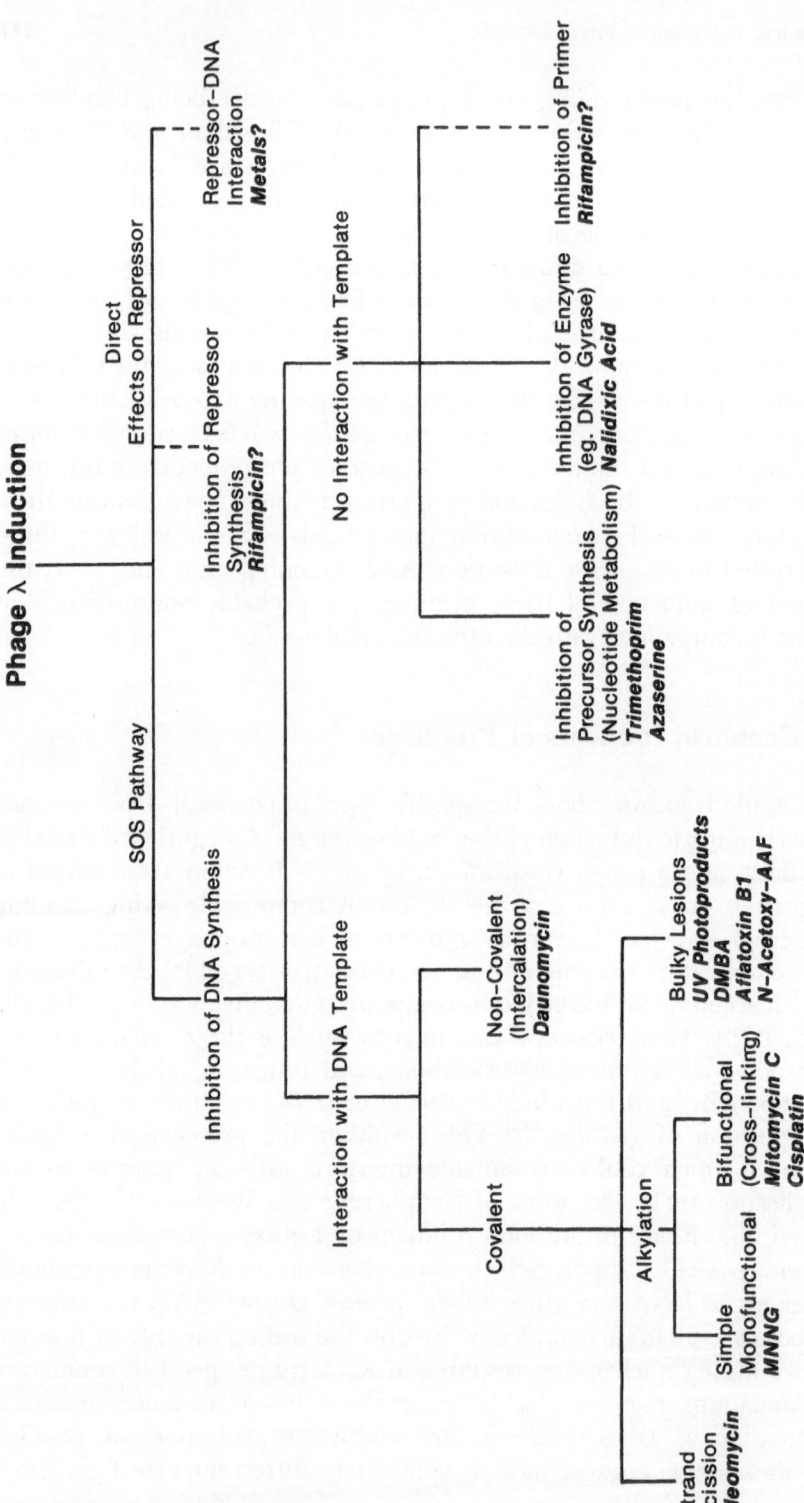

FIGURE 1. Pathways for induction of bacteriophage lambda by diverse chemicals.

conformation. Direct evidence that the presence of "bulky lesions" in the DNA triggers the SOS response is lacking.

Intercalating agents exhibit a wide spectrum of inducing capacity, from no induction (ethidium bromide, aminoacridines) to moderate (daunorubicin) to excellent induction (ICR 170).[19] Because of these differences, it seems clear that intercalation per se is not sufficient for induction; rather, specific types of intercalation or other interactions are necessary. The more chemically complex intercalating agents are the better inducers. Simple alkylating agents, such as nitrosoguanidine, do not cause conformational changes in DNA. The major adduct, N_7-alkyl guanine, is not considered to have major biological consequences, although alkylated DNA is subject to a greater rate of depurination.[29] Lesions considered to be important for mutagenesis and carcinogenesis, alkylations at the O_6 position of guanine, are known to result in the incorporation of incorrect bases. However, no major effect on DNA synthesis would be expected from this lesion. Chemicals that create DNA crosslinks and those that cause strand scissions, such as the mitomycins and bleomycins, respectively, have obvious deleterious effects on DNA synthesis and are excellent inducers.

Induction by metal compounds and ansamycins such as rifampicin is more difficult to rationalize. Because metal ions can affect the activity of proteins, one might speculate that some of the metals might interfere directly with repressor–DNA interactions. Two possibilities for the role of rifampicin in induction are listed in Figure 1, including indirect effects on DNA synthesis via inhibition of RNA primer formation and inhibition of lambda repressor synthesis. Compounds such as these, which are postulated to act directly on the repressor, may have differential activity in inducing SOS functions controlled by different repressors.

4. Prophage Induction versus Mutagenesis

Induction of SOS functions results in mutagenesis via error-prone repair on damaged templates.[50,72] However, a portion of mutations arise by SOS-independent mechanisms.[9,16,30,51] These occur following interactions that directly affect the fidelity of replication; therefore, some divergence in the type of compound detected by the two endpoints, mutagenesis and prophage induction, is expected. Base analogs such as 5-bromodeoxyuridine and 2-aminopurine incorporated into DNA are misread by DNA polymerase, as are the lesions at critical hydrogen bonding positions in DNA, such as alkylations at the O_6 position of guanine resulting from treatment with alkylating agents. Base analogs are mutagenic but do not induce prophage.

On the other hand, some prophage inducers, including bleomycin and streptonigrin, induce the prophage but are not mutagenic in the standard *his* strains of *Salmonella*.[7] Induction at another locus in *Salmonella* was achieved, but the induction levels were only two or three times background.[48] Mitomycin C and cisplatin are not mutagenic in a *uvrB* background. However, several new *Salmonella* strains introduced by Ames and co-workers[34] contain a *his* mutation with an AT target site. Bleomycin reverts this mutation, but only when the *his* genes are placed on a multicopy plasmid. Compounds such as neocarzinostatin and nalidixic acid (D. Levin, personal communication) also are detected as mutagens in this strain, TA102. Although the authors attribute this response to a difference in interaction with plasmid (supercoiled) DNA (D. Levin, personal communication), this explanation does not account for the failure of compounds that induce the SOS response, such as bleomycin, to be detected as mutagens. It is conceivable that the resultant DNA damage (e.g., multiple strand breaks[38,64]) cannot be repaired to allow the formation of mutant colonies in these cases. These compounds may be examples of mutagens whose detection in classical mutagenesis assays is precluded by toxicity. The presence of *his* genes in multiple copies provides a large target for damage that should not interfere with survival. The report of positive results with metal complexes in a prophage induction assay (57) represents a substantial addition to the set of inducers that are nonmutagenic. Additional information on these compounds is necessary prior to an assessment of their mode of action as inducers. As suggested in Section 3, metals could be working by nongenotoxic mechanisms.

5. Phage Induction Assays in Practice

5.1. Advantages

Prophage induction assays, particularly the rapid biochemical versions, offer many potential advantages over other microbial screening methods in widespread use. Derepression of prophage is independent of growth of the host organism, a function that many toxic chemicals, nonsterile samples, S9 mixes, and complex mixtures (e.g., urine) influence either positively or negatively.[59] Assays involving β-galactosidase (*lacZ* gene product) may be accomplished well within 1 day's working time. The availability of several inexpensive substrates for this enzyme allows the assay an unusual versatility.[17–19] The spot test on agar may be used for rapid screening, bioautography (identification of bioactive zones on thin-layer chromatograms), the test of compounds requiring enzymes immobilized in agar, and detection of organisms (as colonies) producing DNA-damaging substances, e.g., antitumor agents.[19] The liquid incubation assay may be quantitated, miniaturized (see Section 7), and controlled, by variation of the expression period, to give the best compromise between sensitivity and speed (reciprocal consequences of expression time) for the problem at hand.[19]

One major advantage is the use of a single bacterial strain in which the entire genome is the target. The *E. coli* and *Bacillus subtilis* DNA repair assays and the Ames *Salmonella* assay all require the use of an ever-increasing number of strains (~6) for the detection of diverse classes of chemicals.[34,43] However, a single lambda lysogen may be used to monitor induction of the SOS response mediated by a great variety of DNA–chemical interactions. There are other practical advantages as well. Once the indicator organism has been grown, the entire phage induction assay may be performed without benefit of sterile technique. It is possible to freeze aliquots of log-phase bacteria, which may be thawed and used as an "instant assay."

5.2. Comparison with Other Assays

Important questions concerning the practical use of prophage induction assays relate to sensitivity, types of chemicals detected, and false negatives (carcinogens not detected in the assay). Prophage induction assays have not been extensively validated, but some comparisons with *Salmonella* mutagenesis and DNA repair tests have been made. It appears, however, that many of these have used a small set of test chemicals, yielding results that are not necessarily representative. A summary of the major literature on induction of prophage by chemicals is provided in Table 1. One widely cited paper is by Speck,[62] in which the Devoret Inductest did not compare favorably with the *polA/pol+* DNA repair assay or the *Salmonella* mutagenesis test. However, this work is flawed by the failure of the workers to detect mitomycin C, a classic prophage inducer, as a positive using Devoret's strain. Because of this result, the other negative prophage induction results are of questionable validity.

A paper with an entirely different assessment of an SOS-type assay is that of Hofnung and co-workers,[49] in which the results using

TABLE 1. Induction of Prophage and Other SOS Functions by DNA-Damaging Chemicals

Authors	Year	Ref.	Assay	Chemicals tested
Epstein and Saporoschetz	1968	22	Classical λ plaque	Analogues of 4-nitroquinoline-*N*-oxide
Mayer et al.	1969	41	Classical λ plaque	Twenty-two antitumor agents, carcinogens, and noncarcinogens not requiring activation
Heinemann	1971	28	Classical λ plaque; review of literature	Eighty-three antitumor agents, teratogens, and chemical carcinogens not requiring metabolic activation
Moreau et al.	1976	44	λ Inductest	Seven chemical carcinogens and noncarcinogens
Ben-Gurion	1978	6	Colicin induction	Chemical carcinogens
Speck et al.	1978	62	λ Inductest, *polA*/*pol*+ test, *Salmonella* mutagenesis	Twenty chemical carcinogens and noncarcinogens
Elespuru and Yarmolinsky	1979	20	λ–*lacZ* fusion phage, β-galactosidase assay	Seven chemical carcinogens
Anderson et al.	1980	1	λ Inductest	Sixteen anthracycline antitumor agents
Rojanapo et al.	1981	56	λ Inductest	Twenty antitumor antibiotics and their derivatives
Dambly et al.	1981	15	λ Zorotest (colony assay)	Forty-two chemical carcinogens and noncarcinogens
Thompson	1981	66	λ Plaque assay	Forty-two chemical carcinogens and noncarcinogens (International Validation Study; blind test)
Ho and Ho	1981	31	λ cI857 endolysin assay	Twenty chemical carcinogens and noncarcinogens
Wheeler et al.	1981	71	Plaque assay in *Salmonella* test strains	Aflatoxin B1
Quillardet et al.	1982	49	SOS Chromotest (*sfiA*–*lacZ* fusion; β-galactosidase assay), Inductest, *Salmonella* mutagenesis	Twenty-four chemical carcinogens and antitumor agents
Elespuru and White	1983	19	λ–*lacZ* fusion phage, β-galactosidase assay	One hundred forty-seven antitumor agents
Mamber et al.	1984	40	λ Inductest, *Salmonella* mutagenesis, *E. coli rec* assay	Forty-six chemical carcinogens and antitumor agents
Rossman et al.	1984	57	WP2 (λ) classical plaque, extended incubation in microtiter	Twenty-one metal compounds of diverse carcinogenicity
Elespuru	1984	18	λ–*lacZ* fusion phage, β-galactosidase assay	One hundred eleven N-nitroso compounds

Salmonella and the *lacZ–sfiA* fusion strain appear to have a near perfect quantitative concordance. However, in this paper the data are presented as nanomoles/assay for the chromotest and nanomoles/plate for the *Salmonella* test, a comparison that is misleading. Because of the differences in volumes of the assay mixtures, the concentration of chemicals in the chromotest is approximately 40 times that in the *Salmonella* assay. On the other hand, the incubation period for the *lacZ–sfiA* bacteria is only 2 hr versus a much longer effective reaction time for *Salmonella*. It is not possible from this analysis to compare the relative sensitivities of the two bacterial systems.

Results obtained with prophage induction assays and other bacterial assays in the blind test of a set of carcinogens and noncarcinogens may provide a more reliable comparison. Such an evaluation was made during the International Study for the Evaluation of Potential Carcinogens organized by de Serres and Ashby. The two lambda prophage induction assays included in this study produced "scores" that were very similar to those obtained with *Salmonella typhimurium* and the DNA repair assays; 50–60% of the carcinogens tested were detected. The prophage induction assays gave significantly fewer false positives (positive results with noncarcinogens) than the DNA repair assays.[3] Although these results indicate that prophage induction assays should be considered as practical screening assays, there are several reasons for caution in this regard. One is that the two prophage induction assays did not have a very high level of agreement on the detection of specific compounds (57%)[3]; a second is that the compounds chosen for test by the committee included a large proportion of non-DNA-interacting carcinogens or noncarcinogens not expected to work in bacterial mutagenesis assays. Thus, the sample of chemicals expected to work via DNA mechanisms was not very large. It is also notable that the prophage induction assays used were not specifically designed for sensitivity in testing chemicals and may not be as useful as Devoret's, Hofnung's, or Yarmolinsky's strains.

A more convincing evaluation of a prophage induction assay has been made by Mamber *et al.*[40] They assessed the activity and relative sensitivity of Devoret's screening strain[44] in comparison with several strains of *Salmonella*[42] and *E. coli* WP100 (*recA uvrA*) versus WP2 (wild type). Using a plate assay for both *Salmonella* and the lambda lysogen, Mamber *et al.* found that *Salmonella* detected approximately tenfold lower concentrations of carcinogen than the *E. coli* lambda lysogen. In addition, the prophage induction assay failed to detect aromatic amines, such as 2-acetylaminofluorene (AAF), which were positive in the other assays.[40] In our laboratory, we have also tested a diverse set of chemical carcinogens and noncarcinogens for induction of the lambda–*lacZ* fusion phage. Our results (unpublished data) agree with those of Mamber *et al.*

6. Mechanism of Prophage Induction Revisited

Results from screening a variety of chemicals as inducers of prophage should contribute to our understanding of the process involved. An assessment of these data, coupled with information on DNA–chemical interactions from the literature, should provide a basis for inferring the types of interactions involved. Moreover, such an analysis should help in the design of superior strains with particular mutations, e.g., in DNA repair functions, that increase the sensitivity of the bacteria to inducing chemicals.

Agents such as mitomycin C, bleomycin, ultraviolet (UV) light, and nalidixic acid are classically known as inducers of prophage. Other well-known compounds that are excellent inducers of prophage include aflatoxin B_1, polypeptide antitumor antibiotics (such as neocarzinostatin), 4-nitroquinoline-*N*-oxide, cisplatin, and ICR 170. It is no coincidence that most of the literature on prophage induction involves these compounds primarily. Some types of DNA-interacting compounds are less efficient inducers of prophage, including nitrosoguanidine, dimethylbenz(*a*)anthracene, daunomycin, and ICR 191. Other DNA-interacting compounds fail to induce the prophage in normal lysogens: base analogues such as 5-bromodeoxyuridine and 2-aminopurine; aromatic amines such as AAF, benzidine, and 2-naphthylamine; alkylating agents such as ethyl methanesulfonate; acridine compounds; and ethidium bromide. All of these compounds are mutagenic.

The quantitative differences in induction efficiency of different DNA-interacting compounds are worth considering. The best prophage inducers (e.g., bleomycin, UV light) are known particularly for their efficiency in causing DNA strand breaks,[24,38,45,47,64] while the noninducers[19,29] are known for other types of DNA interactions (e.g., intercalation, alkylation). In our laboratory we are seeking to determine whether there is a direct relationship between prophage induction and DNA strand breakage. Besides the chemical evidence, there is genetic and biochemical evidence relating these phenomena. Mutations that diminish the capacity of *E. coli* to repair strand breaks, such as *polA* and *lig*$_{ts}$,[25] show greater inducibility than wild-type.[8] Mutants with decreased capacity to generate strand breaks at the site of lesions, such as those with low levels of the RecBC endonuclease, show a diminished capacity for induction.[36,46] The latter mutants are inducible by some chemicals, such as bleomycin, that generate strand breaks by chemical reaction. Nalidixic acid, an excellent inducer, inhibits the strand-rejoining capacity of DNA gyrase, while novobiocin, another gyrase inhibitor with little inducing ability, affects a different function of the enzyme.[24,45] Recent information on the function of RecA protein indicates that single-stranded DNA is required for its activation to a protease, the mediator of the SOS response. A summary of the scenario leading to the generation of single-stranded DNA following a variety of treatments, including mutant conditions, is provided by Roberts and Devoret.[53]

The classical model of the inducing signal, proposed by Radman[50] and recently reiterated by Roberts and Devoret,[53] involves the idea of cessation of DNA replication, envisaged as an impediment to the passage of the replication fork, caused by a bulky lesion or a condition impairing the function of the DNA synthetic apparatus. However, this hypothesis is not necessary to explain the current evidence. It is sufficient to consider single-stranded DNA or, more likely, a single-strand gap as the inducing signal. Sussman and colleagues have provided evidence that RecA protein binds to gaps in DNA along with lambda repressor in the presence of single-strand binding (ssb) protein.[12,52,63]

Information on the prophage-inducing capacity of bacterial mutants is important to our understanding of the induction process. Mutants with different capacities to generate and repair DNA strand breaks have already been mentioned. Other types of mutants have provided results that are more difficult to rationalize in terms of the working model recently presented by Roberts and Devoret.[53] One mutant that has been well studied in our laboratory carries a lexA3 mutation, rendering the LexA protein immune to cleavage by RecA protease. This strain gives enhanced levels of the enzyme β-galactosidase following induction of the lambda–lacZ prophage when induction is carried out in the presence of ampicillin.[18] This strain, in our estimation, is a serious contender as a superior screening strain, given our results with a large variety of chemicals (Ref. 18 and R. K. Elaspuru, S. G. Moore, and S. K. Gonda unpublished data). These data, which cannot be rationalized in terms of our current understanding of Lex-repressed functions, suggest the existence of regulatory elements related to prophage induction that were previously unsuspected. Our work with a lambda cI_{857} (ts repressor) strain has also yielded some puzzling results. This strain is analogous to the one utilized by Ho and Ho. We find, as they did, that this strain may be more sensitive to inducing chemicals. However, in our experience, the sensitivity differences (compared with the cI$^+$) *vary* depending on the inducer used. The greatest enhancement in sensitivity (related to the magnitude of induction achieved and the dose of chemical detected) occurs not with strong inducers such as bleomycin, but rather with weak inducers such as rifampicin, metals and some carcinogens. Induction differences in strains such as cI$_{857}$ versus cI$^+$, which are related to the nature of the repressor–*recA* interaction, should be independent of the inducing chemical. This type of observation suggests that a different induction pathway for weak inducers may exist, or that the lesion itself is somehow involved in the repressor–*recA* interaction.

7. Future Developments in Prophage Induction Assays

There is much room for further development of prophage induction assays, particularly in the area of genetics. An understanding of the critical role of single-stranded DNA in the induction of SOS functions may be utilized in the construction of more sensitive lysogens. The accumulation of strand breaks (and gaps) can be enhanced in general in certain mutants, e.g., *polA* and *lig*$_{ts}$. In addition, there are mutants that may enhance or diminish strand break formation following particular DNA–chemical interactions. For example, *tag* mutants lacking a functional 3-methyladenine glycosylase[32] should have a diminished capacity for prophage induction by nitrosoguanidine, since this enzyme is involved in depurination, a known precursor of DNA strand breaks. Consideration of DNA–chemical interactions and the repair systems for which the lesions are substrates should generate many more ideas on the use of mutants with elevated levels of DNA strand breaks following treatment with chemicals.

In addition, new information on LexA repressor–operator interactions and the genetics and biochemistry of prophage induction is appearing at a rapid rate. These areas of research will yield mutants that are more easily induced by chemicals. Many candidates, such as phi 80 and lambda *ind*^{s},[14,53] already exist; however, only those mutants with low levels of spontaneous induction will be useful. The nature of prophage induction is such that the process can be amplified. Lambda has its own mechanism for amplifying its DNA copy number and thereby increasing the expression of its genes. The results on enhanced induction in a *lexA3* strain suggest that there are other regulatory elements involved in prophage induction that may result in amplified expression of phage DNA. In addition, time-dependent amplification of phage gene expression has been demonstrated.[20,61] It is likely that other manipulations, either genetic or conditional, can be found that increase the magnitude of induction observed following chemical treatment.

Procedural developments also are being made in prophage induction assays. Biochemical assays with colored products can be miniaturized and automated using microtiter plates and Elisa readers. This has been achieved using the lambda–*lacZ* prophage. In our laboratory, we have also succeeded in permeabilizing bacteria to chemicals *in vitro* by a single wash in a simple buffer. This treatment may destroy the colony-forming ability of the bacteria, but it has no effect on their prophage-inducing capacity. This method should alleviate the problems encountered in storing and handling mutants with fragile cell walls. It also offers the convenience of using frozen log-phase bacteria, a difficult proposition with cell wall mutants.

In our laboratory, we are approaching the point where a prophage induction assay may be performed as follows: frozen log-phase bacteria are thawed, washed with permeabilizing buffer, and distributed to microtiter plates containing chemicals; after several hours of incubation, substrate is added for measurement of induced β-galactosidase; 30 min later the enzyme–substrate reaction is terminated and the color absorbance is read by an ELISA spectrophotometer; a computer automatically plots the absorbance readings, which are directly proportional to induced enzyme, to generate dose–response curves. This procedure represents a significant advance over classical prophage induction assays; however, many other advances are anticipated as well. The advantages of prophage induction assays should make them major contenders as screening vehicles. However, more development work is required to overcome the problems of sensitivity and limited spectrum of response to some types of chemicals encountered in the past.

8. References

1. W. A. Anderson, P. L. Moreau, R. Devoret, and R. Maral, Induction of prophage λ by daunorubicin and derivatives: Correlation with antineoplastic activity, *Mutat. Res.* 77, 197–208 (1980).
2. B. N. Ames, W. E. Durston, E. Yamasaki, and F. D. Lee, Carcinogens are mutagens: A simple test system combining liver homogenates for activation and bacteria for detection, *Proc. Natl. Acad. Sci. USA* 70, 2281–2285 (1973).
3. J. Ashby, B. Kilbey, T. Kada, M. Green, M. Mandel, D. Tweats, H. Rosenkraz, C. Dambly, J. Thomson, and B. Rabin, Summary report on the performance of bacterial repair, phage induction, degranulation and nuclear enlargement assays, in: *Progress in Mutation Research, Evaluation of Short-Term Tests for Carcinogens* (F. J. deSerres and J. Ashby, eds.) Vol. 1, pp. 219–223, Elsevier/North-Holland, New York (1981).

4. W. M. Barnes, R. B. Siegel, and W. S. Reznikoff, The construction of λ transducing phages containing deletions defining regulatory elements of the *lac* and *trp* operons in *E. coli, Mol. Gen. Genet.* *129*, 201–215 (1974).

5. J. Baluch, J. W. Chase, and R. Sussman, Synthesis of *recA* protein and induction of bacteriophage lambda in single-strand deoxyribonucleic acid-binding protein mutants of *Escherichia coli, J. Bacteriol.* *144*, 489–498 (1980).

6. R. Ben-Gurion, A simple plate test for screening colicine-inducing substrates as a tool for the detection of potential carcinogens, *Mutat. Res. 54*, 289–295 (1978).

7. W. F. Benedict, M. S. Baker, L. Haroun, and B. N. Ames, Mutagenicity of cancer chemotherapeutic agents in the *Salmonella*/microsome test, *Cancer Res. 37*, 2209–2213 (1977).

8. M. Blanco and L. Pomes, Prophage induction in *Escherichia coli* K12 cells deficient in DNA polymerase I. *Mol. Gen. Genet. 154*, 287–292 (1977).

9. B. A. Bridges, R. P. Mottershead, M. H. L. Green, and W. J. H. Gray, Mutagenicity of dichlorvos and methylmethanesulfonate for *Escherichia coli* WP2 and some derivatives deficient in DNA repair, *Mutat. Res. 19*, 295–303 (1973).

10. R. Brent and M. Ptashne, Mechanism of action of the *lexA* gene product, *Proc. Natl. Acad. Sci. USA 78* 4204–4208 (1981).

11. M. J. Casadaban and S. N. Cohen, Lactose genes fused to exogenous promotors in one step using a *mu-lac* bacteriophage: *In vivo* probe for transcriptional control sequences, *Proc. Natl. Acad. Sci. USA 76*, 4530–4533 (1979).

12. S. P. Cohen, J. Resnick, and R. Sussman, Interaction of single-strand binding protein and *recA* protein at the single-stranded DNA site, *J. Mol. Biol. 167*, 901–909 (1983).

13. N. L. Craig and J. W. Roberts, *E. coli recA* protein-directed cleavage of phage λ repressor requires polynucleotide, *Nature 283*, 26–30 (1980).

14. R. M. Crowl, R. P. Boyce, and H. Echols, Repressor cleavage as a prophage induction mechanism: Hypersensitivity of a mutant λcI protein to RecA-mediated proteolysis, *J. Mol. Biol. 152*, 815–819 (1981).

15. C. Dambly, Z. Toman, and M. Radman, Zorotest, in: *Progress in Mutation Research, Evaluation of Short-Term Tests for Carcinogens* (F. J. de Serres and J. Ashby, eds.), Vol. 1, pp. 219–223, Elsevier/North-Holland, New York (1981).

16. J. W. Drake and R. H. Baltz, The biochemistry of mutagenesis, *in: Annu. Rev. Biochem. 45*, 1n37 (1976).

17. R. K. Elespuru, A biochemical phage-induction assay for carcinogens in: *Topics in Environmental Physiology and Medicine, Short-Term Tests for Chemical Carcinogens* (H. Stich and R. H. C. San, eds.), pp. 1–11, Springer-Verlag, New York (1981).

18. R. K. Elespuru, Induction of bacteriophage lambda by N-nitroso compounds, in: *Topics in Chemical Mutagenesis* Vol. 1, *N-Nitrosamines* (T. K. Rao, W. Lijinsky, and J. L. Epler, eds.), pp. 91–114, Plenum Press, New York (1984).

19. R. K. Elespuru and R. J. White, Biochemical prophage induction assay: A rapid test for antitumor agents that interact with DNA, *Cancer Res. 43*, 2819–2830 (1983).

20. R. K. Elespuru and M. B. Yarmolinsky, A colorimetric assay of lysogenic induction designed for screening potential carcinogenic and carcinostatic agents, *Environ. Mutagen. 1*, 65–78 (1979).

21. H. Endo, M. Ishizawa, T. Kamiya, and S. Sonoda, Relation between tumoricidal and prophage-inducing action, *Nature 198*, 258–260 (1963).

22. S. S. Epstein and I. B. Saporoschetz, On the association between lysogeny and carcinogenicity in nitroquinolines and related compounds, *Experientia 24*, 1245–1248 (1968).

23. W. F. Fleck, Development of microbiological screening methods for detection of new antibiotics, *Postepy. Hig. Med. Dosw. 28*, 479–498 (1974).

24. M. Gellert, K. Mizuuchi, M. H. O'Dea, T. Itoh, and J.-I. Tomizawa, Nalidixic acid resistance: A second genetic character involved in DNA gyrase activity, *Proc. Natl. Acad. Sci. USA 74*, 4772–4776 (1977).

25. M. M. Gottesman, M. L. Hicks, and M. Gellert, Genetics and function of DNA ligase in *Escherichia coli, J. Mol. Biol. 77*, 531–547 (1973).

26. A. Goze, A. Sarasin, Y. Moule, and R. Devoret, Induction and mutagenesis of prophage λ in *E. coli* K12 by metabolites of aflatoxin B1, *Mutat. Res. 28*, 1–7 (1975).

27. D. Grunberger and I. B. Weinstein, Conformational changes in nucleic acids modified by chemical carcinogens, in: *Chemical Carcinogens and DNA* (P. L. Grover, ed.), Vol. II, pp. 59–93, CRC Press, Boca Raton, Florida (1979).

28. B. Heinemann, Prophage induction in lysogenic bacteria as a method of detecting potential mutagenic, carcinogenic, carcinostatic, and teratogenic agents, in: *Chemical Mutagens, Principles and Methods for Their Detection*, Vol. 1 (A. Hollander, ed.), pp. 235–266, Plenum Press, New York (1971).

29. K. Hemminki, Nucleic acid adducts of chemical carcinogens and mutagens, *Arch. Toxicol. 52*, 249–285 (1983).

30. Y. Ishii and S. Kondo, Comparative analysis of deletion and base change mutabilities of *Escherichia coli* B strains differing in DNA repair capacity (wild-type, *uvrA*, *polA*, *recA*) by various mutagens, *Mutat. Res. 27*, 27–44 (1975).

31. Y. L. Ho and S. K. Ho, The screening of carcinogens with the prophage λcIts857 induction test, *Cancer Res. 41*, 532–536 (1981).

32. P. Karran, T. Hjelmgren, and T. Lindahl, Induction of a DNA glycosylase for N-methylated purines is part of the adaptive response to alkylating agents, *Nature 296*, 770–773 (1982).

33. C. J. Kenyon and G. C. Walker, DNA-damaging agents stimulate gene expression at specific loci in *Escherichia coli, Proc. Natl. Acad. Sci. USA 77*, 2819–2823 (1980).

34. D. E. Levin, M. Hollstein, M. F. Christman, E. A. Schwiers, and B. N. Ames, A new *Salmonella* tester strain (TA 102) with A•T base pairs at the site of mutation detects oxidative mutagens, *Proc. Natl. Acad. Sci. USA 79*, 7445–7449 (1982).

35. A. Levine, P. L. Moreau, S. G. Sedgwick, R. Devoret, S. Adhya, M. Gottesman, and A. Das, Expression of a bacterial gene turned on by a potent carcinogen, *Mutat. Res. 50*, 29–35 (1978).

36. J. W. Little and P. C. Hanawalt, Induction of protein X in *Escherichia coli, Mol. Gen. Genet. 150*, 237–248 (1977).

37. J. W. Little, S. H. Edmiston, L. Z. Pacelli, and D. W. Mount, Cleavage of the *Escherichia coli* LexA protein by the RecA protease, *Proc. Natl. Acad. Sci. USA 77*, 3225–3229 (1980).

38. J. D. Love, C. D. Liarakos, and R. E. Moses, Non-specific cleavage of φX174 RFI deoxyribonucleic acid by bleomycin, *Biochemistry 20*, 5331–5336 (1981).

39. A. Lwoff, Lysogeny *Bacteriol. Rev. 17*, 269–337 (1953).

40. S. N. Mamber, V. Bryson, and S. E. Katz, Evaluation of the *Escherichia coli* K12 Inductest for determination of potential chemical carcinogens, *Mutat. Res., 130*, 141–151 (1984).

41. V. W. Mayer, M. G. Gabridge, and R. J. Oswald, Rapid plate test for evaluating phage induction capacity, *Appl. Microbiol. 18*, 697–698 (1969).

42. J. McCann, E. Choi, E. Yamasaki, and B. N. Ames, Detection of carcinogens as mutagens in the *Salmonella*/microsome test: Assay of 300 chemicals, *Proc. Natl. Acad. Sci. USA 72*, 5135–5139 (1975).

43. N. E. McCarroll, B. H. Keech, and C. E. Piper, A microsuspension adaptation of the *Bacillus subtilis* "rec" assay, *Environ. Mutagen. 3*, 607–616 (1981).

44. P. Moreau, A. Bailone, and R. Devoret, Prophage λ induction in *Escherichia coli* K12 *envA uvrB*: A highly sensitive test for potential carcinogens, *Proc. Natl. Acad. Sci. USA* *73*, 3700–3704 (1976).

45. A. Morrison and N. R. Cozzarelli, Site-specific cleavage of DNA by *E. coli* DNA gyrase, *Cell 17*, 175–184 (1979).

46. M. Oishi, R. M. Irbe, and L. M. E. Morin, Molecular mechanism for the induction of "SOS" functions, in: *Progress in Nucleic Acids Research and Molecular Biology* (W. Cohn, ed.), Vol. 26, pp. 281–301, Academic Press, NY (1981).

47. A. O. Olson and K. M. Baird, Single-strand breaks in *Escherichia coli* DNA caused by treatment with nitrosoguanidine, *Biochem. Biophys. Acta 179*, 513–514 (1969).

48. D. M. Podger and G. W. Grigg, Mutagenicity of bleomycins, phleomycins and tallysomycins in *Salmonella typhimurium*, *Mutat. Res. 117*, 9–19 (1983).

49. P. Quillardet, O. Huisman, R. O'Ari, and M. Hofnung, SOS chromotest, a direct assay of induction of an SOS function in *E. coli* K12 to measure genotoxicity, *Proc. Natl. Acad. Sci. USA 79*, 5971–5975 (1982).

50. M. Radman, SOS repair hypothesis, in: *Molecular Mechanisms for Repair of DNA* (P. C. Hanawalt and R. B. Setlow, eds.), Part A, pp. 355–368,.

51. M. Radman, G. Villani, S. Boiteau, M. Defais, P. Caillet-Fauquet, and P. Spadari, On the mechanism and control of mutagenesis due to carcinogenic mutagens, in: *Origins of Human Cancer* (J. P. Watson and H. Hiatt, eds.), pp. 903–922, Cold Spring Harbor Laboratory, Cold Spring Harbor, New York (1983).

52. J. Resnick and R. Sussman, *Escherichia coli* single strand DNA binding protein from wild type and *lexC113* mutant affect *in vitro* proteolytic cleavage of phage/repressor, *Proc. Natl. Acad. Sci USA 79*, 2832–2835 (1982).

53. J. Roberts and R. Devoret, Lysogenic induction, in: *Lambda II* (R. W. Hendrix, J. W. Roberts, F. W. Stahl, and R. A. Weisberg, eds.), pp. 123–144, Cold Spring Harbor Laboratory, Cold Spring Harbor, New York (1983).

54. J. W. Roberts and C. W. Roberts Proteolytic cleavage of bacteriophage lambda repressor in induction, *Proc. Natl. Acad. Sci. USA 72*, 147–151 (1975).

55. J. W. Roberts, C. W. Roberts, and N. L. Craig, *Escherichia coli recA* gene product inactivates phage λ repressor, *Proc. Natl. Acad. Sci. USA 75*, 4714–4718 (1978).

56. W. Rojanapo, M. Nagao, T. Kawachi, and T. Sugimura, Prophage induction test (Inductest) of anti-tumor antibiotics, *Mutat. Res. 88*, 325–335, (1981).

57. T. G. Rossman, M. Molina, and L. W. Meyer, The genetic toxicology of metal compounds: I. Induction of λ prophage in *E. coli* WP2$_s$(λ), *Environ. Mutagen. 6*, 59–69 (1984).

58. G. B. Sancar, A. Sancar, J. W. Little, and W. D. Rupp, The *uvrB* gene of *Escherichia coli* has both *lexA*-repressed and *lexA*-independent promoters, *Cell 28*, 523–530 (1982).

59. J. P. Seiler, Influence of composition and treatment of the growth media on the yield of mutant colonies, in: *Progress in Mutation Research, Progress in Environmental Mutagenesis and Carcinogenesis* (A. Kappas, ed.), Vol. 2, pp. 155–158, Elsevier/North-Holland (1981).

60. C. L. Smith and M. Oishi, The molecular mechanism of virus induction. I. A procedure for the biochemical assay of prophage induction, *Mol. Gen. Genet. 148*, 131–138 (1976).

61. C. L. Smith and M. Oishi, Early events and mechanisms in the induction of bacterial SOS functions: Analysis of the phage repressor inactivation process *in vivo*, *Proc. Natl. Acad. Sci. USA 75*, 1657–1661 (1978).

62. W. T. Speck, R. M. Santella, and H. S. Rosenkranz, An evaluation of the prophage lambda induction (Inductest) for the detection of potential carcinogens, *Mutat. Res. 54*, 101–104 (1978).

63. R. Sussman, J. Resnick, K. Calame, and J. Baluch, Interaction of bacteriophage λ repressor with nonoperator DNA containing single strand gaps, *Proc. Natl. Acad. Sci. USA 75*, 5817–5821 (1978).
64. H. Suzuki, K. Nagai, H. Yamaki, N. Tanaka, and H. Umezawa, On the mechanism of action of bleomycin: Scission of DNA strands *in vitro* and *in vivo, J. Antibiot. 22*, 446–448 (1969).
65. P. A. Swenson, Physiological responses of *Escherichia coli* to far-ultraviolet irradiation, *Photochem. Photobiol. Rev. 1976*, 269–387.
66. J. A. Thomson, Mutagenic activity of 42 coded compounds in the lambda induction assay, in: *Progress in Mutation Research, Evaluation of Short-term Tests for Carcinogens* (F. J. deSerres and J. Ashby, eds.), Vol. 1, pp. 224–235, Elsevier/North-Holland, New York (1981).
67. Z. Toman, C. Dambly, and M. Radman, Induction of stable, heritable epigenetic change by mutagenic carcinogens: A new test system, in: *Molecular and Cellular Aspects of Carcinogen Screening Tests* (IARC Scientific Publications No. 27) (R. Montesano, H. Bartsch, and L. Tomatis, eds.), pp. 243–255, International Agency for Research on Cancer, Lyon (1980).
68. L. D. Vales, J. W. Chase, and J. B. Murphy, Effect of *ssbA1* and *lexC113* mutations on lambda prophage induction, bacteriophage growth, and cell survival, *J. Bacteriol. 143*, 887–896 (1980).
69. G. M. Weinstock and K. McEntee, RecA protein-dependent proteolysis of bacteriophage λ repressor: Characterization of the reaction and stimulation by DNA-binding proteins, *J. Biol. Chem. 256*, 10883–10888 (1981).
70. S. C. West, E. Cassuto, J. Mursalim, and P. Howard-Flanders, Recognition of duplex DNA containing single-stranded regions by RecA protein, *Proc. Natl. Acad. Sci. USA 77*, 2569–2573 (1980).
71. L. Wheeler, M. Halula, and M. Demeo, A comparison of aflatoxin B-1 induced cytotoxicity and prophage induction in *Salmonella typhimurium* mutagen tester strains TA1535, TA1538, TA98 and TA100, *Mutat. Res. 83*, 39–48 (1981).
72. E. M. Witkin, Ultraviolet mutagenesis and inducible DNA repair in *Escherichia coli, Bacteriol. Rev. 40*, 869–907 (1976).
73. M. B. Yarmolinsky and H. Wiesmeyer, Regulation by coliphage lambda of the expression of the capacity to synthesize a sequence of host enzymes, *Proc. Natl. Acad. Sci. USA 46*, 1626–1645 (1960).

CHAPTER 8

The Granuloma Pouch Assay

Peter Maier

1. Introduction

1.1. Significance of *in Vivo* Mutagenicity Tests in Mammals

Chemicals to which man is potentially exposed or that are already distributed in the environment are tested for their mutagenic activity in simple, inexpensive screening tests utilizing prokaryotic and/or eukaryotic cells. These phase I or tier I tests should have a low potential to produce false negative results. This is often achieved by the acceptance of a certain number of false positive results. Therefore chemicals that show positive, contradictory, or inconclusive results in this battery of tier I tests and are still of importance should be further tested in phase II (tier II) tests to provide a broader basis on which risk assessment can be made.

These tests should have the following characteristics:

1. Mutagenic events should be detected in germ cells and/or somatic cells *in vivo* in mammals.
2. Mutations should be recovered at different organizational levels of the genome (gene, chromosome, genome).

Peter Maier • Institute of Toxicology, Swiss Federal Institute of Technology and University of Zurich, Schwerzenbach, Switzerland.

3. The relationship between mutagenicity and toxicity must be defined.
4. The pharmacokinetic profile of the mutagenic metabolites should be characterized.

Ideally, phase II test should cover genetic endpoints comparable to those determined in screening (phase I) tests, and different genetic endpoints should be determined simultaneously in the same cell population. If this is achieved, it should be possible to define whether differences between the response in phase I and phase II tests are due to circumstances related to the *in vivo* test conditions or to the genetic endpoint investigated.

The detection and quantification of the mutagenic activity of chemicals is one important task of genetic toxicology. The other is to define the carcinogenic potential of chemicals. The established correlation between the mutagenic and carcinogenic activity of chemicals suggests a causal relationship between the two events. Most likely during the promotion phase of tumors other important pathways are more critical in carcinogenesis than the initial genotoxic lesions. In order to define the role of mutations in tumor induction and formation, an approach is required in which genetic endpoints are determined *in vivo* in cell populations from which tumors can arise. Mutagenic events might occur at each step of the postulated cascade of events toward neoplastic transformation and tumor formation. Therefore, they should be recovered at defined intervals during promotion and progression of tumors. Furthermore, the simultaneous recovery of different genetic endpoints identifies the most critical types of mutations, which then can be used as an indicator for chemicals with carcinogenic activity.

This double role of a tier II test in the evaluation of the mutagenic and carcinogenic activity of chemicals can be fullfilled to a great extent by the granuloma pouch assay (GPA), a test developed in the past few years in our laboratory. The approach is based on the principle that a rapidly growing defined target tissue in rats is subjected to test compounds *in vivo*, and the genetic alterations are recovered *in vitro* in individual cells. The development of induced tumors can be observed from the same cell population.

1.2. Historical Background

The granuloma pouch technique has been used in pharmacology for many years as a procedure to analyze quantitatively factors that regulate inflammation and wound healing.[32] After injection of a given amount of air into the loose connective subcutaneous tissue of the rat, an ellipsoidal air space is created. Formation of a granulomatous membrane is then induced by injection of a small amount of an irritant (e.g., croton oil, vitamin E) into the air pocket. After pretreatment or simultaneous treatment of the animals with potential antiinflammatory or wound-healing drugs, the development of the inflammatory tissue and the exudate accumulating 4–14 days after pouch formation are investigated. The same animal model has also been used in antibiotic research as a discriminative model for an inflamed body cavity. The biochemical, hematological, and histological changes in tissue and exudates are well characterized and summarized in the reviews by Dalhoff et al.[4,5]

The fact that inflammation processes are associated with rapid proliferation and that the tissue can be isolated easily, prompted us to study genetic endpoints in this animal model.[15,18]

2. Target Tissue and Treatment

2.1. Pouch Formation

On the back, at the midpoint of the scapular area of young adult male rats (Sprague–Dawley, SIV 50, Ivanovas Kisslegg, Federal Republic of Germany, 230–260 g, 54 ± 4 days old), a subcutaneous air pouch is established by injecting 25 ml of germ-free air. The mechanical stretching of the connective tissue that is dorsally loosely connected to the superficial muscles of the skin and ventrally to the fascia of the muscles of the trunk initiates cellular growth at the inside wall of the air pouch. Simultaneous injection of an irritant such as croton oil or vitamin E enhances the inflammatory reaction but is not crucial for the induction of local cell proliferation.

Cells that form this granulation tissue are of mesenchymal origin, mainly fibroblasts, macrophages, and endothelial cells, the latter forming thin-walled capillaries. Most likely the mitogenic stimulus in vivo is similar to that observed in wound repair. It is speculated that macrophages provide signals to stimulate fibroblast DNA synthesis and proliferation.[1,6,11,16]

2.2. Treatment

Mutations are most efficiently fixed in proliferating cells. Therefore it is important to apply chemicals at the time of the highest proliferative activity of target cells. With ^{3}H-thymidine incorporation studies *in vivo* we found the highest percentage of labeled recoverable of fibroblast-like cells attaching to culture dishes 48 hr after pouch formation. At that time after a 6-hr pulse of labeled ^{3}H-thymidine (3 additions of 10 μCi/ml in 0.9% NaCl at intervals of 2 hr), 40.1% of cells surviving in culture dishes are in S-phase.[9] Test compounds are therefore administered 48 hr after pouch formation either systemically by various routes or injected into the pouch. The local application assures a direct contact with the defined target tissue.

So far test chemicals have been applied in a single dose because after the first application of an active test compound membrane structures, enzyme levels, repair mechanisms, and the proliferation kinetics of target cells can be altered. Therefore with a second application we hit target cells that are in an undefined metabolic state, different from those during the first exposure. However, in a situation in which the enzymic activation system in the target cells first has to be stimulated, e.g., with benzo(*a*)pyrene,[19] application 24 and 48 hr after pouch formation is most suitable.

When chemicals are applied locally, problems often arise with water-insoluble chemicals. In the GPA, vegetable oil as a solvent is practicable. It is often a less toxic vehicle than most of the other organic solvents in use.

One of the unique features of the GPA is the easy access to target cells. This permits the performance of pretreatment (e.g., enzyme stimulation, growth stimulation) and post-treatment (e.g., modification of DNA lesions and of repair processes) studies. Furthermore, the defined target tissue allows the testing of substances available only in small amounts. The tissue is dissected out after defined intervals (4 hr to 2 days), depending on the endpoint for which it is investigated. For the demonstration of initial DNA lesions such as alkali-labile sites the granuloma tissue has to be recovered within hours because the lesions can be removed or repaired immediately. Chromosome analysis can be performed within the two cell cycles after treatment. Gene mutations are recoverable after an expression period *in vivo* and *in vitro* of several days.

2.3. Exposure

The exposure of a target tissue *in vivo* after systemic application is determined by the pharmacokinetics of the mutagenic metabolites. Quantitative analysis depends on knowledge of the reactive species and its identification with analytical methods. In the GPA we have the opportunity to determine accurately the exposure of the target tissue cells when test agents are applied locally. This is illustrated with the following examples.

The exposure to dissolved chemicals injected into the pouch can be estimated by the relationship between the total amount of the chemical per weight of target tissue. Two days after pouch formation the inner wall including the dorsal and ventral part of the induced granuloma weighs about 4 g. An optimal distribution of the chemical can be assumed because capillaries are formed in the tissue, which provide an excellent transport system for chemicals. An absorbtion into the circulation is also possible since the granuloma tissue is highly vascularized.

Gaseous chemicals can be injected into the air pouch. With vinylchloride (VC) an aliquot of this gas/air mixture in the pouch was sampled immediately after injection, analyzed in a gas chromatograph, and the exposure calculated.[24] Alternatively, a defined gas volume of a given concentration is applied into the pouch. The volume of the air pouch is approximately 20–22 ml and the temperature 35°C; the actual gas concentration can then be calculated.

With X rays the exposure was determined first during the treatment of the rat and then with the dissected skin between the X-ray source and dosimeter. This procedure allowed us to determine the actual exposure of the granulation tissue.

2.4. Cell Recovery

The dissection procedure of the dorsally located granulation tissue is performed as follows: Rats are sacrificed by cervical dislocation. The area of the pouch is shaved and disinfected with 70% ethanol. Under sterile conditions two longitudinal subcutaneous incisions are made along the bottom of the pouch with as little disruption as possible of the granulation tissue, which still forms an air pouch. The skin is then removed, and the exposed granulation tissue is isolated and washed for 10 sec in distilled water in order to lyze red blood cells. The whole procedure for the recovery of the target tissue from one rat takes approximately 5 min. Subsequently the tissue is minced in 1 ml Dulbecco's phosphate buffer (PBS, pH 7.4) into pieces of less than 1 mm^3. The dissociation into single cells is performed in 10 ml PBS containing collagenase type I (600U) and Dispase (8 U) for 45 min at 37°C while stirring. After 45 min the suspension of tissue fragments is filtered through two layers of Kodak lens paper. Tissue fragments are recovered, minced, and incubated again as described above. At the end of this second incubation the two cell suspensions are pooled, and cells are centrifuged for 5 min at $134 \times g$ and suspended in Dulbecco's modified medium (DMEM). From control animals more than 97% of the granuloma pouch cells (GP cells) are viable, determined by trypan blue exclusion tests. On the average $(30.3 \pm 8.3) \times 10^6$ ($N = 33$ samples) cells are recovered from untreated animals.

In older rats the fat layer in the skin is already enlarged, which results in reduced flexibility. The subcutaneous injected air penetrates into the fat layer, and the recovery of cells able to form clones is drastically reduced.

In earlier studies simultaneously with the air injection a small amount of croton oil was applied into the air pouch. As an irritant the tumor promoter enhanced the total number of nucleated, recoverable cells to $40–50 \times 10^6$ cells. However, the number of cells clonable *in vitro* was not significantly affected. Therefore in standard genetic experiments we initiated growth of the granulation tissue without pretreatment to rule out unpredictable metabolic changes in the target cells.

2.5. Cell Culture

Under rigidly controlled *in vitro* culture conditions (DMEM, 10% FCS, pH 7.4 in the medium adjusted with 12% CO_2) growth was highly reproducible. Within the first 24 hr in culture, the cell cycle of the freshly isolated cells is delayed. With the analysis of BrdUrd-labeled chromosomes, the number of cell cycles after BrdUrd exposure can be determined in metaphases. This investigation demonstrated that cells already in S phase during the last 24 hr before isolation required more than 24 hr to finish their S and G_2 phases *in vitro*.[21] In culture additional cells are pushed into mitosis that were not in cycle 24 hr before isolation. This stimulus is most likely due to protease treatment. Within the consecutive 24 hr *in vitro*, the cell cycle is accelerated to be below 12 hr.

Cells from the granulation tissue show a high clone formation ability in cultures *in vitro*. From control animals the primary cloning efficiency (I CE) is $12.4 \pm 4.4\%$ ($N = 25$). The calculation is based on the total of nucleated living cells recovered, including blood cells. A total of $40–60\%$ of the cells are fibroblast-like cells, which attach to culture dishes within a few hours. The corrected, estimated cloning efficiency based on the seeding efficiency of these cells is therefore $20\%–30\%$. After growth of 3 days, or, with corresponding subcloning, 6 days, *in vitro*, the secondary cloning efficiency is $53 \pm 11.5\%$ ($N =$

25). This value is as high as CE obtained with cells from established cell cultures. The life span of cells *in vitro* is limited, but can be maintained at least for six passages using a dilution factor of 1:3.

Growth was further improved by adapting the culture conditions to the *in vivo* situation. One important physiological parameter that could influence cell proliferation is the pO_2 level. Our measurements *in vivo* showed that the pO_2 level in the subcutaneous tissue was between 30 and 40 mm Hg. This value was adjusted in the culture medium by reduction of the pO_2 with N_2 in incubator atmosphere.[23] Under these culture conditions (pH 7.4) the I CE was increased to 15.8 ± 4.8 ($N = 16$) and the II CE to 73.9 ± 15.4 ($N = 16$). We speculate that the more physiological growth conditions promote cells from additional cell subpopulations to form clones *in vitro*.

The clonability and continuous culture of the freshly isolated cells allows the performance of reconstruction experiments *in vitro*. Cultures obtained from untreated animals can be treated *in vitro* or *in vivo* exposed cells post-treated *in vitro*. In specific cases both concepts could be helpful for the verification of influences due to the *in vivo* situation.

3. Metabolic Activation

3.1. Introduction

In general it is believed that extrahepatic tissues have a low metabolic activation capacity especially poor in cytochrome P-450-dependent and phenobarbital-inducible monooxygenases.[34] This is true when the competence of the tissue as a whole is considered. However, it is possible that a subpopulation of cells exists that contain in individual cells a relatively high intracellular activity. This activity could be sufficient to metabolize chemicals to mutagenic metabolites. Induced genome alterations in these particularly active cells would be recoverable when mutations are checked per individual cells. For tumor initiation such an effect would be of great significance.

3.2. Metabolic Competence of Cells

Because of the expected low level of total enzymes involved in xenobiotic metabolism in extrahepatic cells, one has to use a sensitive enzyme assay that is not disturbed by high amounts of nonenzymic

TABLE 1. Aldrin Epoxidase Activity of a Freshly Isolated Subpopulation from Granulation Tissue[a]

Fraction	Flow rate at 1500 rpm, ml/ min	Cell viability, %	Recovered cells, %	Aldrin epoxidase acitivity[b]	Activity, %
Total	—	92.0 ± 4.6	100.0	0.035 ± 0.008	100
Washing	8.5	88.6 ± 8.3	27.5	0.019 ± 0.010	54.3
F1	19.2	91.7 ± 5.5	46.1	0.067 ± 0.020	191.4
F2	34.5	96.1 ± 6.1	10.6	0.046 ± 0.011	131.4
F3	66.5	96.5 ± 3.7	5.3	0.040 ± 0.009	114.3

[a] Individual cell fractions from the dissociated granulation tissue were obtained by centrifugal elutriation using a Beckman 21 centrifuge with a JE 6 elutriator; rotor and flow rates as indicated.
[b] Picomoles of dieldrin per 10^6 cells per minute.

proteins. As a representative of a phenobarbital (PB)-inducible, P-450 dependent monooxygenase, aldrin epoxidase was assayed.[35] The lipophilic substrate penetrates through cell membranes and facilitates analysis of enzyme activities in intact cells. The method was further improved by using glass-capillary gas chromatography for the detection of the end-product, dieldrin.[10] Enzyme activities were determined in intact granuloma cells as well as in a different subpopulation separated by counterflow centrifugation.[10] Table 1 summarizes the data. Clearly, measurable enzyme levels are present in the granulation tissue cells and this activity is accumulated in specific subpopulations that also differ in their colony-forming capacity. Compared with other extrahepatic tissues, granuloma cells have an activity comparable to bone marrow cells and a three-fold higher activity than peritoneal cells, spleen cells, or thymus lymphocytes.[10]

Pretreatment of animals with aldrin or phenobarbital stimulates the epoxidase activity in freshly isolated, intact granuloma cells by 3-fold and 2.5-fold respectively. However, for technical reasons (hemoglobin and cytochrome oxidase contamination), the cytochrome P-450.7 content was not measurable. Still, part of the epoxidation of aldrin might be mediated by prostaglandin endoperoxide synthetase.[8] When rats are pretreated with phenobarbital, prostaglandin synthetase activity can be stimulated in the liver.[26] Furthermore, granulation tissue transforms arachidonic acid efficiently into hydroxy fatty acids and prostaglandin E_2.[2] In recent experiments, we have found that indo-

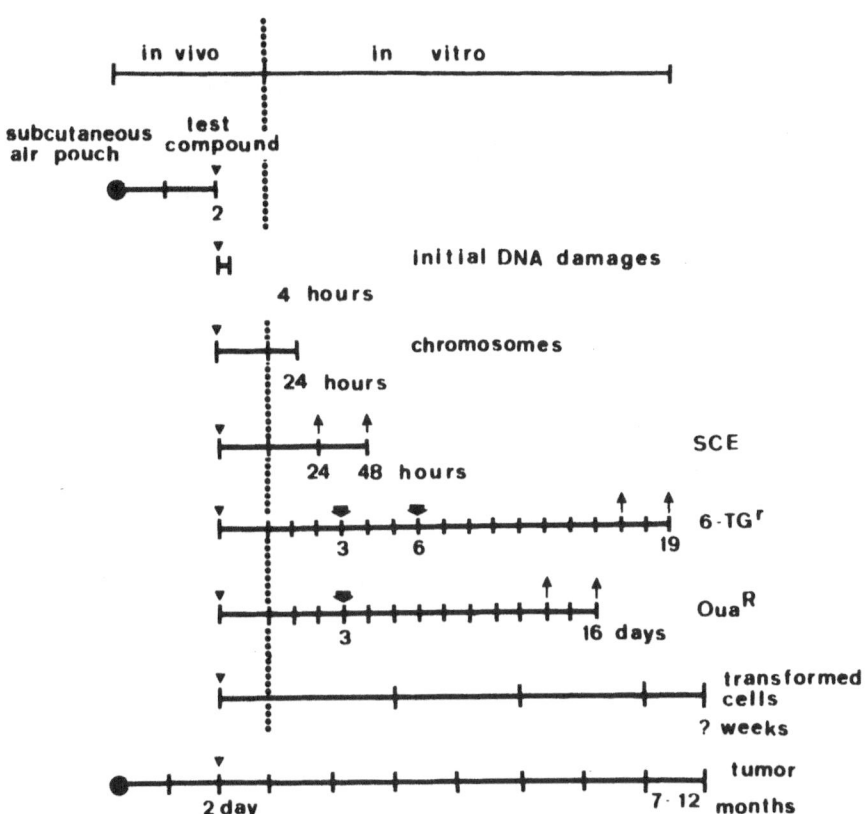

FIGURE 1. Schedule for the granuloma pouch assay (GPA) (transformed cells could not have been detected to date with *in vitro* methods). (●) Pouch formation, (▼) application of the test compound, (···) dissection of the granuloma tissue, (⬇) subcloning, (↑) fixation, staining.

methacin decreased, whereas arachidonic acid increased the aldrin epoxidase activity when added to isolated granuloma cells *in vitro*.[14]

4. Genetic Endpoints

In phase II tests different genetic endpoints should be detectable ideally in a combined approach. In the GPA this is possible within the same tissue of different animals. Gene mutations and chromosome aberrations can be recovered simultaneously in the same cell population from an animal (Figure 1).

4.1. Primary DNA Lesions

The primary interaction of chemicals with DNA from GP cells, such as covalent binding to DNA or the induction of alkaline-labile DNA lesions, can be studied as in most other tissues in the body. These primary interactions are indirect evidence for a mutagenic effect and cannot be related to individual living cells or cell types. This handicap can be reduced in the GPA. The elution profile or radioactivity found of chemicals covalently bound to DNA can be calibrated with the number of recoverable cells mutated at specific loci. So far the GPA has demonstrated the appearance of alkali-labile lesions.[15] Using a standard elution protocol with labeled DNA, it was found that monofunctional alkylating agents are highly active, whereas bifunctional chemicals are not. For DNA-binding studies, the amount of the totally extractable DNA from recoverable cells might be a limiting factor.

4.2. Cytogenetic Analysis

Cytogenetic changes detectable in cells exposed to mutagenic agents are: alterations in chromosome structure (chromosome aberrations), in chromosome number (genome mutations), and enhancements in sister chromatid exchange frequencies. The majority of clastogenic chemicals induce lesions during the G_1/S phase, resulting in chromatid type aberrations that are detectable in the subsequent metaphase. Therefore, only an acute treatment schedule assures the efficient recovery of substances which induce chromosomal changes.

4.2.1. Chromosome Analysis

Chromosome analysis in metaphase is possible in several dividing cell population *in vivo*. However, the accumulation of a sufficient number of metaphases with high resolution of the chromosomes is often the limiting factor.

In the GPA, cells are isolated from the tissue 72 hr after pouch formation or 24 hr after application of the test compound (Figure 1). Metaphases are accumulated *in vitro* on glass coverslips *in situ*. This procedure assures an excellent resolution of the chromsomes. From cell cycle analysis using differentially labeled BrdUrd substituted chromosomes, we concluded that nearly 60% of metaphases accumulated within 24 hr *in vitro* are from cells that performed not more than one cycle *in vivo* 24 hr immediately before isolation.[21] Most of them are from cells that started DNA synthesis *in vitro*. From untreated animals the mitotic index of freshly isolated cells after 24 hr was 3–4% and after the subsequent 24 hr 6%. This situation can be altered drastically by the use of cytostatic test chemicals.

Chromosomes from rats are well defined in size and centromere position. Cells have a stable diploid, in rare cases a tetraploid, chromosome set even after prolonged culture *in vitro*. Therefore, unlike cells from established cell lines, GP cells are extremely suitable for genome mutation analysis. This has special significance, since chemicals classified as tumor promoters (e.g., DES, asbestos) are also able to induce chromosome aberrations and aneuploidy.[29,33]

In vivo accumulated metaphases cannot be recovered efficiently in the GPA. Exposure to proteases during the dissociation of the granulation tissue will preferentially destroy mitotic cells in metaphase. Furthermore, metaphase arrest *in vivo* with a spindle poison will exclude any other combined studies with another genetic endpoint.

4.2.2. Sister Chromatid Exchanges

In the GPA, SCEs can be analyzed after cells are exposed to BrdUrd (1) *in vivo*, (2) *in vivo* and *in vitro*, or (3) *in vitro* only.

In vivo BrdUrd tablets were implanted s.c. in the lumbar region of the rats. The subcutaneous rat tissue absorbed the nucleotide analogue much more slowly than in mice.[21] Still, a peak concentration of 10 µg/ml blood is obtained within 5 hr after tablet implantation. This is sufficient to double the spontaneous mutation frequency at the 6-TG^r locus. Further reduction of the BrdUrd absorption during 24 hr and without peak concentration can be achieved by the use of agar-coated tablets.[21] Using this technique, it is possible to combine a gene mutation test with SCE analysis.

The spontaneous SCE values found were between 5.66 and 16.10 exchanges per metaphase (mean 11.47). In bone marrow cells this value dropped to 2.6 SCE/metaphase. Most likely this high control value is an inherent characteristique of the cells because this value also was obtained when BrdUrd was applied *in vivo* only.

With cyclophosphamide (CP) and mitomycin C (MMC) *in vivo* induced lesions are long-lived and were recovered in cells exposed to BrdUrd *in vitro* only. Values obtained were similar to those found after continuous exposure to BrdUrd during application of the test compound *in vivo* and the recovery period *in vitro*.

With MMC we compared the systemic versus the local application routes. After local injection into the air pouch with increasing dose a pronounced plateau effect was obtained.[22] This is in contrast to the systemic application routes. The longer presence of an active chemical after local injection is more cytotoxic than a short pulse that is in general achieved with a systemic application route. Therefore, after local exposure it is possible that relatively more cells enter mitosis *in vitro*, and are in an insensitive stage in the cell cycle at the time of treatment *in vivo*.[22]

4.3. Gene Mutations

The clonability of the GP cells makes them suitable for growth studies. Therefore the drug resistance and clone formation capacity of mutated cells in selective medium can be determined. The knowledge accumulated over the past decade with established cell cultures in the detection of specific locus mutations can be used.

The total number of cells recovered from the granulation tissue is sufficient to perform tests at different loci simultaneously. So far the two loci Oua^R (base-pair substitutions) and $6\text{-}TG^r$ (broad spectrum, including small deletions) have been used.[18,19] Critical steps in culture conditions are similar to what is known from *in vitro* tissue culture systems (expression period, cell densities). However, important differences exist:

1. The freshly isolated cell population from the granulation tissue is heterogeneous, as pointed out earlier. During *in vitro* culture a certain homogeneity can be obtained because all blood cells are dying out within 48 hr. By visual inspection among clone-forming cells, including mutant cells, we distinguish three classes that differ in size, structure, and growth pattern of their clones.

2. The expression period can be separated into an *in vivo* and an *in vitro* part. The optimal schedule was found to be 1–2 days *in vivo* and 3 days *in vitro* for the Oua^R locus and 1–2 days *in vivo* and 2–3 days for the $6\text{-}TG^r$ locus. In fact the expression period depends on the type of mutagen used and on the dose to which cells are exposed. With high doses and strong clastogenic compound, however, we found no further improvement of mutation frequencies at the $6\text{-}TG^r$ locus after the second 3 days of expression *in vitro*.

3. The optimal concentration of 6-TG was evaluated. In standard tests dialyzed fetal calf serum (dFCS) tested for its influence on the cloning efficiency was used. With increasing 6-TG concentration we did not find a stable baseline spontaneous mutation frequency.[19] Clearly, mutants from subpopulations of cells might differ in their response to the toxicity of the base analogue. With a concentration of 10 μg/ml in the culture medium a spontaneous mutation frequency of 5.3×10^{-6}

was reached. This is comparable to frequencies obtained in cell culture systems. In a recent series of experiments we enhanced the concentration to 15 μg 6-TG/ml. The spontaneous mutation frequency dropped to 2.1×10^{-6} (reduction factor 0.4), whereas the induced frequencies were reduced by a factor of 0.7.

4. The different morphology of clones can be helpful for deciding whether or not 6-TG-resistant clones that grow close together arose from colony splitting.

To date X rays and 17 chemicals have been tested in the gene mutation assay (Table 2), including alkylating agents, one nucleotide (BrdUrd), one frameshift mutagen (9-aminoacridine), and the so-far nongenotoxic carcinogen clofibrate. A chemical was considered to be a mutagen in the GPA when at least two exposures induced a significant increase over the spontaneous mutation frequencies using the test by Kastenbaum and Bowman[13] and when the two frequencies showed an exposure/effect relationship. To satisfy these requirements the optimal exposure range had to be determined in pilot experiments. With the exception of the frameshift mutagens, all the substances (including gaseous VC) induced 6-TG^r clones after local and/or systemic application. This demonstrates that a wide spectrum of lesions can be recovered at this locus. ENU, procarbazine, and MNNG were the most powerful agents and are most suitable to serve as positive controls.

Cells isolated from clones resistant to 6-TG^r were subcloned in HAT medium. Cells from 92% of the clones tested did not survive, but all remained resistant in 6-TG medium. Induced 6-TG^r in cells isolated from treated animals were detectable even after subcloning them over five passages (dilution factor 1:3).

4.4. Combined Approaches

Genetic endpoints induced by a given chemical cover a wide spectrum. Using a 24 hr *in vivo* expression period in the gene mutation assay, SCE analysis can be performed simultaneously. Comparative analysis shows that MMC and CP induce SCEs efficiently[22] but are extremely low inducers of gene mutations (P. Maier, unpublished results). By contrast, MNNG is an extremely powerful gene-mutation inducer (Figure 5), but only slightly enhances the SCE values (P. Maier, unpublished results). These data prove that *in vivo* initial lesions caused by a mutagen and its metabolites are compound specific and lead to different consequences at different organizational levels of the genome. It is necessary to know these relationships because each requires different risk estimations.

TABLE 2. Gene Mutation Test at the 6-TG Locus in the GPA

	Application[a]			
	Systemic		Local	
Test compound	Exposure range, mg/kg	6-TG	Exposure range, mg/GP	6-TG
MNNG (N-Methyl-N'-nitro-N-nitrosoguanidine)	2.5–50	+ +	0.1–1.8	+ +
EMS (Ethyl methanesulfonate)	20–50	+	5	+
MMS (Methyl methanesulfonate)	20–100	+ +	5–20	+ +
ENU (1-Ethyl-1-nitrosourea)	20–80	+ +	0.06–1.33	+ +
DR (Doxorubicin)	1–10	+ +	—	NT
MIH (Procarbazine)	20–300	+ +	1–75	+ +
AF-2 (Furylfuramide)	50–200	+/–	5	+
MMC (Mitomycin-C)	1–5	+/–	0.001–0.025	+
B(a)P [Benzo(a)pyrene]	—	NT	0.03–0.25	+ +
CP (Cyclophosphamide)	2–50	+/–	1–5	+
DMN (Dimethylnitrosamine)	20–40	+/–	0.25–1	+ +
VC (Vinylchloride)	—	NT	1–4 ml	+ +
AFB$_1$ (Aflatoxin B$_1$)	0.25	–	0.01–0.04	+ +
2-AAF (2-Acetyl aminofluorene)	10–100	+ +	—	NT
BrdUrd (5-Bromo-2'-deoxyuridine)	10	+	—	NT
CF (Clofibrate)	—	NT	1–4	+ +
AA (9-Aminoacridine)	40	–	0.2–10	–
X-rays	—		200–800 r	+ +

[a] + +, Dose-dependent effect; with two exposures mutation frequencies were induced in individual animals that are significantly different from untreated animals. The two exposures showed an exposure–effect relationship when results when 3–4 animals are pooled. +, Positive at one dose; only one exposure was tested, or from the exposures chosen only one has resulted in a significant increase of mutation frequencies. +/–, Inconsistent results; mutation frequencies recovered individually from less than one-half of the treated animals showed no significantly enhanced values. –, No effect; mutation frequencies from all animals tested were not significantly different from control values. NT, Not tested. Significance was determined according to Kastenbaum and Bowman.[13]

4.5. Evaluation of the Pharmacokinetic Profile of Mutagenic Metabolites

In the evaluation of the GPA model, compounds were used that differ widely in their induction of different types of DNA lesion and in their activation/detoxification pathways. Among substances already tested, we identified chemicals that are active after local and systemic application (group I) and chemicals that are active only after local application or showed a drastically reduced activity after systemic

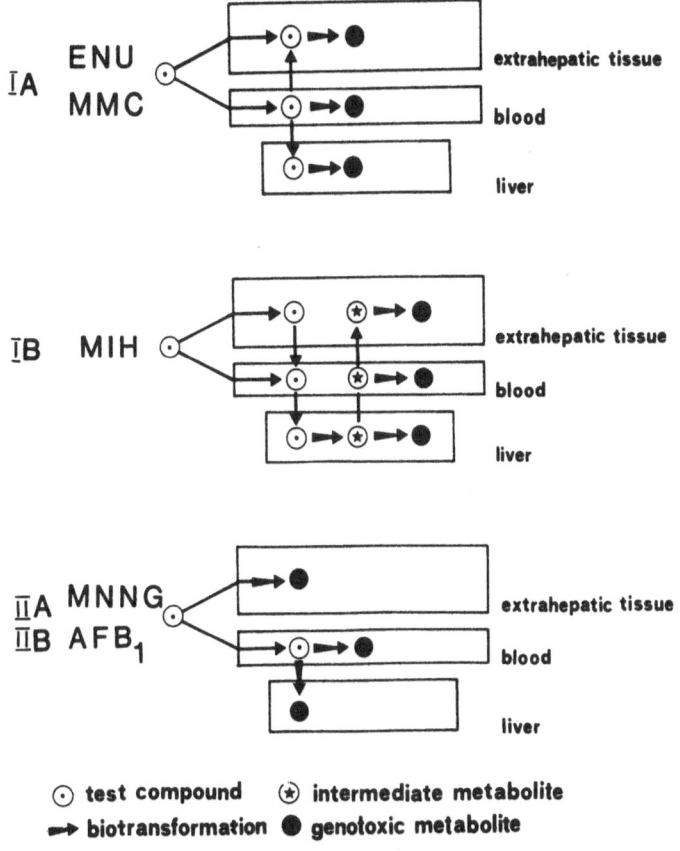

FIGURE 2. The pharmacokinetics of mutagenic metabolites evaluated in the GPA (granuloma tissue = extrahepatic tissue). Chemicals tested can be classified into four groups (IA, IB, IIA, IIB); see Section 4.5.

application (group II). Each of the two groups can be further subdivided (Figure 2).

In group IA the mutagenic response of target tissue is related to the exposure per weight of target tissue after systemic and local application routes. With MMC (Figure 3), using SCE as an endpoint, a close tissue-exposure relationship was obtained especially at the lowest exposure level.[22] With higher exposure of locally applied MMC, the cytotoxic and cell cycle-inhibiting activity disturbs the efficient expression of SCE in lethally damaged cells. This selection for cells that were in an insensitive stage at the time of tissue application disturbs the characteristic exposure–tissue effect relationship usually obtained after

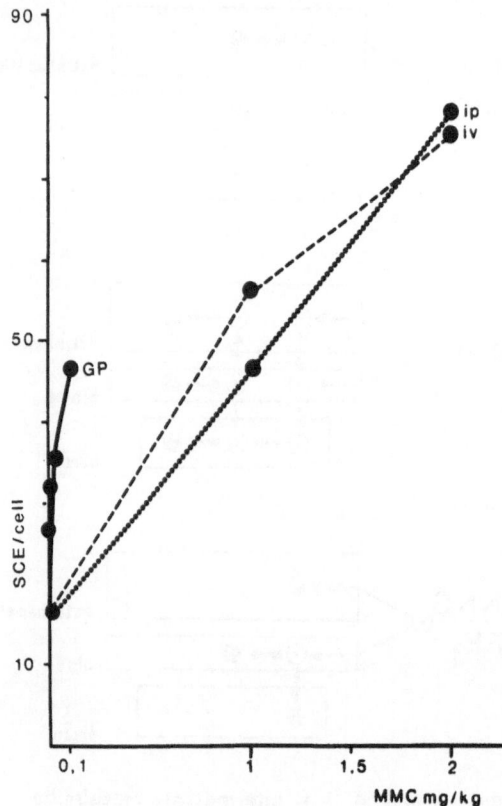

FIGURE 3. SCE induction with mitomycin C. Dose–effect response curve.

exposure to a corresponding amount of the chemical when applied systemically.

In group IB (Figure 4) an identical response toward an exposure is obtained whether the chemical is applied locally or systemically. With procarbazine the dose–response curves obtained after oral, intravenous, or intraperitoneal administration are nearly identical.[20] We conclude that after passage through the liver, stable, still unidentified metabolites are formed, which, after release into the circulation, reach the granulation tissue. There, as at other body sites, the mutagenic species might be formed in a second intracellular activation/degradation step. In a standard bacterial screening test the chemical is inactive,[3] most likely because bacteria cannot perform the second activation step.

In group IIA (Figure 2) we find chemicals that are active in most of the target cells and are detoxified before entering the blood system. An example is MNNG. This chemical, which is highly mutagenic in *in*

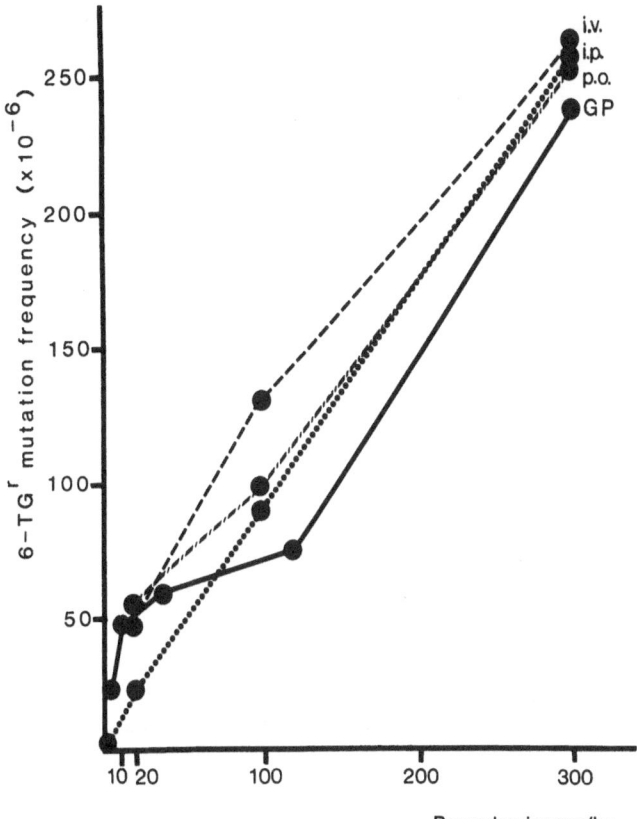

FIGURE 4. The 6-TG^r mutation frequency induced with procarbazine.

vitro tests,[30] shows no clastogenic response in bone marrow cells[25] or in germ cells.[28] In the GPA, MNNG is a powerful mutagen and induces gene mutations after local application, but showed only a reduced gene mutation activity after systemic applications (Figure 5).

Aflatoxin B$_1$ (Figure 6), previously reported to be negative in the GPA after systemic application,[17] is a typical representant of a carcinogen with an organ-specific or cell-type-specific single-step activation (group IIB). After systemic application no mutant cells were detectable in the granulation tissue. However, after local application a dose-dependent increase was obtained. The dose range was critical. Cells carrying mutations are those that contain the enzymes involved in the activation of AFB$_1$, mainly accumulated in liver cells but also in subpopulations of the granulation tissue.[10] It seems that the genotoxic activity of reactive metabolites is restricted to those cells in which they

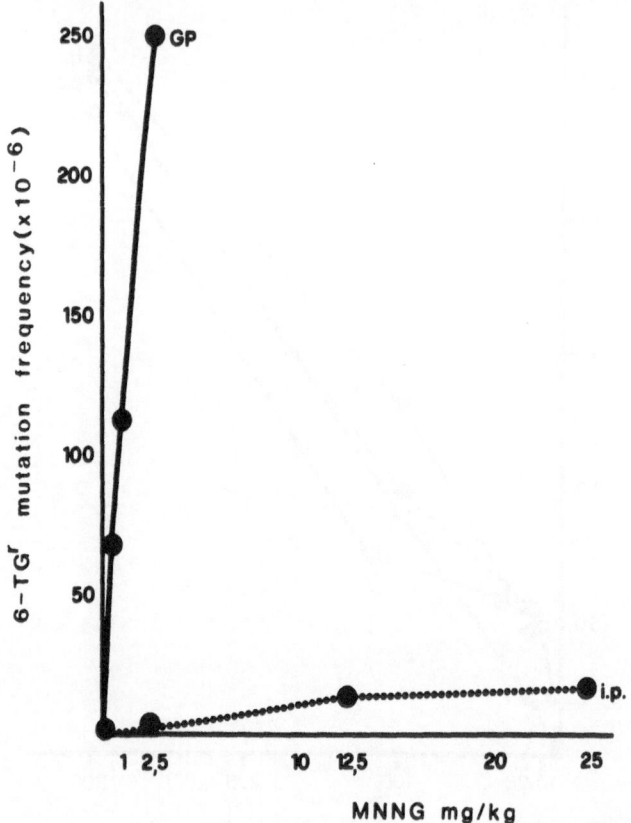

FIGURE 5. The 6-TG^r mutation frequency induced with N-methyl-N'-nitro-N-nitroso-guanidine.

are created. An excessive dose is lethal or cytostatic toward the cells of the subpopulation carrying the activation enzymes.[24] Because cells carrying the oxidative activation enzymes reflect less than 40% of the total number of recoverable cells, their susceptibility is neither expressed in the total of viable cells nor in the primary cloning efficiency as observed with MNNG, a chemical that affects most of the cells. The relatively low maximal inducible mutation frequency is characteristic for the optimal response of a small subpopulation. Further studies are necessary to evaluate whether the epoxidation of AFB_1 is due to cytochrome P-450-dependent monooxygenases or to a prostaglandin synthetase-dependent pathway or both.

The fact that a dose-dependent mutation frequency can be detected in a tissue with a 3500 times lower total PB-inducible enzyme activity

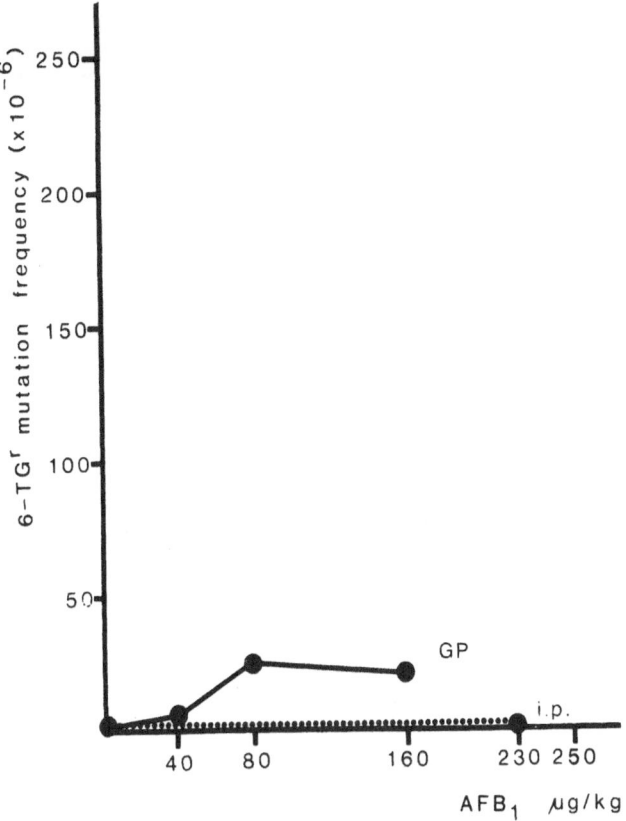

FIGURE 6. The 6-TG^r mutation frequency induced with aflatoxin B$_1$.

than in liver (comparison based on mg proteins) demonstrates the potential and sensitivity of mutagenicity tests in which individual mutant cells are recovered by selective procedures. Furthermore, we have been able to see that hepatocarcinogens can induce gene mutations in an extrahepatic tissue. This means that the intracellular level of epoxidase activity in individual cells from subpopulations of an extrahepatic tissue is sufficient to cause mutant cells. However, in the case of AFB$_1$ the locally effective exposure calculated to a corresponding whole-body exposure is approximately 2.5 mg/kg, a lethal dose for the whole animal. This systemic lethal effect will not allow the exposure of the extrahepatic tissue to a mutagenic dose level via a systemic exposure. Therefore the risk that AFB$_1$ will induce gene mutations in extrahepatic tissue will be from local exposure of the target tissue, e.g., through inhalation, ingestion, or skin contact.

TABLE 3. Comparison of the Effective Exposure Range in the Mouse
Specific-Locus Test (SLT) and the Granuloma Pouch Assay (GPA)[a]

	Systemically applied effective exposures, mg/kg	
Substance	GPA (rat)	SLT (mouse)
EMS	10, 50	300
ENU	20, 80	250
MMS	20, 50	20
MMC	1, 2.5, 5.0	5.25
MIH	20, 100, 300	400, 600

[a] The genetic endpoint investigated was 6-TG^r.

So far, negative results with whole-mammal mutagenicity tests
obtained with chemicals active in phase I tests are often limited to their
clastogenic activity in bone marrow cells. Whether the reactive metabolites reach the target tissue or the mutagenic metabolites induce only
gene mutations cannot be decided in such cases. With the GPA this can
be verified with a combined gene and chromosome mutation test after
local and systemic application of the test compound.

5. Risk for Germ Cells

Data on mutagens tested to date in the GPA indicate that compounds active in the granulation tissue after systemic application
represent a high risk for germ cells. The GPA might therefore be used
as an efficient indicator system to identify chemicals with a potential
risk for germ cells.

This statement can be substantiated when data from the GPA are
compared with those obtained in the mouse specific-locus test (SLT), a
direct test for gene mutations in germ cells.[7] It is problematic to
compare the rate of transmissible gene mutations induced in a specific
cell type during spermatogenesis in mice with mutation frequencies at
a different locus in somatic mesenchymal cells from rats. But as an
approximation we compare the effective exposure range of chemicals.
The available data from five chemicals tested in both systems are
summarized in Table 3. Clearly chemicals mutagenic in the granulation
tissue after systemic application represent a high risk for mutation
induction in germ cells, at least for substances for which no blood–
testicular barrier exists.

6. Relationship between Mutagenic and Carcinogenic Events

In standard long-term carcinogenicity tests only the biological consequences of the carcinogen are described but its mode of action remains unknown. In order to define priorities in elimination, protection, and regulation of a potential carcinogen, the initiating versus promoting activity of a chemical should be defined. This can be achieved in the GPA with mesenchymal cells. Mutagenic and carcinogenic events can be determined in a combined approach in the same cell population, within an organ, and in an intact animal. The *in vivo* situation is mandatory not only because of the complexity of absorption, distribution, activation, detoxification, and excretion of chemicals, but also because preneoplastic tissues and tumors are definitely characterized by their morphology and growth behaviour within the surrounding tissue.

For the assessment of carcinogenic effects in the GPA, rats are kept alive after an initial single or multiple dose of the test substance. The injected air that forms the pouch is absorbed within 2 weeks. Fibrosarcomas developing in the area of the granuloma pouch can be palpated and time of appearance and growth can be observed.[37,38]

6.1. Tumor Formation

With MNNG and procarbazine the induction of specific-locus mutations in the granulation tissue (Figure 4 and 5) and the development of tumors was determined using identical exposures.[37,38] In carcinogenicity studies MNNG was injected in single exposures of 0.1 and 0.6 mg/GP and procarbazine of 20 and 300 mg/kg into the pouch, orally and intraperitoneally. A small amount of croton oil was administered at the time of pouch formation. After a mean latency period between 35.5 and 60.0 weeks, fibrosarcomas of various histopathological types developed at site of pouch formation and other body sites. These results demonstrate that lesions induced with a single dose of MNNG or procarbazine are sufficient to transform cells capable of growing out to tumors.

With s.c. injection a much weaker carcinogenic effect was noted.[37,38] This proves that application of an air pouch alone[9] or combined with a tumor promoter enhances drastically the carcinogenic activity of a chemical. However, croton oil itself was not mutagenic in the gene mutation test and the use of croton oil as an additional growth stimulant did not enhance significantly MNNG-induced 6-TG^r mutants.

TABLE 4. Comparison of the Mutagenic and Carcinogenic Activity of
MNNG and Procarbazine in the GPA

Chemical exposure	Application	Animals with fibrosarcoma at site of GP	Tumors at other sites	Mutation frequency (6-TG^r × 10^{-6})
MNNG				
2.4 mg/kg	Into GP	86.7%	13.6%	251.3^a
25 mg/kg	IP	NT^b	NT^b	16.8
Control	—	0	6.7%	2.1
Procarbazine				
300 mg/kg	Into GP	15.0%	30.0%	236.3
300 mg/kg	IP + PO	12.5%	42.5%	252.4
Control	—	0	16.7%	5.3

^{a} Exact dose per kg calculated per animal weight was 2.7 mg/kg.
^{b} NT Not tested.

With both compounds the carcinogenic metabolites behave similar
to the mutagenic metabolites found in mutagenicity tests. With MNNG
mainly local sarcomas at the site of GP formation were found, whereas
with procarbazine tumors appeared predominantly at other body sites
(Table 4).

From growing tumors, cells can be dissociated and cultured *in vitro*.
These *in vivo* transformed cell populations serve as valuable reference
cultures for the evaluation and definition of markers for preneoplastic
or transformed cells derived from treated, freshly isolated granulation
tissue cells.[12]

6.2. Correlation and Mechanisms

The overall correlation existing between the mutagenic and carcinogenic activity of chemicals does not prove causality between the two
events. The degree of correlation depends on the test compound, the
mutagenic endpoint examined, the target organ, and the species investigated. Most likely we are dealing with a multistep process, probably a
sequence of mutagenic and epigenetic events.

Tumor initiation is considered to be a mutagenic event depending
on the type of DNA damage, on error-prone repair, and/or replication.
Cell growth necessary for tumor formation might be initiated by a
second mutation, e.g., at a regulatory locus, that leads directly to the
loss of cellular growth control under tumor-promoting conditions, or
in genes that alter receptors on cell membranes, leading to an enhanced

susceptibility to exogenous or endogenous tumor promoters. Alternatively, c-*onc* genes are activated pathologically by their rearrangement in the genome.[31] Comparing the mutagenicity and carcinogenicity data obtained in the GPA with MNNG and procarbazine we observe a close correlation between the dose applied, the tumor incidence, and the gene mutation frequency (Table 4).

The types of genetic events involved can be described as follows: At the time of treatment the granulation tissue *in situ* contains on an average at least 60×10^6 fibroblast-like cells. With 0.6 mg MNNG/kg, the most efficient carcinogenic dose, we created, therefore, 15×10^3 cells carrying a mutation at the 6-TG^r locus or 4.3×10^3 cells carrying a base pair substitution leading to Qua^R. Based on the total number of structural genes in a mammalian genome (5×10^4?[27]), every cell carries at least one mutation. Assuming that the initial event in tumor formation follows the same kinetics as the induction of a gene mutation and that the tumor has a clonal origin, we conclude that the critical initial DNA alteration can be highly specific and might occur in a single fibroblast-like cell.

With higher doses of MNNG, mutation frequencies decreased due to cytotoxic effects, which can be detected by a reduced primary cloning efficiency of freshly isolated cells 2 days after treatment *in vivo*.[19] Correspondingly, tumor frequency decreased in a dose-dependent way when MNNG was applied in cytotoxic doses ranging from 1.25 to 5 mg per pouch.[36]

This direct relationship even in the cytotoxic/cytostatic dose range can be obtained in the GPA because MNNG is a compound that is active mainly locally in the exposed GP cell population. Furthermore, both the genetic and mutagenic approaches detect individual cells that survive treatment and perform several cell divisions before being recovered either as a mutant cell forming a clone or as an *in vivo* transformed cell grown into a fibrosarcoma.

The decreased tumor incidence after treatment with cytostatic/cytotoxic exposures suggests that only cells from specific subpopulations (stem cells?) have the ability to escape the immunological control of the host and act as tumor progenitors. These cells are more susceptible to the test compound than the cells surrounding them and therefore the tumor incidence is decreased with higher exposures. The enhanced carcinogenic effect after the simultaneous application of croton oil during pouch formation before the application of MNNG might be due to an enlargement of this tumor progenitor-cell population during growth of the subcutaneous granulation tissue. It is likely that the tumor promoter alters cells irreversibly, inducing a state in which specific gene mutations are sufficient to initiate tumor growth.

In order to establish a quantitative relationship between the mutagenic and carcinogenic activity of a chemical one should be able to distinguish between mutations induced in the hypothetical tumor progenitor cell population and those in the other *in vitro* clonable fibroblast-like cells.

7. Technical Problems, Reproducibility, Use of Other Species

The GPA can be performed in laboratories equipped for cell culture work. An animal facility is necessary to keep the rats under standardized conditions. Inherent to the concept is its flexibility with respect to application schedule, test compounds, and mutagenic endpoints. This assures the optimal testing of each chemical. Further development of the methodology can be expected from advances in the rapidly progressing field of cell biology. Furthermore, the *in vivo–in vitro* concept stimulates methods that give insight into the culture conditions of freshly isolated cells.

Reproducibility is within the range acceptable for *in vivo* tests. The main source of variability is the handling of cells *in vitro*. This variability can be minimized with rigid laboratory practice protocols. Cell recovery, trypan blue exclusion, and platting efficiencies after cell isolation, as well as growth densities after each 3-day expression period and primary and secondary cloning efficiencies are recorded and compared to standardized values. For one data point in general 3–4 animals are sufficient to give statistically significant data. Further improvement could be obtained by the use of inbred rats. However, the risk of specific responses toward specific chemicals has to be considered.

In principle in the GPA other strains and species can be used. The limitation is that a loose connection must exist between skin and muscles. With Wistar rats, e.g., this connection is a little tighter toward the side of the skin, but the tissue is still easy to dissect out. With Chinese hamsters and mice we found that this junction between the superficial muscles of the skin and the fascia of the muscle is too loose in order to define a target tissue by an air pouch. Although cells are recoverable and culturable *in vitro*, locally enhanced cell proliferation cannot be achieved, and chemicals applied into the pouch will be distributed over the whole body. It might well be that other strains of mice or hamster are suitable for the GPA.

8. Significance of the Test and Future Developments

In contrast to other dividing cell populations *in vivo* (e.g., bone marrow) suitable for studying genetic endpoints, in the GPA single cells are obtained after dissociation of the target tissue. This effort is justified because the animal model offers several advantages. These are:

1. Different mutagenic events, including gene mutations induced *in vivo*, can be measured within the same target cell population.
2. Target cells can be exposed directly or systemically to test compounds, thus giving useful information on the pharmacokinetic behavior of genotoxic metabolites.

   The data collected with the chemicals so far tested clearly demonstrate the significance of the information that can be obtained.
3. Volatile chemicals can be tested.
4. Chemicals available only in small amounts can be tested *in vivo*.
5. A large number of diploid metaphases with an excellent resolution of the chromosomes at different intervals after treatment can be recovered.
6. Mutagenic events occur in a tissue from which tumors can arise.
7. The accessibility of the tissue to local and systemic treatments allows one to test possible synergistic/protective effects between chemicals or between chemicals and the animals diets.

Cells treated *in vivo* in the GPA respond in a qualitatively similar way to those in mammalian cell culture system when the comparison is based on the same genetic endpoints. This gives the assay a key role in evaluating the relevance of *in vitro* mammalian cell culture mutagenicity tests. On the other hand, the GPA shows that quantitative results are influenced dramatically *in vivo* by the pharmacokinetics of the mutagenic metabolites and therefore can only be obtained in whole-animal tests. Additional chemicals should be tested to further define the potential and limitations of the GPA.

Developments in the future need to focus on three aspects: (1) the recovery of mutant cells from granulation tissue that cannot be cloned on a substrate in petri dishes *in vitro*, (2) the use of the GPA in the detection of solid carcinogens, such as asbestos, and (3) the detection of *in vivo* induced transformed or preneoplastic cells with selective methods *in vitro*.

The data accumulated so far indicate, however, that the GPA is already a valuable phase II mutagenicity test in its present form.

ACKNOWLEDGMENTS

The author wishes to express his gratitude to Prof. G. Zbinden for his support of this work and for stimulating discussions. Further, he gratefully acknowledges the contributions of Dr. K. Frei, B. Weibel, and H. P. Schawalder, who are studying different aspects of the GPA.

This research was supported by the Swiss National Science Foundation, contracts 3.819-0.79 and 3.935-0.82.

9. References

1. S. Aho, P. Lethinen, and E. Kulonen, Penetration of macrophage ribonucleases into fibroblasts and the effects on nucleic acid and collagen metabolism, *Acta Pathol. Microbiol. Immunol. Scand. C 90* 147–154 (1982).
2. P. C. Bragt and I. L. Bonta, In vivo metabolism of [1-^{14}C] arachidonic acid during different phases of granuloma development in the rat, *Biochem. Pharmacol. 28*: 1581–1586 (1979).
3. G. Bronzetti, E. Zeiger, and H. V. Malling, Genetic toxicity of procarbazine in bacteria and yeast, *Mutat. Res. 68*, 51–58 (1979).
4. A. Dalhoff, G. Frank, and G. Luckhaus, The granuloma pouch—an in vivo model for pharmacokinetic and chemotherapeutical investigations. I. Biochemical and histological characterization, *Infection 10*, 354–360 (1982).
5. A. Dalhoff, G. Frank, and G. Luckhaus, The granuloma pouch—an in vivo model for pharmacokinetic and chemotherapeutical investigations, II. Microbiological characterization, *Infection 11*, 41–46 (1983).
6. R. F. Diegelmann, I. K. Cohen, and A. M. Kaplan, Effect of macrophages on fibroblast DNA synthesis and proliferation, *Proc. Soc. Exp. Biol. Med. 169*, 445–451 (1982).
7. U. H. Ehling, Specific-locus mutations in mice, in: *Chemical Mutagens, Principles and Methods for Their Detection* (F. J. de Serres and A. Hollaender, eds.), Vol. 5, pp. 233–256, Plenum Press, New York (1978).
8. T. E. Eling, J. A. Boyd, G. A. Reed, R. P. Mason, and K. Sivarajah, Xenobiotic metabolism by prostaglandin endoperoxide synthetase, *Drug Metab. Rev. 14* (5), 1023–1053 (1983).
9. B. G. Flückiger, Wachstum and chemisch induzierte maligne Transformation von Zellen eines subcutanen Granuloms der Ratte unter dem Einfluss von Stimulatoren der Zellproliferation, Dissertation ETH Nr. 7292 Swiss Federal Institute of Technology, Zurich, Switzerland (1983).
10. K. Frei, P. Maier and G. Zbinden, Aldrin epoxidase activity in freshly isolated cells from extrahepatic rat tissues, (submitted) (1984)
11. B. Helpap and H. Cremer, Autoradiographische Untersuchungen am Granulationsgewebe mit radioaktiv markiertem und unmarkiertem Thymidin, *Virchows Arch. Abt. B Zellpathol. 10*, 145–151 (1972).
12. C. Holzer, P. Maier and G. Zbinden, Mesenchymal rat cells stimulate growth of their transformed counterpart in vivo and in vitro, Experientia, (in press) (1984)
13. M. A. Kastenbaum and K. O. Bowman, Tables for determining the statistical significance of mutation frequencies, *Mutat. Res. 9*, 527–549 (1970).

14. B. Lang, P. Maier and G. Zbinden, Metabolic activation of aldrin in extrahepatic cells, Experientia, (in press) (1984).

15. I. P. Lee and G. Zbinden, Differential DNA damage induced by chemical mutagens in cells growing in a modified Selye's granuloma pouch, *Exp. Cell Biol. 47*, 92–106 (1979).

16. S. J. Leibovich and R. Ross, A macrophage dependent factor that stimulates the proliferation of fibroblasts *in vivo*, *Am. J. Pathol. 84*, 501–513, (1976).

17. P. Maier, The granuloma pouch assay for mutagenmicity testing, *Arch. Toxicol. 46*, 151–157, (1980).

18. P. Maier, P. Manser, and G. Zbinden, Granuloma pouch assay. I. Induction of ouabain resistance by MNNG *in vivo*, *Mutat. Res. 57*, 159–165 (1978).

19. P. Maier, P. Manser, and G. Zbinden, II. Induction of 6-thioguanine resistance by MNNG and benzo[a]pyrene *in vivo*, *Mutat. Res. 77*, 165–173 (1980).

20. P. Maier and G. Zbinden, Specific locus mutations induced in somatic cells of rats by orally and parenterally administered procarbazine, *Science 109*, 299–301 (1980).

21. P. Maier, B. Weibel, and G. Zbinden, The mutagenic activity of BrdU *in vivo,Environ. Mutagene. 5*, 695–703 (1983).

22. P. Maier, K. Frei, B. Weibel, and G. Zbinden, Granuloma pouch assay IV. Induction of sister chromatid exchanges *in vivo*, *Mutat. Res. 97*, 349–357 (1982).

23. P. Maier, B. Lang, H. P. Schawalder, and G. Zbinden, Influence of physiological pO$_2$ *in vitro* on growth, enzymic activity and mutation frequency in freshly isolated mesenchymal rat cells, (submitted).

24. P. Maier, H. P. Schawalder, and G. Zbinden, The mutagenic activity of heptocarcinogens in an extrahepatic tissue *in vivo* in rats, (submitted).

25. B. E. Matter and J. Grauwiler, Micronuclei in mouse bone-marrow cells. A simple *in vivo* model for the evaluation of drug induced chromosomal aberrations. *Mutat. Res. 23*, 239–249 (1974).

26. S. Murato and I. Morita, Prostaglandin-synthesizing systems in rat liver: Changes with aging and various stimuli, in: *Advances in Prostaglandin and Thromboxane Research* (B. Samuelsson, P. W. Ramwell and R. Paoletti, eds.), Vol. 8, pp. 1495–1506, Raven Press, New York (1980).

27. J. V. Neel, a consideration of two biochemical approachest approaches to monitoring human populations for a change in germ cell mutation rates, in: *Genetic damages in Man Caused by Environmental Agents* (K. Berg, ed.), pp. 29–62, Academic Press, New York (1979).

28. R. Parkin, H. B. Waynforth, and P. N. Magee, The activity of some nitroso compounds in the mouse dominant lethal assay, *Mutat. Res. 21*, 155–161 (1973).

29. J. M. Parry, E. M. Parry, and J. C. Barrett, Tumor promoters induce mitotic aneuploidy in yeast, *Nature 294*, 263–265 (1981).

30. A. R. Peterson, D. F. Krahn, H. Peterson, C. Heidelberger, B. K. Bhuyan, and L. H. Li, The influence of serum components on the growth and mutation of Chinese hamster cells in medium containing 8-azaguanine, *Mutat. Res. 36*, 345–356 (1976).

31. M. Radman, P. Jeggo, and R. Wagner, Chromosomal rearrangement and carcinogenesis, *Mutat, Res. 98*, 249–264 (1982).

32. H. Selye, Use of granuloma pouch technique in the study of antiphlogistic corticoids, *Proc. Soc. Exp. Biol. Med. 82*, 328–333 (1952).

33. T. Tsutsui, T. H. Maizumi, J. A. McLachlan, and J. C. Barrett, Aneuploidy induction and cell transformation by diethylstilbestrol: A possible chromosomal mechanism in carcinogenesis, *Cancer Res. 43(8)*, 3814–3821 (1983).

34. H. Vainio and E. Hietanen, Role of extrahepatic metabolism, in: *Concepts in Drug Metabolism* (P. Jenner and B. Testa, eds.), Part A, pp. 251–284, Marcel Dekker, New York (1963).
35. T. Wolff, E. Deml, and H. Wanders, Aldrin epoxidation a highly sensitive indicator specific for cytochrome P-450 dependent monooxygenase activities, *Drug Metab. Disp.* 7, 301–305 (1979).
36. G. Zbinden and C. Schlatter, New approaches to mutagenicity and carcinogenicity testing *in vivo* in mammalian systems, in: *Chemical Toxicology of Food* (C. L. Galli, R. Oaoletti, and G. Vettorazzi, eds.), pp. 153–166, Elsevier, Amsterdam (1978).
37. G. Zbinden, P. Maier, and S. Alder, Granuloma pouch assay, III. Enhancement of the carcinogenic effect of MNNG, *Arch. Toxicol.* 45, 227–232 (1980).
38. G. Zbinden and P. Maier, Single dose carcinogenicity of procarbazine in rats, *Cancer Lett.* 21, 155–161 (1983).

CHAPTER 9

The Use of Multiply Marked *Escherichia coli* K12 Strains in the Host-Mediated Assay

G. R. Mohn, P. R. M. Kerklaan, and P. A. van Elburg

1. Introduction

Due to the development in the last decade of rapid and sensitive bacterial genetic test systems for detecting potential mutagens and carcinogens[1-3] there has been a recrudescence of testing of environmental chemicals for genotoxicity. Experiments performed in particular with the *Salmonella*/mammalian microsome test,[4,5] in which the ability of chemicals to cause reversion in auxotrophic *Salmonella* strains is determined on selective agar medium under the influence of various mammalian organ fractions, have yielded results on a variety of environmental compounds that show definite mutagenic activity in this assay system and are therefore to be considered as potentially mutagenic and carcinogenic in animals (see also Refs. 6 and 7).*

In spite of a striking *qualitative* correlation between the ability of chemicals to induce genotoxic effects in bacteria and their mutagenic

G. R. Mohn, P. R. M. Kerklaan, and P. A. van Elburg • Department of Radiation Genetics and Chemical Mutagenesis, State University of Leiden, 2333 AL Leiden, The Netherlands.

and tumor-initiating properties in animals,[3,8–10] attempts at *quantitatively* and directly relating the mutagenic potency of chemicals in bacterial plate tests with their genotoxic (including tumor-initiating) potency in exposed animals[11,12] have raised an active discussion about the feasibility of this approach.[13–15] The general conclusion has been that, depending on the chemical class investigated, results of mutagenicity plate tests using exclusively DNA repair-deficient tester strains may lead to gross overestimations or underestimations of the actual genotoxic potency of chemicals in animals (for review see Refs. 6 and 7); the main reason for the shortcomings of *in vitro* tests is that the distribution and metabolism of the test compounds are not adequately represented, which may lead to large misinterpretations of the actual *dose* required to produce a given genetic effect *in vivo*.

On the other hand, comparative mutagenesis studies performed in various organisms indicate that the available bacterial genetic test systems can be employed for obtaining more quantitative information about the spectrum of genetic changes induced by chemicals at the gene and the nucleotide level, the influence of DNA (mis)repair in this process, and the probability of DNA alterations leading to heritable changes (mutagenic potency). In addition to the high empirical correlation between mutagenesis in bacteria and in mammalian cells mentioned above (see also Refs. 16 and 17), this assumption is further supported by the findings that in some cases a good correlation is observed between alkylating activity (with DNA or the model nucleophilic chemical nitrobenzylpyridine), mutagenicity in bacteria, and mutagenicity in higher organisms, including mammalian cells,[21–23,52–57,74,75] and by the observations that some genetic changes, including gene mutations, can be causally related to the initiation of neoplastic changes in animal and human cells.[18–20]

To obtain, therefore, quantitative information on the presence, amount, and persistence of genotoxic factors in treated animals it would appear necessary to expose the bacterial test systems directly to the metabolism of these animals. First attempts at doing this were made in the "host-mediated assay" developed by Gabridge *et al.*,[24] in which indicator bacteria (*Salmonella typhimurium* cells) were injected into the peritoneal cavity of mice that were subsequently treated with the substance under test and from which the bacteria were recovered and tested for induced genetic effects (back mutations from auxotrophy). At present, the procedure has been improved, both with regard to the genetic endpoints that can be assayed for (various gene mutations and differential DNA repair) and the organs of treated animals in which genotoxic effects can be determined (liver, spleen, lungs, kidneys, pancreas, intestine, and the blood stream). It is the purpose of the present review to describe genetic endpoints and strains of *E. coli* K12 that are now available for use in different host-mediated assay procedures.

2. Genetic Endpoints

Escherichia coli was chosen as the indicator organism because it is best characterized genetically and physiologically and because of its symbiotic adaptation to mammalian species, thereby allowing various types of host-mediated assays (intrasanguineous, intraintestinal) to be performed. Furthermore, in *E. coli* K12, several genetic endpoints can be easily scored, such as gene mutations and differential DNA repair, which have been shown repeatedly as being sensitive to the action of mutagens and carcinogens (see, for example, Refs. 1, 3, 9, 10, and 25). For economic reasons and also to allow an accurate comparison between different genetic endpoints, all strains described here are derivatives of *E. coli* K12/343/113, which has already been used in various mutagenicity test procedures, especially in host-mediated assays.[26–29,76]

2.1. Gene Mutations

The *E. coli* K-12 strain 343/113 is a multipurpose indicator strain specifically constructed for the simultaneous determination of genetic effects induced in various genes and gene loci. Among the forward and back mutation systems developed,[27,30] a few have been retained that exhibit high sensitivity to various genotoxic agents. These include the NAL^{res}, VAL^{res}, and MTR^{res} forward mutation systems: the growth of wild-type *E. coli* K-12 is inhibited, among others, by the antimetabolites nalidixic acid (NAL), 1-valine (VAL), and 5-methyl-DL-tryptophan (MTR), and mutations in a number of genes can lead to resistance against them. Some of the genes involved are *trpR*, *trpO*, *trpE*, *mtr*, and *aroP* in the MTR^{res} system, and it is likely that even deletions (in the *trpR* gene) will lead to viable MTR^{res} mutants (see Ref. 30). In the NAL^{res} system, it has been shown that the genes *nalA* (*gyrA*), *nalB*, *nalC*, and possibly *sloB* are involved, while mutations in several of the *ilv* genes as well as permeability genes have been indicated as productive of the VAL^{res} phenotype.[31–33] The mutational systems NAL^{res}, VAL^{res}, and MTR^{res} have been calibrated with a variety of mutagens; the spectrum of detectable changes includes both base-pair substitution and frameshift-type mutational events.[23,26–30]

2.2. Differential DNA Repair

The isolation and characterization of *E. coli* mutants defective in the repair of damage induced by chemical and physical mutagens[34-39] has permitted not only the elucidation of various DNA repair pathways, including those responsible for induced mutagenesis (for reviews see Refs. 40 and 41), but also has led to the development of tests for "differential DNA repair," in which genotoxic agents are detected by their increased toxicity for the DNA repair-deficient strains as compared to the repair proficient parents.[42-44] Extensive calibration with chemicals of known mutagenicity and carcinogenicity has shown that agents that are active in the differential DNA repair test will also exhibit mutagenic and probably tumor-initiating properties.[45] Because different DNA lesions are substrates for various types of repair enzymes,[40,41] one can assume that a combination of different repair-deficient strains or that the use of strains marked with several repair deficiencies will minimize the probability that particular genotoxic agents may be overlooked. Furthermore, since the recovery of bacterial cells may vary between individual animals, a prerequisite for successfully performing differential DNA repair tests in the host-mediated assay will be that the strains differing in DNA repair capacity be injected simultaneously into mice and recovered from various organs, from which the individual survival of each strain can be determined separately; therefore, the strains were additionally marked with auxotrophic growth requirements, in analogy to derivatives of *Bacillus subtilis* previously used in an intraperitoneal host-mediated assay.[43]

2.3. Suitable *E. coli* Strains

The derivatives of *E. coli* K-12 strain 343/113 developed and calibrated for use in host-mediated assays are listed in Table 1. Strain 343/113 itself has been repeatedly used in host-mediated assay procedures[26,28-30] and the penicillin- and streptomycin-resistant substrain 343/540 has been isolated as indicator for intraintestinal host-mediated assays.

For performing differential DNA repair tests, a set of derivatives was constructed that are isogenic to 343/113 but have, in addition, mutational deficiencies in the genes *uvrB*, *recA*, *polA*, or *dam*-3. Furthermore, a double mutant, *uvrB/recA*, was isolated after conjugation of the *recA*₁₃ allele into the Δ*uvrB*₃₀₁ deletion-carrying derivative. To allow determination of the viable cell titer of each individual strain when mixtures of the various strains are being used, auxotrophic

TABLE 1. Derivatives of *E. coli* K12 Strain 343/113 for Use in Host-
Mediated Assays

Strain	Relevant genetic markers	Genetic endpoints scored
343/113[a]	$galR^s{}_{18}$, arg_{56}, nad_{113}, MTR^{sens}, NAL^{sens}, VAL^{sens}	Forward and back mutations in various genes
343/540	Same as 343/113, PEN^{res}, STR^{res}	Same as 343/113
343/358	Same as 343/113, *trp*$^-$	Differential DNA repair
343/636	Same as 343/113, Δ(*lac*$^-$, *pro*$^-$)	Differential DNA repair
343/673	Same as 343/113, Δ(*uvrB*, *bio*), *phe*$^-$	Differential DNA repair
343/415	Same as 343/113, *his*$^-$, *recA*	Differential DNA repair
343/447	Same as 343/113, *ile*$^-$, *polA*	Differential DNA repair
343/435	Same as 343/113, *dam*-3, *thr*$^-$	Differential DNA repair
343/591	Same as 343/113, Δ(*uvrB*, *bio*), *recA*, *lys*$^-$	Differential DNA repair

[a] For further description of *E. coli* strain 343/113 and the genetic endpoints involved, see Refs. 48, 65 and 76.

markers were finally introduced that enable only one particular strain to form colonies on a specific growth agar medium. The auxotrophic growth requirements of the strains are *Trp*$^-$, Phe$^-$/Bio$^-$, His$^-$,Ile$^-$, Thr$^-$, and Lys$^-$/Bio$^-$ of the wild-type, *uvrB*, *recA*, *polA*, *dam*-3, and *uvrB*/*recA* derivatives, respectively. Furthermore, a Δ(*lac*/*pro*) deletion was introduced into the wild-type DNA repair strain 343/113, so that mixtures of this strain with, for example, the *uvrB*/*recA* derivative (which still carries the *lac*$^+$ allele) will produce, when plated on MacConkey-lactose agar, both lactose-nonfermenting white colonies originating from the *lac*$^-$ DNA repair-proficient strain and lactose-fermenting red colonies arising from the *lac*$^+$, *uvrB*/*recA* strain.

In order to enhance the permeability of the strains to large molecules, such as polycyclic hydrocarbons and aromatic amines, further derivatives were isolated as spontaneous or UV-induced mutants resistant to phage T7 (and T3). Some of the mutants obtained this way are characterized by a deficiency in the cell-wall lipolysaccharide layer (LPS) and the phage T7 receptors,[46] which also leads to greatly enhanced sensitivity to the lethal action of crystal violet. These mutants remain to be further characterized, but are probably allelic to *rfa* or *lps* mutants of *E. coli* and *Salmonella*.[46,47]

3. Animal-Mediated Assays

In the original host-mediated assay devised by Legator and colleagues, the indicator bacteria (*Salmonella*) were injected into the peritoneal cavity of mice, from which they were recovered and tested for induced genetic effects; this procedure allowed the demonstration that dimethylnitrosamine, among other carcinogenic substances, is activated inside the animal body to factors that are mutagenic for the implanted bacterial cells.[24] Subsequently, modifications of this procedure were applied in order to bring the indicator bacteria in closer contact to various organs of the treated animals, e.g., after intravenous injection of *E. coli* cells into mice.[26] This procedure, called the intrasanguineous host-mediated assay, and a newly developed test in which the bacteria are allowed to establish themselves in the intestinal tract of rodents (intraintestinal host-mediated assay) are described below.

3.1. Intrasanguineous Type

After intravenous injection of bacterial cells into mice or rats, these undergo a blood clearance and accumulate in organs of the reticulo endothelial system, among others,[58,59] and can be found especially in liver and spleen in quantities large enough for the performance of mutational tests. This was first demonstrated using *E. coli* K12 strain 343/113,[26] in a procedure which is still employed, with some modifications: approximately 5×10^9 viable, stationary cells suspended in 0.2 ml of buffered saline are injected into a lateral tail vein of mice, to which the substance under test is then administered. Depending on the type of experiment, the test substance can, of course, also be administered before the injection of the bacteria. After a certain period of time (usually up to 4–5 hr), the animals are killed by cervical dislocation, and the liver, spleen, kidneys, lungs, pancreas, and 10 μl of blood are suspended in buffered saline. The bacterial cells present are processed further as described below for the determination of induced gene mutations or differential DNA repair. The titer of viable cells in this type of experiment 120 min after i.v. injection of *E. coli* K12/343/113 into mice is shown in Table 2. The table shows that sufficient numbers of *E. coli* cells can be recovered from liver, spleen, lungs, kidneys, pancreas, and the blood stream for the performance of differential DNA repair tests (for which, in principle, $\sim 10^3$ viable cells of the wild-type strain are enough). For the determination of gene mutation frequencies with accuracy, it appears that only the liver and the spleen can be used as target organs, i.e., those in which more than 10^7 cell accumulate. The further processing of the cells present in the various organs is performed as follows.

TABLE 2. Survival of *E. coli* K12 Cells in Various Organs of Mice 120 min after Intravenous Injection[a]

Organ	Titer per ml	Titer per organ
Liver	9.3×10^7	4.7×10^8
Spleen	1.2×10^7	6.0×10^7
Lungs	2.4×10^5	1.2×10^6
Kidney	6.0×10^3	3.0×10^4
Pancreas	3.0×10^3	1.5×10^4
Blood (10 μl)	5.0×10^2	1.0×10^5

[a] A total of 0.2 ml of stationary phase cells (titer ~5 × 10^9 per ml) of strain 343/113 was injected into a lateral tail vein of Swiss albino mice. After 2 hr, the mice were killed by cervical dislocation and the various organs removed and suspended in 5 ml of buffered phosphate saline, pH 7.0. The suspensions were homogenized in a Potter tube with Teflon pestle, and viable cell counts were performed by plating diluted aliquots of the suspensions over supplemented growth agar. The titer of the bacteria in the blood stream was determined by assuming a total blood volume of 2 ml in mice weighing ~20 g.

3.1.1 Induction of Gene Mutations

The liver and spleen suspended in buffered saline are homogenized in a Potter tube (or with an Ultra-Turrax homogenizer) and centrifuged at $200 \times g$ to eliminate larger cell debris. The supernatant is then centrifuged for 10 min at $7500 \times g$ and 4°C and the pellet containing the *E. coli* cells is suspended in nutrient broth or in fully supplemented minimal growth medium. The bacterial suspension, with an original viable cell titer of $\sim 10^8$ per ml, is then incubated overnight (~ 12 hr) at 37°C in the dark in a rotary shaker. During this incubation period, the viable cell titer reaches a value of $\sim 10^9$ per ml, which is equivalent to a mean number of 3–4 cell divisions and is necessary and sufficient for the genetic fixation and phenotypic expression of eventually induced premutations (see Ref. 48). After this posttreatment incubation, the bacterial cells are washed free of excess nutrients and resuspended in buffered saline (viable cell titer $\sim 10^9$ per ml). Aliquots of the undiluted bacterial suspension are then spread over selective media for the determination of mutations in various genes and/or gene loci, such as NAL[res], VAL[res], and MTR[res], as described earlier (see Section 2.1).

3.1.2. Differential DNA Repair

In this assay, a mixture of wild-type repair-proficient strain and one or more DNA-repair-deficient derivatives is injected i.v. into mice,

TABLE 3. Recovery of *E. coli* K12 Strain 343/358 (Repair-Proficient) and
of Different DNA-Repair-Deficient Substrains from Various Organs of Mice
after Intravenous Injection

Organ	Experiment number	Wild type (343/113)	Relative recovery,[a] % wild type				
			uvrB	*recA*	*polA*	*dam*-3	*uvrB/recA*
Liver	1	100	103	72	70	84	—
	2	100	92	69	106	112	84
Spleen	1	100	87	80	83	76	—
	2	100	89	78	80	92	76
Lung	3	100	—	—	—	—	88
Kidney	3	100	—	—	—	—	96
Blood (10 μl)	1	100	116	63	96	93	—
	3	—	—	—	—	—	82

[a] The experiments were performed essentially as described in the footnote to Table 2. In experiment 3, mixtures of strain 343/636 (wild-type DNA repair, Δ*lac*⁻) and of the *uvrB/recA* (Lac⁺) derivative were employed and the survival of the individual strains was determined by plating aliquots of the diluted samples over McConkey-lactose agar medium, on which the Lac⁺ strain gave red colonies upon incubation for 24 hr at 37°C. Survival of the various strains employed in experiments 1 and 2 was determined on growth media supplemented with speficic growth factors required, as mentioned in Table 1.

which are subsequently treated with the test substance and from which various organs are then removed and suspended in buffered saline. The suspensions are then homogenized in a Potter tube, or, better, with an Ultra-Turrax homogenizer. The survival of the individual strains is determined by plating aliquots of appropriately diluted samples over selective growth media on which only one particular strain can grow and form colonies, due to the presence in this medium of the specific growth factors required. The recovery of *E. coli* K12/343/113 and of various DNA-repair-deficient substrains from various organs of otherwise untreated mice 120 min after i.v. injection of the bacteria is shown in Table 3. The results demonstrate that mixtures containing five different strains (343/113 and the derivatives *uvrB*, *recA*, *polA*, and *dam*-3) can be used and the survival of each strain determined accurately; furthermore, it appears that the recovery of the various strains in liver, spleen, and the blood stream is not substantially different from that of the wild-type strain, with the possible exception of *recA* cells, which seem to undergo a more pronounced clearance rate from these organs, a phenomenon that remains to be elucidated and had already been observed by Kada *et al.*[43] in their experiments using mixtures of wild-type and *rec*⁻ strains of *Bacillus subtilis*. The difference in survival, however, is not of such amplitude as to require discarding the use of *recA* strains in this type of host-mediated assay.

3.2. Intraintestinal Type

This type of host-mediated assay was devised to broaden the range of organs under study and to include the intestinal tract as an organ, at least the large bowel, which is a preferred site for the occurrence of malignant tumors induced by various chemical carcinogens.[49] Again, *E. coli* appeared as an organism of choice in such studies, since it is a natural component of the microbial intestinal flora of mammalian species and is likely to establish itself quite rapidly in various parts of the intestinal tract upon oral administration into mice or rats. A prerequisite for a successful implantation is that the normal *E. coli* flora of the host animal be replaced as quantitatively as possible with the desired indicator strain and that those microbial species that might lead to infections on the selective mutation media be reduced to a minimum. To reach this, a derivative of strain 343/113 was isolated (strain 343/540) upon sequential selection on media containing streptomycin (50 μg/ml) and penicillin G (50 μg/ml).

The feasibility of intraintestinal host-mediated assays using indicator strain *E. coli* 343/540 has been assayed with rats as host animals, in the following procedure: female Wistar rats, SPF, weight ~200 g, received orally 0.5 ml of an aqueous solution of penicillin G and streptomycin sulfate to reach a final concentration of 5 mg of each antibiotic per rat. After 4 hr, 0.5 ml of a suspension of stationary cells (titer ~10^9 per ml) of strain 343/540 in buffered saline containing 50 μg of streptomycin sulfate and 50 μg of penicillin G per ml were applied again orally to the rats. The establishment of the indicator bacteria in the intestinal tract was then determined by assessing the titer of viable cells of strain 343/540 in the feces of the animals and in contents of the small intestine and the colon. Results of an experiment are shown in Table 4. They demonstrate that, indeed, the penicillin- and streptomycin-resistant *E. coli* cells spread out and can be recovered from this organ and from collected feces in amounts sufficient for performing mutation frequency determinations. The results also show that under the conditions of the present experimental setup, i.e., with only one high initial dose of penicillin and streptomycin to the host animals, the *E. coli* indicator cells remain established up to 3 days after administration before they become gradually replaced by other members of the microbial flora.

TABLE 4. Establishment of Cells of *E. coli* K12/343/540 in the Intestinal Tract of Mice

Time schedule, hr	0	4	24	28	48	72
	Administration[a]	Administration[b]	Recovery of viable *E. coli* 343/540 cells[c]			
Small intestine			ND	≤10,000	<100	0
Colon			ND	1.6×10^9	5.5×10^6	1.2×10^6
Feces			4.7×10^8/g	5.1×10^8/g	3.7×10^8/g	2.5×10^7/g

[a] Oral administration of 0.5 ml of a solution of penicillin G and of streptomycin sulfate in phosphate buffer (5 mg per rat).

[b] Oral administration of 0.5 ml of a suspension of stationary cells of strain 343/540 in phosphate-buffered saline containing 50 μg of penicillin G and of streptomycin sulfate per ml (5×10^9 cells per rat).

[c] Titer of viable cells of strain 343/540 on growth medium containing penicillin G (50 μg/ml) and streptomycin sulfate (50 μg/ml). At the same periods indicated, the contents of the small intestine, the colon, and of ~100 mg feces were suspended in buffered saline and the viable cell titer determined. The values given are number of viable cells present in the organ (intestine, colon) or in 1 g feces. From 25 colonies grown, all had the phenotype PEN^rs, STR^r, Arg⁻, Nad⁻ and thus represented descendents of the originally inoculated 343/540 cells. ND, Not determined.

The sensitivity of this type of host-mediated assay to mutagens and carcinogens remains to be assessed more systematically, as well as the possible use of lipolysaccharide-deficient *E. coli* mutants as indicators; it is, however, likely that this assay may display analogous sensitivity to genotoxic chemicals as the intraintestinal tests previously described, in which *Salmonella* or yeast as indicators were used[50,51]; corresponding experiments to test this are presently underway.

4. Some Experimental Results

Shortly after the introduction of the intrasanguineous host-mediated assay it was recognized that this procedure, in which the indicator bacteria are removed from the livers of treated mice,[26] is more sensitive than the traditional assay. This is obviously due to a closer contact of the bacteria to the organ in which most xenobiotic metabolism is known to occur, namely the liver.[27,58] Several recent studies using various microbial indicators have also indicated that the intrasanguineous test can be more sensitive than tests performed *in vitro* with homogenates of rodent livers.[59–61] In the following, some further examples of the superiority of animal-mediated assays over *in vitro* tests are presented; furthermore, the use of intrasanguineous tests to determine the presence and amount of genotoxic (DNA-repair-eliciting) factors in various organs of treated mice are shown.

4.1. Comparison of *in Vitro* Tests with Host-Mediated Assays

4.1.1. Gene Mutations

This study was specifically undertaken to determine whether the mutagenic activity of selected dialkylnitrosamines with known carcinogenicity would be detected more efficiently in the intrasanguineous host-mediated assay than in *in vitro* tests using the same indicator bacteria and organ homogenates (S9) from the same strain of mice. The reason for performing these comparisons is that some of the nitrosamines are known to be carcinogenic but have not been detectable as mutagens in standard *in vitro* mutagenicity plate tests. The results, which have been reported *in extenso* elsewhere[28] and are summarized in Table 5, do indeed indicate that bacteria present in the livers of mice treated with dimethylnitrosamine (DMNA), methylethylnitrosamine (MENA), and diethylnitrosamine (DENA) are mutated at higher frequencies (and at lower chemical exposure levels) than bacteria treated

TABLE 5. Comparison of the Mutagenic Activity of Different
Dialkylnitrosamines in *in Vitro* Tests and in Animal-Mediated Assays,
Using the Induction of VAL[res] Mutations in *E. coli* K12 Strain 343/113 as
Indicator of Genetic Effects

Chemical	Mutagenicity tests "*in vitro*"[a]		Intrasanguineous host-mediated assay[b]	
	Concentration, mmole/liter	VAL[res] mutants per plate	Concentration, mmole/kg	VAL[res] mutants per plate
Dimethylnitrosamine (DMNA)	5.0	295	0.81	589
Methylethylnitrosamine (MENA)	5.3	217	0.68	273
Diethylnitrosamine (DENA)	14.9	112	0.59	682
Diethanolnitrosamine (DELNA)	149	5.8	0.45	84.2
Methyl-*n*-propylnitrosamine (MnPNA)	196	11.0	1.18	44.8
Diisopropylnitrosamine (DiPNA)	154	6.0	1.38	14.4
Control	—	6.5	—	10.3

[a] Stationary cells were exposed for 180 min at 37°C in the dark to the chemicals at the indicated concentrations, then washed and tested for induced VAL[res] mutants as described elsewhere.[28,48] The experiments were performed in the presence of S9 liver homogenates from Swiss albino mice from the same strain as those used for the host-mediated assays.
[b] The bacteria were isolated from the livers of treated animals 180 min after injection (i.p.) of the chemicals.

in vitro with the same chemicals in the presence of S9 liver homogenates from the same mouse strain. In addition, diethanolnitrosamine (DELNA), methyl-*n*-propylnitrosamine (MnPNA), and perhaps diisopropylnitrosamine (DiPNA) are mutagenically active in the host-mediated assay procedure, whereas significant mutagenic effects are not observed *in vitro*, with the possible exception of MnPNA. Whether the higher sensitivity of the host-mediated assay for dialkylnitrosamines is also representative of the carcinogenic potency of these chemicals in rodents cannot be established accurately because of lack of adequate carcinogenicity data in the same strain of mice. However, results from experiments using rats[62-64] indicate that the two strongest mutagens in the host-mediated assay, namely DMNA and DENA, are also the strongest carcinogens in rats, followed by MENA and DELNA, again in parallel with the mutagenicity data. The two weak carcinogens MnPNA and DiPNA are also the weakest mutagens in the host-mediated assay. Noteworthy is the fact that this parallelism is not observed when results from the *in vitro* tests are examined. A preliminary conclusion from these findings is that host-mediated assays, in addition to being more sensitive to the mutagenic action of carcinogenic nitrosamines

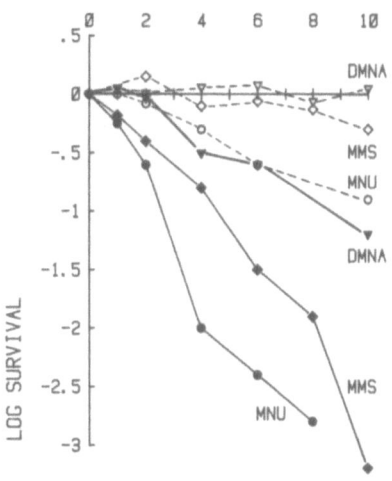

SUBSTANCE CONC. (mM)

FIGURE 1. Survival of wild-type and *recA* derivative of *E. coli* K12/343/113 after treatment *"in vitro"* with methyl methanesulfonate (MMS), methylnitrosourea (MNU), and dimethylnitrosamine (DMNA). Mixtures of stationary cells of the two strains were incubated for 120 min at 37°C in buffered saline, pH 7.0, under rotary shaking with the chemicals at the indicated concentrations. The individual survival of the wild type (dashed lines) and of the *recA* mutant (solid) was determined on minimal agar medium supplemented with the specifically required growth factors of the strains.

than *in vitro* tests, are also better representative of the *in vivo* (genotoxic) carcinogenicity of these compounds in animals.

4.1.2. Differential DNA Repair

In this study, the bacterial genetic endpoint used was differential killing of *recA* and wild-type *E. coli* cells (DNA repair test) exposed to the monofunctional methylating agents methyl methanesulfonate (MMS), methylnitrosourea (MNU), and dimethylnitrosamine (DMNA). The experiments were performed *in vitro*, either without (MMS and MNU) or with (DMNA) mouse liver S9 homogenates, and in the host-mediated assay using mice from the same population as host animals. The results are summarized in Figures 1 and 2, and are being published *in extenso* elsewhere.[65] In Figure 1, the effectiveness of the three methylating agents to induce *recA*-mediated repairable DNA damage in liquid suspension tests is shown, indicating a ranking of relative genotoxic potency in the decreasing order MNU>MMS>DMNA (+S9) under the present experimental conditions; a similar relative effectiveness is obtained when a *polA*/wild-type pair of *E. coli* strains is used (data not shown). When a host-mediated assay is performed, however, in which the indicator bacteria are removed from the livers of mice treated with the same chemicals, as shown in Figure 2, the ranking of

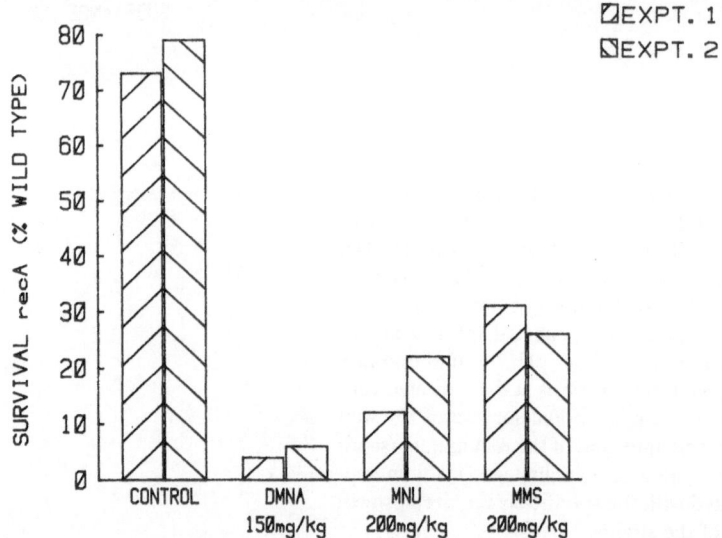

FIGURE 2. Survival of the *recA* derivative relative to the *E. coli* K12 wild type in the livers of mice treated with DMNA, MNU, or MMS. Mixtures (0.2 ml) of stationary cells of the two strains (~10^{10} viable cells/ml) were injected intravenously into Swiss albino mice (females, weight 25 g), which received immediately thereafter intraperitoneal injections (0.2 ml) of aqueous solutions of DMNA, MNU, or MMS to reach the concentrations indicated. Control mice remained untreated. Two hours after injection of the bacteria, the mice were killed by cervical dislocation and the livers removed and suspended in buffered saline after gentle homogenization in a Potter tube. The titers of viable cells of the *recA* derivative and the wild type were determined by plating aliquots of the diluted suspensions on minimal agar media containing specific growth factors for each strain.

genotoxic potency of the three chemicals is quite different; since the exposure levels of 200 mg/kg body weight (MMS and MNU) and 150 mg/kg (DMNA) represent roughly equimolar concentrations of 0.47– 0.5 mmole/kg, respectively, the genotoxic potency *in vivo* decreases in the order DMNA>MNU>MMS. Interestingly, this is similar to the relative carcinogenic potency of the three chemicals in rodents,[66–69] and obviously also reflects the degree of methylation of liver DNA in rats treated with these compounds.[70] It is probable that the low activity of DMNA in *in vitro* tests is due to its nonoptimal liver S9-mediated activation, as has been shown in the case of gene mutations.[28] Furthermore, it is very likely that the strong genotoxic agents *in vitro*, MMS and MNU, do undergo a rapid deactivation in the animal body before they reach the livers of the treated mice, as previously shown in rats,[71] while the relatively long half-life (5–6 hr) of DMNA[72] may prolong its availability for being metabolized in the liver and for inducing

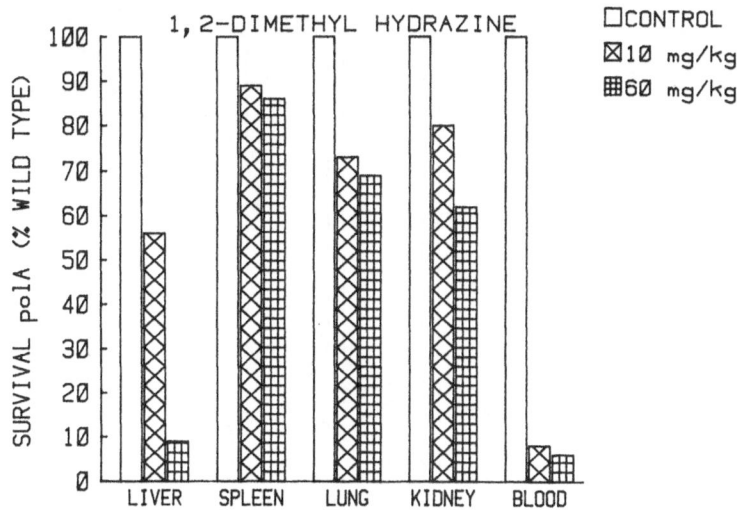

FIGURE 3. Differential survival of the *polA* derivative compared to the wild-type *E. coli* K12 parent in various organs of mice treated with 1,2-dimethylhydrazine (SDMH). Mixtures of both strains were injected intravenously into mice, which were subsequently treated with SDMH to reach the concentrations indicated (control mice were not treated). Three hours after injection of the bacteria, the mice were killed and the livers removed and suspended in buffered saline. These suspensions were then gently homogenized in a Potter tube. Individual survival of both strains in the mixtures was determined by plating on minimal agar containing specific growth factors required for each strain.

genotoxic effects in the bacteria present in that organ. These results again indicate that host-mediated assays are better representative of the presence and amount of genotoxic factors in livers of treated rodents than are *in vitro* tests using homogenates of that organ.

4.2. Organ-Specific Genotoxic Effects

A particular feature of the action of chemical carcinogens is the pronounced selectivity for inducing tumors in specific mammalian and human organs[73]; this effect is strongly dependent on the type of reactive species produced (and of the dose) and is probably the result of various processes inside the animal body, including, among others, the toxicokinetics of the compound and its metabolites and the probability of DNA adducts formed to lead to premutagenic (and tumor-initiating?) alterations, the so-called genotoxic potency. While host-mediated assays in which the indicator bacteria are recovered from the liver are now relatively well established, as demonstrated before, the accurate determination of genotoxic effects in other rodent organs was,

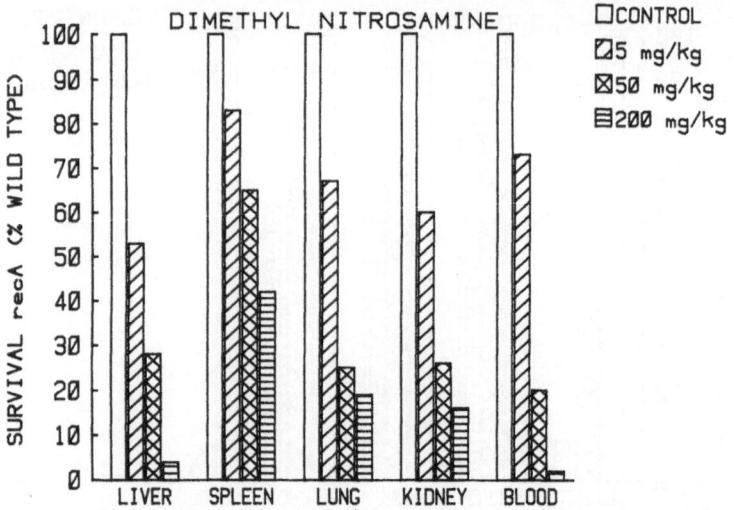

FIGURE 4. Differential survival of the *recA* derivative compared to the DNA-repair-proficient *E. coli* K12 parent in various organs of mice treated with dimethylnitrosamine (DMNA). The experiments were performed as described in legend of Figure 3.

up till now, difficult to realize due to the low recovery of viable cells from these organs and the thereby encountered difficulty in adequately measuring (induced) mutation frequencies.

The introduction of the differential DNA repair test in animal-mediated procedures is likely to overcome this difficulty, because a few hundred viable bacterial cells are sufficient for the determination of survival of both wild-type and DNA-repair-deficient derivatives. As demonstrated in Table 3, the recovery of intravenously injected wild-type and DNA-repair-deficient *E. coli* K-12 cells is such that differential survival assays can be performed with the bacteria present in liver, spleen, lungs, kidneys, pancreas, and the blood stream, among others (testes, bone marrow). First experiments were performed with known mutagens/carcinogens to determine the amounts of genotoxic factors (as measured by differential survival) present in liver, spleen, lungs, kidneys, and the blood stream of treated mice. The results are summarized in Figures 3–5 after treatment of mice with dimethylnitrosamine (DMNA), 1,2-dimethylhydrazine (SDMH), mitomycin C (MMC), 4-nitroquinoline-1-oxide (4-NQO), and *cis*-dichlorodiammineplatinum-II (*cis*-Pt), respectively. The results presented in Figures 3 and 4 were obtained with the *recA* derivative and the corresponding wild type (strain 343/358), while in Figure 5, results of differential survival of the

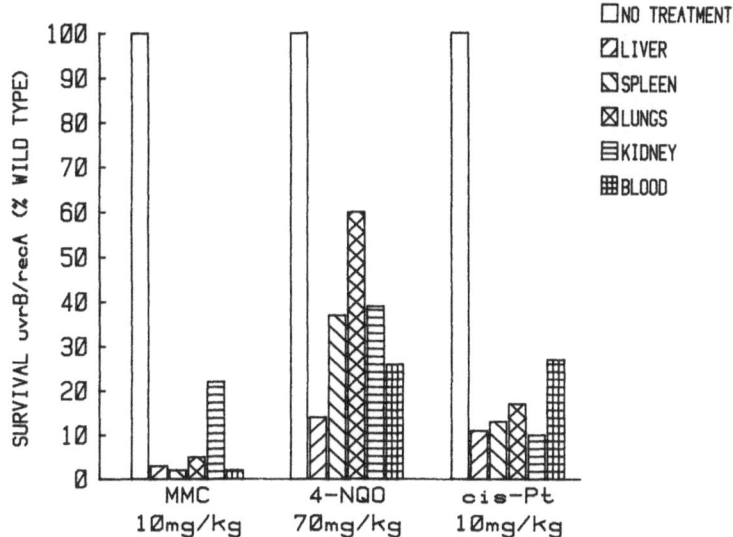

FIGURE 5. Differential survival of the *uvrB/recA* derivative compared to the wild-type *E. coli* K12 parent in various organs of mice treated with mitomycin C (MMC), 4-nitroquinoline-1-oxide (4-NQO), or *cis*-dichlorodiammineplatinum II (*cis*-Pt). Mixtures of both strains were injected intravenously into mice, which were subsequently treated with the chemicals to reach the concentrations indicated. Two hours after injection of the bacteria, the livers were removed and the viable cell titers of each strain were determined by plating dilutions of the suspensions on McConkey-lactose agar on which the *uvrB/recA* (Lac$^+$) strain produces red colonies, while colonies of the Lac$^-$ DNA-repair-proficient strain remain white.

double DNA-repair-deficient *uvrB/recA* strain relative to the 343/636 repair-proficient wild type are shown. Although preliminary in nature and based on only one or two mice per experimental point, the present results clearly indicate, first, that all tested chemicals are active in this host-mediated assay; second, both dose-dependent (Figures 3 and 4) and organ-specific (Figure 5) effects are evident, the strongest organ specificity being observed after treatment with those compounds that require bioactivation, e.g., in the liver, namely, DMNA and SDMH. The directly acting chemical *cis*-Pt or those that can be activated inside bacterial cells, MMC and 4-NQO, exhibit less pronounced organ-specific effects. Before being able to adequately assess the value of this type of host-mediated assay for determining genotoxic factors in different organs of treated animals, it remains to be seen whether the effects found in the bacterial indicator cells are representative of analogous DNA reactions in the cells of the various organs investigated; corresponding experiments are presently under way.

ACKNOWLEDGMENTS

This work was supported by grants from the Koningin Wilhelmina Fonds of the Netherlands, project no. IKW-81,92 and from the Commission of the European Communities, contract no. 139-77 ENV N. The authors are very grateful to Drs. B. W. Glickman, I. Mattern, P. Starlinger, and P. van de Putte for providing bacterial strains. The competent technical assistance of S. Bouter and P. de Knijff is gratefully acknowledged.

5. References

1. B. J. Kilbey, M. Legator, W. Nichols, and C. Ramel, (eds.), *Handbook of Mutagenicity Test Procedures*, Elsevier, Amsterdam (1984).
2. K. H. Norpoth, and R. C. Garner, (eds.), *Short-Term Test Systems for Detecting Carcinogens*, Springer, Berlin (1980).
3. F. J. de Serres and J. Ashby, (eds.), *Evaluation of Short-Term Tests for Carcinogens*, Elsevier/North-Holland, Amsterdam (1981).
4. B. N. Ames, J. McCann, and E. Yamasaki, Methods for detecting carcinogens and mutagens with the *Salmonella*/mammalian microsome mutagenicity test, *Mutat. Res.* 31, 347–364 (1975).
5. D. M. Maron and B. N. Ames, Revised methods for the *Salmonella* mutagenicity test, *Mutat. Res. 113*, 173–215 (1983).
6. ICPEMC, International Commission for Protection against Environmental Mutagens and Carcinogens, Committee 1 Final Report, *Mutat. Res. 114*, 117–177 (1983).
7. ICPEMC, *International Commission for Protection against Environmental Mutagens and Carcinogens, Committee 2 Final Report*, Elsevier Biomedical Press, Amsterdam (1982).
8. J. McCann, E. Choi, E. Yamasaki, and B. N. Ames, Detection of carcinogens as mutagens in the *Salmonella*/microsome test: Assay of 300 chemicals, *Proc. Natl. Acad. Sci. USA 72*, 4135–5139 (1975).
9. F. J. de Serres and M. D. Shelby, (eds.), *Comparative Chemical Mutagenesis*, Plenum Press, New York (1981).
10. IARC, *International Agency for Research on Cancer Monographs, Long-Term and Short-Term Screening Assays for Carcinogenesis: A Critical Appraisal*, IARC, Lyon (1980).
11. J. McCann and B. N. Ames, The *Salmonella*/microsome mutagenicity test: Predictive value for animal carcinogenicity, in: *Origins of Human Cancer* (H. H. Hiatt *et al.*, eds.), Book C, pp. 1431–1450, Cold Spring Harbor Laboratory, Cold Spring Harbor, New York (1977).
12. M. Meselson and K. Russell, Comparisons of carcinogenic and mutagenic potency, in: *Origins of Human Cancer* (H. H Hiatt *et al.*, eds.), Book C, pp. 1473–1481, Cold Spring Harbor Laboratory, Cold Spring Harbor, New York (1977).
13. B. N. Ames and K. Hooper, Does carcinogenic potency correlate with mutagenic potency in the Ames assay? *Nature 274*, 19–20 (1978).
14. J. Ashby and J. A. Styles, Factors affecting mutagenic potency *in vitro*, *Nature 274*, 20–22 (1978).

15. J. Ashby, Implications of carcinogenicity, in: *Mutagenesis in Sub-mammalian Systems. Status and Significance* (F. E. Paget, ed.), pp. 165–184, MIT Press, Cambridge, Massachusetts (1979).

16. G. R. Mohn, Bacterial systems for carcinogenicity testing, *Mutat. Res. 87*, 191–210 (1981).

17. G. R. Mohn, On the correlation between mutagenicity and carcinogenicity, in: *Genetic Origins of Tumor Cells* (F. J. Cleton and J. W. I. M. Simons, eds.), pp. 11–24, Martinus Nijhoff, The Hague (1980).

18. E. P. Reddy, R. K. Reynolds, E. Santos, and M. Barbacid, A point mutation is responsible for the acquisition of transforming properties by the T24 human bladder carcinoma oncogene, *Nature 300*, 149–152 (1982).

19. G. Mardon and H. E. Varmus, Frameshift and intragenic suppressor mutations in a Rous sarcoma provirus suggest *src* encodes two proteins, *Cell 32*, 871–879 (1983).

20. E. Taparowsky, Y. Suard, O. Fasano, K. Shimizu, M. Goldfarb, and M. Wigler, Activation of the T24 bladder carcinoma transforming gene is linked to a single amino acid change, *Nature 300*, 762–765 (1982).

21. G. R. Mohn, A. A. van Zeeland, and B. W. Glickman, Influence of experimental conditions and DNA repair ability on EMS-induced mutagenesis and DNA binding in *Escherichia coli* K-12. Comparison with mammalian cell mutagenesis, *Mutat. Res. 92*, 15–27 (1982).

22. C. S. Aaron, A. A. van Zeeland, G. R. Mohn, A. T. Natarajan, A. G. A. C. Knaap, A. D. Tates, and B. W. Glickman, Molecular dosimetry of the chemical mutagen ethyl methanesulfonate. Quantitative comparison of mutation induction in *Escherichia coli*, V79 Chinese hamster cells and L5178Y mouse lymphoma cells, and some cytological results *in vitro* and *in vivo*, *Mutat. Res. 69*, 201–216 (1980).

23. G. R. Mohn, T.-M. Ong, D. F. Callen, P. G. N. Kramers, and C. S. Aaron, Comparison of the genetic activity of 5-nitroimidazole derivatives in *Escherichia coli, Neurospora crassa, Saccharomyces cerevisiae,* and *Drosophila melanogaster, J. Environ. Pathol. Toxicol. 2*, 657–670 (1979).

24. M. G. Gabridge, A. DeNunzio, and M. S. Legator, Microbial mutagenicity of streptozotocin in animal-mediated assays, *Nature 221*, 68–70 (1969).

25. G. M. Williams, R. Kroes, H. W. Waaijers, and K. W. van de Poll, eds., *Predictive Value of Short-Term Screening Tests in Carcinogenicity Evaluation*, Elsevier/North-Holland, Amsterdam (1980).

26. G. Mohn and J. Ellenberger, Mammalian blood-mediated mutagenicity tests using a multipurpose strain of *Escherichia coli* K-12, *Mutat. Res. 19*, 257–260 (1973).

27. G. Mohn, J. Ellenberger, D. McGregor, and H.-J. Merker, Mutagenicity studies in microorganisms *in vitro*, with extracts of mammalian organs, and with the host-mediated assay, *Mutat. Res. 29*, 221–233 (1975).

28. P. Kerklaan, G. Mohn, and S. Bouter, Comparison of the mutagenic activity of dialkylnitrosamines in animal-mediated and *in vitro* assays using an *Escherichia coli* indicator, *Carcinogenesis 2*, 909–914 (1981).

29. P. Kerklaan, S. Bouter, and G. Mohn, Mutagenic activity of three isomeric *N*-nitroso-*N*-methylamino-pyridines towards *Escherichia coli* K-12 *in vitro* and animal-mediated assays, *Carcinogenesis 3*, 415–421 (1982).

30. G. R. Mohn and J. Ellenberger, The use of *Escherichia coli* K-12/343/113(λ) as a multipurpose indicator strain in various mutagenicity testing procedures, in: *Handbook of Mutagenicity Test Procedures* (B. Kilbey *et al.*, eds.), pp. 95–118, Elsevier/North-Holland, Amsterdam (1977).

31. B. J. Bachmann and K. B. Low, Linkage map of *Escherichia coli* K-12, edition 6, *Microbiol. Rev. 44*, 1–56 (1980).

32. R. Favre, A. Wiater, S. Puppo, M. Iaccarino, R. Noelle, and M. Freundlich, Expression of a valine-resistant acetolactate synthase activity mediated by the *ilvO* and *ilvG* genes of *Escherichia coli* K-12, *Mol. Gen. Genet. 143*, 243–252 (1976).

33. M. W. Hane and T. H. Wood, *Escherichia coli* K-12 mutants resistant to nalidixic acid: Genetic mapping and dominance studies, *J. Bacteriol. 99*, 238–241 (1969).

34. P. Howard-Flanders and L. Theriot, Mutants of *Escherichia coli* K-12 defective in DNA repair and in genetic recombination, *Genetics 53*, 1137–1150 (1966).

35. T. Kato and S. Kondo, Two types of x-ray sensitive mutants of *Escherichia coli* B: Their phenotypic characters compared with UV sensitive mutants, *Mutat. Res. 4*, 253–263 (1967).

36. A. Rörsch, P. van de Putte, I. E. Mattern, and H. Zwenk, in: *Radiation Research. Proceeding of the Third International Congress of Radiation Research* (G. Silini, ed.), pp. 771–789, Elsevier/North-Holland, Amsterdam (1967).

37. R. P. Boyce and P. Howard-Flanders, Genetic control of DNA breakdown and repair in *E. coli* K-12 treated with mitomycin C or ultraviolet light, *Z. Vererbungsl. 95*, 345–350 (1964).

38. K. W. Kohn, H. H. Steigbigel, and C. L. Spears, Cross-linking and repair of DNA in sensitive and resistant strains of *E. coli* treated with nitrogen mustard, *Proc. Natl. Acad. Sci. USA 53*, 1154–1161 (1965).

39. P. de Lucia and J. Cairns, Isolation of an *E. coli* strain with a mutation affecting DNA polymerase, *Nature 224*, 1164–1166 (1969).

40. P. E. Hartman, Bacterial mutagenesis: Review of new insights, *Environ. Mutagen. 2*, 3–16 (1980).

41. P. C. Hanawalt, P. K. Cooper, A. K. Ganesan, and C. A. Smith, DNA repair in bacteria and mammalian cells, *Annu. Rev. Biochem. 48*, 783–836 (1979).

42. E. E. Slater, M. D. Anderson, and H. S. Rosenkranz, Rapid detection of mutagens and carcinogens, *Cancer Res. 31*, 970–973 (1971).

43. T. Kada, K. Tutikawa, and J. Sadaie, *In vitro* and host-mediated "rec-assay" procedures for screening chemical mutagens; and phloxine, a mutagenic red dye detected, *Mutat. Res. 16*, 165–174 (1972).

44. M. Nagao and T. Sugimura, Sensitivity of repair-deficient mutants and similar mutants to *N*-nitroquinoline-l-oxide, *N*-nitropyridine-l-oxide, and their derivatives, *Cancer Res. 31*, 2369–2374 (1972).

45. Z. Leifer, T. Kada, M. Mandel, E. Zeiger, R. Stafford, and H. S. Rosenkranz, An evaluation of tests using DNA repair-deficient bacteria for predicting genotoxicity and carcinogenicity, A report of the US EPA's Gene-Tox program, *Mutat. Res. 87*, 211–297 (1981).

46. R. E. W. Hancock and P. Reeves, Lipolysaccharide-deficient bacteriophage-resistant mutants of *Escherichia coli* K-12, *J. Bacteriol. 127*, 98–108 (1980).

47. B. N. Ames, F. D. Lee, and W. E. Durston, An improved bacterial test system for the detection and classification of mutagens and carcinogens, *Proc. Natl. Acad. Sci. USA 70*, 782–786.

48. G. Mohn, P. Kerklaan, P. de Knijff, and S. Bouter, Influence of phenotypic expression lag and division delay on apparent frequencies of induced mutations in *Escherichia coli* K-12, *Mutat. Res. 91*, 419–425 (1981).

49. M. Lipkin and R. Good, eds., *Gastro-Intestinal Tract Cancer and Carcinogenesis*, Plenum Press, New York (1978).

50. L. A. Wheeler, J. H. Carter, F. B. Soderberg, and P. Goldman, Association of *Salmonella* mutants with germfree rats: Site specific model to detect carcinogens as mutagens, *Proc. Natl. Acad. Sci. USA 72*, 4607–4611 (1975).

51. R. Barale, D. Zucconi, M. Romano, and N. Loprieno, The intragastric host-mediated assay for the assessment of the formation of direct mutagens *in vivo*, *Mutat. Res. 113*, 21–32 (1983).

52. E. Balbinder, C. I. Reich, D. Shugarts, J. Keogh, R. Fibiger, T. Jones, and A. Banks, Relative mutagenicity of some urinary metabolites of the anti-tumor drug cyclophosphamide, *Cancer Res. 41*, 2967–2972 (1981).

53. K. Yano and M. Isobe, Mutagenicity of *N*-methyl-*N'*-aryl-*N*-nitrosoureas and *N*-methyl-*N'*-methyl-*N*-nitrosoureas in relation to their alkylating activity, *Cancer Res. 39*, 5147–5149 (1979).

54. K. Hemminki and K. Falck, Correlation of mutagenicity and 4-(*p*-nitrobenzyl)-pyridine alkylation by epoxides, *Toxicol. Lett. 4*, 103–106 (1979).

55. E. Eder, T. Neudecker, D. Lutz, and D. Henschler, Mutagenic potential of allyl and allylic compounds. Structure–activity relationship as determined by alkylating and direct *in vitro* mutagenic properties, *Biochem. Pharmacol. 29*, 993–998 (1980).

56. A. W. Wood, R. L. Chang, W. Levin, D. E. Ryan, P. E. Thomas, R. E. Lehr, S. Kumar, M. Schaefer-Ridder, U. Engelhardt, H. Yagi, D. M. Jerina, and A. H. Conney, Mutagenicity of diolepoxides and tetrahydroepoxides of benz(*a*)acridine and benz(*c*)acridine in bacteria and in mammalian cells, *Cancer Res. 43*, 1656–1662 (1983).

57. R. Majumbar, S. C. Mathur, and K. Roy, Mutagenicity and K-region reactivity of monomethyl derivatives of benz(α)anthracene in a self-consistent-field molecular orbital theory, *Biochem. Biophys. Res. Commun. 106*, 836–841 (1982).

58. R. Hauser and B. Matter, Localization of *E. coli* K-12 in livers of mice used for an intrasanguineous host-mediated assay, *Mutat. Res. 46*, 45–48 (1977).

59. P. Arni, The microbial host-mediated assay in comparison with *in vitro* systems: Problems and evaluation, predictive value, and practical application, in: *Short-Term Test Systems for Detecting Carcinogens* (K. H. Norpoth and R. C. Garner, eds.), pp. 190–198, Springer, Berlin (1980).

60. D. Frezza, B. Smith, and E. Zeiger, The intrasanguineous host-mediated assay procedure using *Saccharomyces cerevisiae*: Comparison with two other metabolic activation systems, *Mutat. Res. 108*, 161–168 (1983).

61. W.-Z. Whong and T. Ong, Mediated mutagenesis of dimethylnitrosamine in *Neurospora crassa* by various metabolic activation systems, *Cancer Res. 39*, 1525–1528 (1979).

62. H. Druckrey, R. Preussmann, S. Ivankovic, and D. Schmähl, Organotrope carcinogene Wirkungen bei 65 verschiedenen *N*-Nitroso Verbindungen an BD Ratten, *Z. Kerbsforsch. 69*, 103–201 (1967).

63. W. Lijinsky, M. D. Reuber, and W. B. Manning, Potent carcinogenicity of nitrosodiethanolamine in rats, *Nature 288*, 289–590 (1980).

64. G. Reznik, U. Mohr, and F. W. Krüger, Carcinogenic effect of di-*n*-propyl-nitrosamine, β-hydroxy-propyl-*n*-propylnitrosamine, and methyl-*n*-propyl-nitrosamine on Sprague Dawley rats, *J. Natl. Cancer Inst. 54*, 937–943 (1975).

65. G. R. Mohn, P. R. M. Kerklaan, W. P. C. ten Bokkum-Coenradi, and T. E. M. ten Hulscher, A differential DNA repair test using mixtures of strains of *E. coli* K-12 in liquid suspension and animal-mediated assays, *Mutat. Res. 113*, 404–415 (1983).

66. P. N. Magee and J. M. Barnes, Carcinogenic nitroso compounds, *Adv. Cancer Res. 10*, 163–246 (1967).

67. A. E. Pegg and J. W. Nicoll, Nitrosamine carcinogenesis: The importance of the persistence in DNA of alkylated bases in the organotropism of tumor induction, in: *Screening Tests in Chemical Carcinogenesis* (R. Montesano, H. Bartsch, and L. Tomatis, eds.), pp. 571–590, IARC, Lyon (1976).

68. IARC, *Monographs on the Evaluation of the Carcinogenic Risk of Chemicals to Humans*, Vol. 17, *Some Nitroso Compounds*, IARC, Lyon (1978).

69. IARC, *Monographs on the Evaluation of the Carcinogenic Risk of Chemicals to Humans*, Vol. 7, *Some Anti-thyroid and Related Substances, Nitrofurans and Industrial Chemicals*, IARC, Lyon (1974).

70. B. F. Swann and P. N. Magee, Nitrosamine induced carcinogenesis. The alkylation of nucleic acids of the rat by N-methyl-N-nitrosourea, dimethyl nitrosamine, and methyl methanesulfonate, *Biochem. J. 110*, 39–47 (1968).

71. B. F. Swann, The rate of breakdown of methyl methanesulphonate, dimethyl sulfate and N-methyl-N-nitrosourea in the rat, *Biochem. J, 110*, 49–52 (1968).

72. D. F. Heath, The decomposition and toxicity of dialkylnitrosamines in rats, *Biochem. J. 85*, 72–91 (1962).

73. IARC, *Monographs on the Evaluation of the Carcinogenic Risk of Chemicals to Humans*, Supplement 4, *Chemicals, Industrial Processes and Industries Associated with Cancer in Humans*, IARC, Lyon (1982).

74. J. Ellenberger and G. Mohn, Mutagenic activity of major mammalian metabolites of cyclophosphamide toward several genes of *Escherichia coli, J. Toxicol. Environ. Health 3*, 585–599 (1977).

75. J. J. Wong and D. P. H. Hsieh, Mutagenicity of aflatoxins related to their metabolism and carcinogenic potential, *Proc. Natl. Acad. Sci. USA 73*, 2241–2244 (1976).

76. G. R. Mohn, P. R. M. Kerklaan, A. A. van Zeeland, J. Ellenberger, R. A. Baan, P. H. M. Lohman, and F. W. Pons, Methodologies for the determination of various genetic effects in permeable strains of *E. coli* K-12 differing in DNA repair capacity. Quantification of DNA adduct formation, experiments with organ homogenates and hepatocytes, and animal-mediated assays, *Mutation Res. 125*, 153–184 (1984).

CHAPTER 10

The Detection of Mutagens in Human Feces as an Approach to the Discovery of Causes of Colon Cancer

H. F. Mower

Colon cancer is a major health problem. It is the second most prevalent organ-site cancer for both sexes in "Western societies"[35] (in the U. S. and Western Europe women get more breast cancer and men get more lung cancer than colon cancer). In the U. S. over 100,000 cases of colon cancer can be expected each year in the 1980s, and about 45% of the cases will be fatal.[19] The incidence of the disease in the U. S. appears to have stabilized since the 1930s. It is slowly increasing among white males and slowly decreasing among white females.[3] About 3% of the population will be affected some time during their lives. As the population of the U. S. shifts to larger proportions of older people, the number of colon cancer cases will steadily increase, and the many adverse effects of this disease on our society will multiply.

Colon cancer is a member of the triad of major Western lifestyle cancers, which also includes female breast cancer and prostate cancer.[5] These cancers have low incidence in rural undeveloped areas of Japan and other Asian and African countries. The incidence of these cancers is about 5- to 10-fold lower in these areas than in Western countries.

H. F. Mower • Department of Biochemistry and Biophysics and Cancer Center of Hawaii, University of Hawaii, Honolulu, Hawaii 96822.

Colon cancer, like the other Western lifestyle cancers, appears to be dependent on the presence of environmental agents or lifestyle factors that are persistent and unnoticed in the everyday activities of individuals. This is most clearly demonstrated by the study of populations migrating from countries of low colon cancer incidence to countries of high colon cancer incidence. These populations, such as Japanese living in Hawaii and California, and blacks moving from Nigeria to the U. S., acquire the higher colon cancer incidence of the host countries within two generations.[32] It appears, therefore, that at least in the majority of colon cancer cases, environmental influences are the primary determining factor, and that most people are equally susceptible to these factors. A genetic predisposition to the cancer is clearly evident, however, in some cases. For example some families get colon cancer more often than expected[14] and in the uncommon disease familial polyposis, colon cancer often develops at an early age.[15]

When one considers the possible environmental factors that could cause these cancers, one is inevitably drawn to the composition of the colon contents, and since the direct sampling of this material is not possible, the study of the composition of the feces is considered as a suitable surrogate.

Cancer of the large bowel is a complex disease and different causes may be responsible for the cancers that form in the ascending colon, the descending and sigmoid colon, or in the rectum. The nature of the compounds found in feces is probably directly related to cancer of the sigmoid colon. This may be true since feces reside in the rectum for only a short time, and usually several hours elapse between the time material leaves the cecum and ascending colon and the time it is passed as fecal material from the body. Hence it is uncertain whether any of the observed fecal mutagens were ever present in the contents of the right-sided portions of the colon.

As more detailed epidemiological studies within the U. S. took place, the complexities of the etiology of colon cancer began to emerge. At first it was thought that the primary determinant in Western society that increased colon cancer was fat, principally saturated fat in the diet, and increased consumption of meat, principally beef.[1] Eating these materials, it was thought, caused increased bile acid synthesis. The bile acids were postulated to be precursors of carcinogens that caused colon cancer.[10] Despite extensive efforts, the demonstration of the existence of the bile acid-derived carcinogens remains unproven.

Although colon cancer incidence of a population correlates well with the average individual fat intake within the particular population, a given person's fat consumption and serum lipid levels are poor predicators of colon cancer risk.[26,27]

For example, Mormons living in Utah, eating a high-fat, high-beef diet, have a colon cancer risk about 60% of the U. S. national average, a risk that is about midway between the U. S. and Japan.[16] Seventh Day Adventists, who eat vegetarian diets, have an overall colon cancer risk similar to the Mormons.[22] Thus the Mormons have a lower cancer risk and the Seventh Day Adventists have a higher colon cancer risk than would be expected if animal fat intake were the sole determinant of colon cancer risk.

Finally, it must be pointed out that the carcinogens producing colon cancer may arrive at the sensitive cells of colon mucosa not in the fecal stream but as blood constituents. This is probably true for other environmental cancers, such as prostate cancer, and perhaps for female breast cancer as well. Animal experiments show this to be true for some colon carcinogens, such as dimethylhydrazine and azoxymethane. When these compounds are administered orally to rats, they reach the colon mucosa through the blood, as shown by the work of Rubio et al.,[25] who isolated colonic segments by surgical ligation of the distal and proximal ends of a portion of a rat colon. The segments were otherwise unaffected and remained attached to their normal blood supply. When ip. administration of 1,2-dimethylhydrazine was given to these rats, tumors were formed in the mucosal surfaces of the isolated colon segments, which were devoid of intestinal contents.

A rational approach to understanding some of the causes of colon cancer may lie in the study of mutagens contained in human feces. The validity of this approach rests upon the hypothesis that transformation of normal cells to the neoplastic state proceeds through a series of somatic mutations. This theory is supported by the observation that most carcinogens are mutagens, and the importance of the short-term mutagen assays rests on the high probability that compounds with mutagenic properties are likely to be carcinogenic if exposed to a suitable host for a sufficient length of time and at a high but subtoxic dose. Recent studies on the characterization of human oncogenes show the importance of point mutations and chromosomal translocation in human carcinogenesis. These studies[24,30] of the T24 human bladder carcinoma cells have shown, by restriction enzyme digestion of the cellular DNA, that these cells contain a dominant DNA-transforming oncogene. This gene is a 4.6-kilobase segment of DNA, and by transfection assays this DNA will induce the transformation of NIH 3T3 cells. These studies have further demonstrated that this bladder carcinoma oncogene is closely related to the oncogene of murine sarcoma viruses and to human DNA sequences in normal human cells. A difference exists between the oncogene sequence in bladder cancer cells and that in normal human cells. These two gene sequences differ from one another by a single point mutation, the change of guanosine into thymidine. This substitution results in the incorporation of valine instead of glycine as the 12th amino acid residue of the T24 oncogene P21 protein. This simple amino acid substitution appears to be sufficient to bring about some key change in normal cells that may be involved in their conversion to neoplastic cells.

Another oncogene of some viruses, called the *myc* gene, is also found in human DNA sequences of normal cells. These sequences are located on chromosome 8, and recently they have been associated with Burkitt's lymphoma.[21] Tumor cells from these cancers show the presence of aberrant chromosomes formed by an exchange of segments between chromosomes 8 and 14, 8 and 2, and 8 and 22. In each of these exchange processes, the *myc* gene is moved from its normal location and inserted in regions of DNA coding for some portion of the constant region of antibody heavy-chain protein on chromosomes 14, 2, and 22. Frequently an incomplete *myc* gene, shortened by 0.4– 0.5 kilobases, is inserted in the DNA sequence of antibody protein. In its new location, presumably under the influence of the gene promoter for the heavy chain, increased expression of the *myc* gene occurs. This translocation process appears to be important in the genesis of these lymphomas by causing either increased *myc* gene expression and/or altered *myc* gene product formation.

Thus, the chemically caused point mutations, which are measured in many short-term bacterial mutagen assay procedures, and chromosomal translocations, which are measured in chromosomal aberration or sister chromatid exchange assays, appear to be fundamental to the process of oncogenesis. The importance of genotoxic chemicals present in the environment to the causation of human cancer seems to be reaffirmed by these recent developments in biomolecular oncology.

The study of mutagens in human feces has some unique uncertainties, as described in detail in a recent excellent review by Venitt.[31] For example, the presence of some mutagens in feces may be artifactual because of the anaerobic condition existing within the colon. Upon exposure of feces to oxygen, oxidation of numerous reduced substances undoubtedly takes place, and this process could, in some circumstances, result in the formation of mutagenic compounds that were never present within the colon.

Another concern centers on the presence of inhibitors of mutagenesis in the feces. Hayatsu has shown that oleic acid and heme products prevent the mutagenic activity of certain substances.[8] This raises the complication that the lack of mutagenicity in certain fecal samples may be due to inhibitors that may mask the presence of genotoxic material.

Dietary fiber and other fecal contents have been postulated to provide protection against colon cancer by adsorbing mutagen/carcinogens that would otherwise be able to interact with the DNA of mucosal cells. This situation may also make mutagen analysis more complicated, as solvent extraction of feces may not transfer those mutagens adsorbed on fiber to the extracting phase.

Comutagens, antimutagens, inhibitors, and promoters of DNA repair and a host of other factors may increase, suppress, or completely mask the action of fecal mutagens and add a special challenge to the study of these important substances.

The analysis of human feces reaches its greatest potential relevant to an understanding of colon cancer when feces from a high-risk population and a low-risk population are analyzed simultaneously and compared as to frequency of mutagens detected by different short-term mutagen-testing techniques.

While simple in conception, these studies are difficult to carry out. Ideally the two populations should be of the same race, have similar demographic parameters, such as age distribution, numbers of each sex, and occupation. Moreover, the collection and handling of the feces should be the same, as should the storage time before analysis. When fecal samples are transported for long distances, it is often not possible to prevent them from warming to ambient temperature even in a container in which small amounts of dry ice are still present. The value of these studies is increased many fold if they can be carried out within existing long-range prospective health studies in which many other parameters of the populations are compiled as well, such as diet history, occupation, and medical records of the fecal donors and their relatives. A further benefit would accrue in those studies in which the fecal donors can be monitored until their death. In this way the presence of fecal mutagens in an individual and the eventual development of colon cancer in that individual can be assessed.

Three groups have attempted studies of this kind,[6,18,23] but none has been carried out in an ideal manner. The most complete was the study by the Japanese–Hawaii Cancer Study based at Kuakini Hospital in Honolulu, Hawaii.[18] Feces were analyzed for mutagens in populations of high-risk Japanese living in Hawaii and low-risk Japanese living in Japan. The Hawaiian Japenese group is part of a long-range multivariant prospective study in which the cohort is being continually monitored. Fewer data and less followup study, however, are possible with the low-risk group of this study.

A second study measured mutagens in feces of whites and blacks living in various regions of South Africa.[6] A third, more complex study involved three groups: Finns living in Finland, an unexpectedly low-risk group, Seventh Day Adventists living in New York at low risk for colon cancer, and other New Yorkers.[23] In the last two studies very few epidemiologic data were taken, and no followup observations are planned. These three studies all used similar fecal mutagen assay techniques, which involved solvent extraction (ether or methylene chloride) of the feces, usually after drying the fecal samples. The organic extract was then evaporated to dryness and the residue was redissolved in dimethylsulfoxide and was examined by the Ames pour plate method of mutagen detection. The frameshift (TA98) and base-substitution (TA100) strains were used with and without rat liver microsomes. All three groups reported "direct-acting" mutagens detected by either TA98 or TA100 in some of the fecal samples. These mutagens were found in 20–25% of the feces samples of the high-risk populations and were present less often in the low-risk populations. In one study the mutagens were more frequently found in the females of the high-risk populations, and smaller differences were observed between the males of the high- and low-risk populations.[18] Some mutagens that require rat liver microsomes for activation were also observed in these fecal samples, and they were also present more often in the high-risk group. The Hawaiian studies included a study of water-soluble mutagens as well as those that were soluble in organic solvents. The water-soluble mutagens were measured in aqueous extracts of freeze-dried feces, which were first treated with ether or methylene chloride to remove substances soluble in these solvents. Mutagenic substances soluble in dilute pH 7.5 buffer solutions were detected by a DNA-repair-deficient strain of *E. coli* by the procedure of Ichinotsubo *et al.*[11] This study found more mutagens in the aqueous fecal extracts of the high-risk population than in the low-risk population. When mutagens were studied in feces from some individuals over an extended period of time, it was found that mutagens were not consistently present in sequential fecal samples. Some individuals seem never to have mutagens in their feces, while others could be described as mutagen "producers" and 30–40% of their fecal samples contained mutagens.

Why some individuals appear to be mutagen producers and others do not is a mystery, as is the relationship of this to the development of subsequent large bowel cancer. One group reported that an individual who was a consistent mutagen producer was married to a person who never produced fecal mutagens.[34] The two presumably largely ate the same diet, yet were completely opposite in fecal mutagen formation. Needless to say, much more needs to be learned in this interesting area of environmental carcinogenesis.

In only one of these studies was an attempt made to detrmine if the inhibitors of mutagen detection were present in equal amounts in the high- and low-risk populations. Hayatsu et al.[9] studied a small subset of the high-risk Hawaii population and the low-risk Japanese population and found similar amounts of the inhibitory substances in the feces of both populations.

Fecal material is a complex mixture of many components, some of which are toxic to the mutagen-testing organisms. In response to this situation, some investigators, such as Bruce et al.,[2] have devised simple purification schemes in which some of the inhibitory substances are removed by solvent partition steps. Others have used different mutagen assay procedures, such as the fluctuation technique[13,31] or mammalian cell cultures/chromosome aberration procedures,[28] which seem to be less sensitive to toxins. The use of these latter procedures gives an entirely different view of human fecal mutagens. These studies have demonstrated that virtually every fecal sample contains mutagens, and it is the exceptional fecal sample that is void of mutagens. These mutagens are present in the feces of humans who are strict vegetarians as well as those on a high meat and fat diet. They are also present in animal feces of herbivores, carnivores, and omnivores.[29] The emerging view of the human situation from these studies using the fluctuation test is in contrast to the studies using the Ames plate incorporation procedure. The fluctuation test shows that most people contain some mutagens in their feces and there are only a few who do not produce fecal mutagens. This individual mutagen pattern can persist over a long period of time, as shown in carefully controlled dietary studies. When meat is eaten, mutagens tend to increase in the feces, but by only a modest amount.[13,30]

The fluctuation tests have been carried out using several different endpoints, such as the Salmonella reversion assays,[13] or the E. coli strains of Venitt.[31] Measurements have been made of mutagens in aqueous extracts[13] or ether extracts.[31]

These results of the above studies raise the possibility that all feces contain low levels of mutagenic and hence cancer-initiating substances and that the incidence of colon cancer may depend on the variable concentration of tumor-promoter substances present in the feces or blood. Animal model experiments support this possibility. For example, tumors produced by rectal doses of low levels of MNNG are increased by subsequent rectal installation of bile acids.[20] suggesting that bile acids are tumor promoters. While this hypothesis is supported by some epidemiological evidence, it is far from established whether bile acids are important tumor-promoter substances in colon cancer, but this certainly merits further consideration, as does a well-planned and well-executed study to determine the presence of other tumor-promoter substances in feces.

The relationship between the mutagens detected by the fluctuation tests and those detected by the pour plate assay of Ames is unclear. It is possible that both techniques detect identical groups of compounds, but the pour plate procedure is only positive with fecal extracts that contain unusually large amounts of these mutagens or those that contain unusually low amounts of inhibitory substances.

On the other hand, the fluctuation test may be detecting different mutagens. This is because the fluctuation test is less influenced by toxic substances than the pour plate procedure and hence is capable of detecting the mutagenicity of substances that would otherwise only be recognized as bactericidal agents.

The greatly increased sensitivity of the fluctuation test compared to the pour plate procedure is, however, tempered by some disadvantages. First, it is cumbersome to carry out, and even a small study will require the scoring of tens of thousands of culture tubes. In addition, it is very sensitive to contamination. Fifty contaminating auxotrophic bacterial cells in 1 ml of fecal extract could, after the 5-day incubation used in a fecal extract assay, cause the appearance of a positive result in nearly every test vessel. None of the studies using the fluctuation assay considered checking the bacteria grown in positive test vessels for the characteristics of the revertent tester strain.

Stich and his colleagues[28] have shown by a chromosome aberration assay that human feces contain substances that are potent chromosome-damaging agents. The clastogenic effect of these substances is enhanced by the transition metals Cu^{2+} and Mn^{2+} and is greatly reduced by catalase. This suggests that the extracts contain reducing agents that upon exposure to air form active oxygen species, including H_2O_2, which are likely to be responsible for the chromosome damage. This work of Dr. Stich describes one broad group of genotoxic substances to which the colon mucosa is exposed, and the possible role of these materials in colon carcinogenesis should be investigated further.

Despite the existence of the body of evidence suggesting that a plethora of mutagens exists in feces, a considerable effort has been made to determine the structure of one of the mutagens detected by the Ames mutagen test. The rationale for choosing the particular mutagen was rather weak, and the role of this mutagen in colon cancer is yet to be established, although some indirect association has been found.[2] The mutagen was first discovered by Bruce, who was among the first to become interested in fecal mutagens. The mutagen was detected by Ames tester strain TA100 in the absence of rat liver microsomes in extracts of freeze-dried feces. These extracts were partitioned between ether and an aqueous basic solution. Bruce and his colleagues made some progress in determining the structure of the mutagen, and showed that the material had a characteristic UV absorption spectrum with maxima at 320, 340, and 365 nm. Further work was possible only when a greater supply of the material could be found. Wilkins *et al.* were able to solve this problem by incubating in an anaerobic environment for several days feces from a donor who was usually a consistent mutagen producer.[33] By this procedure the mutagen material in the feces was increased many fold, presumably by bacterial action. Despite this advancement, the assignment of a structure to this material proved to be a difficult task. The material on purification became increasingly unstable and sensitive to light, oxygen, and acid. Finally the following structure was proposed for the mutagen, and it has been named fecapentaene[7]: The structure is a conjugated pentaene vinyl ether of glycerol. It is closely related to the common class of lipids called plasmalogens, which are simple, vinyl ethers of glycerol containing only one double bond. Multiple conjugated double-bonded substances such as carotenoids also are commonly found in most living systems. Neither carotenoids nor plasmalogens has been shown to be

(S)-3-(1,3,5,7-dodecapentaenyloxy-1,2- propanediol (fecapentaene) (I)

mutagenic, and it is difficult to understand why structure I would be mutagenic. The synthesis of I is eagerly awaited in hopes that these questions can be answered.

In a recent publication Dion and Bruce[4] describe a newly designed fraction procedure based upon acetone extraction of wet human feces, followed by SepPak cartage (Walters, Inc., Waltham, Massachusetts) absorption and elution and normal phase HPLC. The results of the examination of 24 fecal samples showed that about one-third contain mutagens, all of which appeared to be fecapentaenes. These results suggest that these polyene structures may be the predominant mutagenic substances in human feces.

The study of mutagenic substances found in feces has just begun, and further work can be expected to discover and elucidate the structures of many new mutagenic compounds. This fact alone should encourage investigators to continue their studies. An important caveat is the problem of toxic and bacteriostatic compounds in feces that may mask the detection of mutagens. It is difficult to speculate how best to solve this problem, but probably screening of fecal mutagens should be preceded by simple, rapid, chemical fraction procedures to separate the mutagens from those substances that interfere with their detection. These separation procedures depend upon the chemical nature of the mutagens being sought, and premutagen testing purification procedures can be expected to vary from one study to another.

The anaerobic incubation procedure devised by Wilkins to increase the amount of fecapentaene could also be examined as a possible way to increase the amount of other fecal mutagens, such as those detected by TA98 or TA100, which are clearly different from fecapentaene, or those that require metabolic activation. This simple anaerobic incubation procedure may be very important in the study of fecal mutagens. The increase in mutagens upon anaerobic incubation of feces demonstrates the bacterial origin of the mutagens and indicates that exposure of feces to oxygen tends to destroy mutagens rather than to create them. Diet affects the bacterial population of the colon in complex ways, and the amounts of bacterial metabolites, both mutagenic and nonmutagenic, can be expected to change with variation in food intake.

An added impetus for the study of fecal mutagens is their possible role in colon cancer. The complex nature of this disease suggests that many chemical agents, each responding to a variety of environmental, genetic, and metabolic factors, may cause the disease. An optimistic view of this situation is that many important discoveries will be made in this area, each contributing to the lowering of colon cancer in one or many populations.

The key to unraveling the relationship of mutagens to colon cancer may involve the study of diet and mutagen formation in humans as well as epidemiological studies of world populations linking mutagen presence to colon cancer incidence. This approach has been employed by Bruce, who has shown that fecapentaene levels in feces respond to diet, declining with high dietary intake of ascorbic acid, α-tocopherol, and dietary fiber, but are unresponsive to added fat. Through studies of this kind colon cancer may be the first of the "environmental Western lifestyle cancers" to be controlled by preventive measures.

References

1. B. Armstrong and R. Doll, Environmental factors and cancer incidence and mortality in different countries, with special reference to dietary practices, *Int. J. Cancer 15*, 617–631 (1975).
2. W. R. Bruce, A. J. Varghese, P. C. Land, and J. J. F. Krepinsky, Properties of a mutagen isolated from feces, in: *Banbury Report 7, Gastrointestinal Cancer: Endogenous Factors* (W. R. Bruce, P. Correa, M. Lipkin, S. R. Tannenbaum, and T. D. Wilkins, eds.), pp. 227–234, Cold Spring Harbor Laboratory, Cold Spring Harbor, New York (1981).
3. F. Burbank, Patterns in cancer mortality in the United States, *Natl. Cancer Inst. Monogr. 33*, 1950–1967 (1971).
4. P. Dion and W. R. Bruce, Mutagenicity of different fractions of extracts of human feces, *Mutat. Res. 119*, 151–160 (1983).
5. R. Doll and R. Peto, The causes of cancer: Quantitative estimates of avoidable risks of cancer in the United States today, *J. Natl. Cancer Inst. 66*, 1193–1308 (1981).
6. M. Ehrich, J. E. Aswell, R. L. Van Tassell, A. R. P. Walker, N. J. Richardson, and T. D. Wilkins, Mutagens in the feces of 3 South African populations at different levels of risk for colon cancer, *Mutat. Res. 64*, 231–240 (1979).
7. I. Gupta, J. Baptista, W. R. Bruce, C. T. Che, R. Furrer, J. S. Gingerish, A. A. Grey, L. Marai, P. Yates, and J. J. Krepinsky, Structures of fecapentaenes, the mutagens of bacterial origin isolated from human feces, *Biochemistry 22*, 241–245 (1983).
8. H. Hayatsu, personal communication.
9. H. Hayatsu, S. Arimoto, K. Togawa, and M. Makita, Inhibitory effect of the ether extract of human feces on activities of mutagens; inhibition by oleic and linoleic acids, *Mutat. Res. 36*, 287–293 (1981).
10. M. J. Hill, B. S. Drasar, R. E. O. Williams, T. W. Meade, A. G. Cox, J. E. P. Simpson, and B. C. Morson, Fecal bile acids and clostridia in patients with cancer of the large bowel, *Lancet 1*, 535–538 (1975).
11. D. Ichinotsubo, J. Setliff, H. F. Mower, and M. Mandel, The use of rec⁻ bacteria for testing of carcinogenic substances, *Mutat. Res. 46*, 53–62 (1977).
12. U. Kuhnlein, D. Bergstrom, and H. Kuhnlein, Mutagens in feces from vegetarians and non-vegetarians, *Mutat. Res. 85*, 1–12 (1981).
13. H. Kuhnlein, U. Kuhnlein, and P. A. Bell, The effect of short-term dietary modification on human fecal mutagenic activity, *Mutat. Res. 113*, 1–12 (1983).
14. I. P. Law, R. B. Herberman, and R. K. Oldham, Familial occurrence of colon and uterine carcinoma and of lymphoproliferative malignancies, clinical description, *Cancer 39*, 1224–1228 (1977).
15. M. Lipkin, P. Sherlock, and J. J. De Cosse, Identification of risk factors and preventive measures in the control of cancer of the large intestine, *Curr. Problems Cancer 4*, 4–57 (1980).
16. J. L. Lyon, J. Gardner, M. R. Klauber, and C. R. Smart, Cancer incidence in Mormons and non-Mormons in Utah 1966–1970, *N. Engl. J. Med. 294*, 129–133 (1976).
17. T. H. Maugh, Potent mutagen from human feces identified, *Science 218*, 363 (1982).
18. H. F. Mower, D. Ichinotsubo, L. W. Wang, M. Mandel, G. Stemmermann, A. Nomura, L. Heilbrun, S. Kamiyama, and A. Schimada, Fecal mutagens in two Japanese populations with different colon cancer risks, *Cancer Res. 42*, 1164–1169 (1982).

19. M. H. Myers and B. F. Hanky, Cancer patient survival in the United States, in: *Cancer Epidemiology and Prevention* (D. Schottenfeld and J. F. Fraumeni, eds), pp. 166–178, Saunders, Philadelphia, Pennsylvania (1982).

20. T. Narisawa, N. E. Magadia, J. H. Weisburger, and E. L. Wynder, Promoting effect of bile acid on colon carcinogenesis after intrarectal installations of N-methyl-N'-nitro-N-nitrosoguanidine in rats, *J. Natl. Cancer Inst. 53*, 1093–1097 (1974).

21. P. Newmark, Still more about oncogenes, *Nature 301*, 111 (1983).

22. R. L. Phillips, Role of life-style and dietary habits in risk of cancer among Seventh-Day Adventists, *Cancer Res. 35*, 3403–3406 (1975).

23. B. S. Reddy, C. Sharma, L. Darby, K. Laakso, and E. L. Wynder, Metabolic epidemiology of large bowel cancer; fecal mutagens in high and low-risk populations for colon cancer, *Mutat. Res. 72*, 511–522 (1980).

24. E. P. Reddy, R. K. Reynolds, E. Santos, and M. Barbaad, A point mutation is responsible for the acquisition of transforming properties by the T24 human bladder carcinoma oncogene, *Nature 300*, 149–152 (1982).

25. C. A. Rubio, G. Nylander, and M. Santos, Experimental colon cancer in the absence of intestinal contents in Sprague-Dawley rats, *J. Natl. Cancer Inst. 64*, 569–571 (1980).

26. P. D. Sorbie and N. Feinleib, The serum cholesterol–cancer relationship: An analysis of time trends in the Framingham study, *J. Natl. Cancer Inst. 69*, 689–696 (1982).

27. G. N. Stemmermann, A. M. Y. Nomura, L. K. Heilbrun, E. S. Pollack, and A. Kagan, Serum cholesterol and colon cancer incidence in Hawaiian Japanese men, *J. Natl. Cancer Inst. 67*, 1179–1182 (1981).

28. H. F. Stich and U. Kuhnlein, Chromosome breaking activity of human feces and its enhancement by transition metals, *Int. J. Cancer 24*, 284–287 (1979).

29. H. F. Stich, W. Stich, and A. B. Acton, Mutagenicity of fecal extracts from carnivorous and herbivorous animals, *Mutat. Res. 78*, 105–112 (1980).

30. C. J. Tabin, S. M. Bradley, C. I. Bargmann, R. A. Weinberg, A. G. Papageorge, E. M. Scolnick, R. Dahr, D. R. Lowy, and E. H. Chang, Mechanism of activation of a human oncogene, *Nature 300*, 143–149 (1982).

31. S. Venitt, Mutagens in human faeces: Are they relevant to cancer of the large bowel?, *Mutat. Res. 98*, 265–286 (1982).

32. J. Waterhouse, C. Muir, and P. Correa, Eds., *Cancer Incidence in Five Continents*, Vol. III, IARC Scientific Publications No. 15, International Agency for Research on Cancer, Lyon, France (1976).

33. T. D. Wilkins, M. Lederman, and R. L. Van Tassell, Isolation of a mutagen produced in the human colon by bacterial action, in: *Banbury Report 7, Gastrointestinal Cancer: Endogenous Factors* (W. R. Bruce, P. Correa, M. Lipkin, S. R. Tannenbaum, and T. D. Wilkins, eds.), pp. 205–212, Cold Spring Harbor Laboratory, Cold Spring Harbor, New York (1981).

34. T. D. Wilkins, M. Lederman, R. L. Van Tassell, D. G. I. Kingston, and J. Henion, Characterization of a mutagenic bacterial product in human feces, *Am. J. Clin. Nutr. 33*, 2513–2520 (1980).

35. J. L. Young and E. S. Pollack, The incidence of cancer in the United States, in: *Cancer Epidemiology and Prevention* (D. Schottenfeld and J. F. Fraumeni, eds.), pp. 138–165, Saunders, Philadelphia, Pennsylvania (1982).

Index